Conference Proceedings of the Society for Experimental Mechanics Series

Series Editor
Tom Proulx
Society for Experimental Mechanics, Inc.,
Bethel, CT, USA

For further volumes:
http://www.springer.com/series/8922

Todd Simmermacher • Scott Cogan • Babak Moaveni
Costas Papadimitriou
Editors

Topics in Model Validation and Uncertainty Quantification, Volume 5

Proceedings of the 31st IMAC, A Conference on Structural Dynamics, 2013

Springer

Editors
Todd Simmermacher
Sandia National Laboratories
Albuquerque, NM, USA

Scott Cogan
CNRS
University of Franche-Comte
Besançon, France

Babak Moaveni
Department of Civil and Environmental Engineering
Tufts University
Medford, MA, USA

Costas Papadimitriou
Department of Mechanical and Industrial Engineering
University of Thessaly
Volos, Greece

ISSN 2191-5644 ISSN 2191-5652 (electronic)
ISBN 978-1-4614-6563-8 ISBN 978-1-4614-6564-5 (eBook)
DOI 10.1007/978-1-4614-6564-5
Springer New York Heidelberg Dordrecht London

Library of Congress Control Number: 2013936727

Printed on acid-free paper

Springer is part of Springer Science+Business Media (www.springer.com)

Preface

Topics in Model Validation and Uncertainty Quantification represents one of seven volumes of technical papers presented at the 31st IMAC, A Conference and Exposition on Structural Dynamics, 2013 organized by the Society for Experimental Mechanics, and held in Garden Grove, California February 11–14, 2013. The full proceedings also include volumes on Nonlinear Dynamics; Experimental Dynamics Substructuring; Dynamics of Bridges; Dynamics of Civil Structures; Special Topics in Structural Dynamics; and, Modal Analysis.

Each collection presents early findings from experimental and computational investigations on an important area within Structural Dynamics. Model Validation and Uncertainty Quantification is one of these areas.

Model Verification and Validation (V&V), refers to technology developed to assess the accuracy of numerical predictions and quantify the uncertainty and confidence with which predictions are made. V&V includes activities such as code and calculation verification, the design of validation experiments, test-analysis correlation, uncertainty propagation and quantification for both models and experiments, and finite element model updating.

The organizers would like to thank the authors, presenters, session organizers, and session chairs for their participation in this track.

Albuquerque, NM, USA — Todd Simmermacher
Besançon, France — Scott Cogan
Medford, MA, USA — Babak Moaveni
Volos, Greece — Costas Papadimitriou

Contents

1 **Optimal Inequalities to Bound a Performance Probability** 1
François M. Hemez and Christopher J. Stull

2 **Remaining Fatigue Life Predictions Considering Load and Model Parameters Uncertainty** 17
Maurizio Gobbato, Joel P. Conte, and John B. Kosmatka

3 **Fast Computing Techniques for Bayesian Uncertainty Quantification in Structural Dynamics** 25
Costas Papadimitriou and Dimitra-Christina Papadioti

4 **Bayesian Uncertainty Quantification and Propagation in Nonlinear Structural Dynamics** 33
Dimitrios Giagopoulos, Dimitra-Christina Papadioti, Costas Papadimitriou, and Sotirios Natsiavas

5 **Probabilistic Damage Identification of the Dowling Hall Footbridge Using Bayesian FE Model Updating** ... 43
Iman Behmanesh and Babak Moaveni

6 **Considering Wave Passage Effects in Blind Identification of Long-Span Bridges** 53
S. Farid Ghahari, M. Ali Ghannad, James Norman, Adam Crewe, Fariba Abazarsa, and Ertugrul Taciroglu

7 **Quantification of Parametric Model Uncertainties in Finite Element Model Updating Problem via Fuzzy Numbers** 67
Yildirim Serhat Erdogan, Mustafa Gul, F. Necati Catbas, and Pelin Gundes Bakir

8 **Quantifying Maximum Achievable Accuracy of Identified Modal Parameters from Noise Contaminated Free Vibration Data** 75
Eric M. Hernandez

9 **Using P-Box and PiFE to Express Uncertainty in Model Updating** 81
Ramin Madarshahian, Juan M. Caicedo, and Boris A. Zárate

10 **Robust Model Calibration with Load Uncertainties** 89
D. Pereiro, S. Cogan, E. Sadoulet-Reboul, and F. Martinez

11 **Simulating the Dynamics of the CX-100 Wind Turbine Blade: Model Selection Using a Robustness Criterion** 99
Kendra L. Van Buren, Sez Atamturktur, and François M. Hemez

12 **Defining Coverage of a Domain Using a Modified Nearest-Neighbor Metric** 113
Matthew C. Egeberg, Sez Atamturktur, and François M. Hemez

13 **Orthogonality for Modal Vector Correlation: The Effects of Removing Degrees-of-Freedom** 123
Michael L. Mains

14 **CAE Model Correlation Metrics for Automotive Noise and Vibration Analysis** 135
Qijun Zhang, Shawn Hui, and Kurt Schneider

15 **Damage Localization Using a Statistical Test on Residuals from the SDDLV Approach** 143
L. Marin, M. Döhler, D. Bernal, and L. Mevel

16 Robust Tolerance Design in Structural Dynamics ... 153
Chaoping Zang, Jun Yang, and M.I. Friswell

17 Uncertainty Propagation in Floating Raft System by FRF-Based Substructuring Method for Elastic Coupling ... 165
Huang Xiuchang, Hua Hongxing, Chen Feng, and Xu Shiyin

18 Crossing and Veering Phenomena in Crank Mechanism Dynamics ... 175
Elvio Bonisoli, Gabriele Marcuccio, and Carlo Rosso

19 Validating Low-Level Footfall-Induced Vibration Predictions in Steel and Concrete Structures ... 189
Michael J. Wesolowsky, Julia M. Graham, J. Shayne Love, Jon K. Galsworthy, and John C. Swallow

20 Finite Element Model Updating of an Assembled Aero-Engine Casing ... 199
Chaoping Zang, Shuangchao Ma, and M.I. Friswell

21 Experimental Modal Analysis and Modelling of an Agricultural Tire ... 213
F. Braghin, F. Cheli, S. Melzi, S. Negrini, and E. Sabbioni

22 International Space Station Modal Correlation Analysis ... 221
Kristin Fitzpatrick, Michael Grygier, Michael Laible, and Sujatha Sugavanam

23 Numerical Modeling of Vibration Induced Atomization of Liquids ... 243
Jesi Ehrhorn and William Semke

24 Dynamical Modeling and Verification of Underwater Acoustic System ... 255
Ahmet Levent Avşar, İstek Tatar, and Cihangir Duran

Chapter 1
Optimal Inequalities to Bound a Performance Probability

François M. Hemez and Christopher J. Stull

Abstract A challenging problem encountered in engineering applications is the estimation of a probability-of-failure based on incomplete knowledge of the sources of uncertainty and/or limited sampling. Theories formulated to derive upper probability bounds offer an attractive alternative because first, they avoid postulating the probability laws that are often unknown and second, they substitute numerical optimization for statistical sampling. A critical assessment of one such technique is presented. It derives upper probability bounds from the McDiarmid concentration-of-measure theory, which postulates that fluctuations of a function are more-or-less concentrated about its mean value. Two applications of this theory are presented. The first application analyzes a "toy" polynomial function defined in two dimensions. The upper bounds of probability are calculated and compared to sampling-based estimates of the true-but-unknown probabilities. For this function, the upper bounds obtained are too broad to be useful. These results are confirmed by conducting a similar analysis on a real engineering system, where upper bounds of probability associated with resonant frequencies of a structural system are estimated. A high-fidelity finite element model, previously validated using vibration measurements, is used to predict the frequencies. In this application, the uncertainty is introduced by way of material properties and the effective preload of a beam-to-column connection, modeled explicitly. These applications suggest that the theory not only leads to upper bounds that are inefficient but that can also be sub-optimal if their numerical estimation is based on too few model runs. It is concluded that this particular theory, while mathematically attractive, may not be well suited for engineering applications.

Keywords Concentration-of-measure • Probability-of-failure • McDiarmid diameter • Parametric uncertainty • Model validation (Approved for unlimited, public release on September 15, 2012, LA-UR-12-24769, Unclassified.)

1.1 Introduction

An analysis of system reliability estimates a probability-of-failure, denoted as P_F, where "failure" is defined as a condition where the "demand" exceeds the "capacity" that an engineered system is capable of providing. Demand, for example, refers to a peak stress resulting from an applied load, while capacity refers to a material yield stress criterion. Another application would be to require that a performance metric, that defines the capacity C, meets a user requirement, or demand D. The probability-of-failure can be written in a generic sense as:

$$P_F = \text{Prob}\left[D \geq C\right], \tag{1.1}$$

F.M. Hemez (✉)
Technical Staff Member, Los Alamos National Laboratory, X-Theoretical Design Division (XTD-3),
PO Box 1663, Mail Stop T087, Los Alamos, NM 87545, USA
e-mail: hemez@lanl.gov

C.J. Stull
Technical Staff Member, Los Alamos National Laboratory, Applied Engineering and Technology Division (AET-6),
PO Box 1663, Mail Stop P915, Los Alamos, NM 87545, USA
e-mail: stull@lanl.gov

T. Simmermacher et al. (eds.), *Topics in Model Validation and Uncertainty Quantification, Volume 5,*
Conference Proceedings of the Society for Experimental Mechanics Series 41,
DOI 10.1007/978-1-4614-6564-5_1,

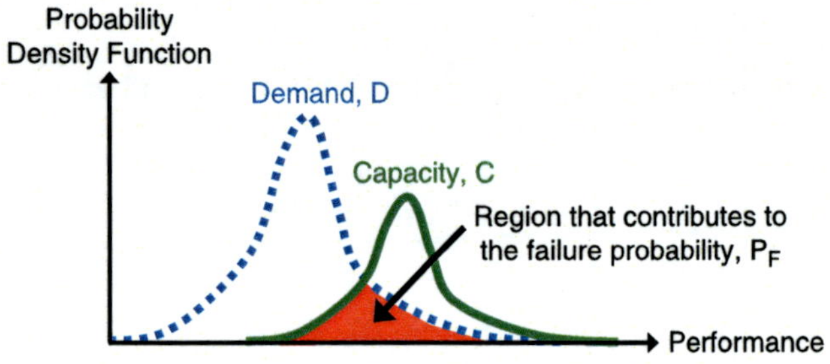

Fig. 1.1 Generic definition of a probability-of-failure P_F

where the symbol Prob[•] denotes a probability function. The probability-of-failure P_F is shown graphically in Fig. 1.1 as the shaded area, where the Probability Density Function (PDF) of encountering a demand that exceeds capacity is non-zero. The failure probability integrates the area under the curve and the reliability R of the engineered system is, by definition, equal to:

$$R = 1 - P_F. \tag{1.2}$$

Equations (1.1) and (1.2) indicate that the probability laws of the random variables (C; D) must be known analytically or, at a minimum, can be sampled a sufficient number of times, in order to estimate P_F with reasonable accuracy. In many applications, however, this is not possible. These include cases where the datasets available to estimate the pair (C; D), and their uncertainties, are too sparse to be useful for numerical integration. Another commonly encountered situation is one where the capacity C is estimated numerically by analyzing the output of a numerical simulation. The large computational expense of running the model may prevent the accumulation of enough samples to yield satisfactorily converged statistics. Also, the high dimensionality of an input space may make it difficult to numerically integrate the probability-of-failure (1.1).

This report evaluates another approach to bypass the challenge of sampling a high-dimensional space. This alternative approach searches for an upper bound of the failure probability. Put forth by the Predictive Science Academic Alliance Program (PSASP) Center at Caltech in Pasadena, California, it is a technique that relies on concentration-of-measure inequalities [1]. The method derives an upper bound U_P of the failure probability:

$$P_F \leq U_P, \tag{1.3}$$

without having to assume probability laws for the demand and capacity terms of Eq. (1.1). Such an upper bound can be derived from a variety of theories. The technique of Ref. [1] is based on the concentration-of-measure theory that postulates that "fluctuations" of a function are more-or-less concentrated about the mean value of the function. This assumption is taken advantage of to derive the upper bound of probability U_P.

The technique is evaluated because of the great potential that it offers. As alluded to, the upper bound is by construction, valid irrespective of the probability laws of the variables (D; C). This is an attractive property in cases where the datasets analyzed are so sparse that probabilities cannot be inferred reliably. It also is useful in situations where sampling the function is computationally expensive. The examples discussed herein evaluate the upper bounds when applied to nonlinear functions that are defined with two or three variables and fluctuate significantly within their design spaces. Nonlinearity, moderate dimensionality, and large fluctuations characterize the class of function properties of interest for engineering applications. The material modeling discussed in Ref. [2], for example, involves 7 parameters while the simulations of Refs. [3–5] are defined with 20 variables, approximately. Higher-dimensional problems are, at this time, considered to be out-of-reach for practical applications.

Unfortunately, the upper probability bounds derived using concentration-of-measure inequalities do not live up to their promises, at least not for the examples considered in this work. The main two observations are that: first, the upper bounds are too broad to be useful for sensible decision-making; and second, the computational cost of their estimation is prohibitive even for the relatively low-dimensional problems considered. Furthermore, it is observed that the true-but-unknown failure probability P_F can exceed the upper bound U_P if its estimation relies on too few samples or iterations, which can lead to a dangerous, false sense of confidence.

After briefly overviewing the derivation of upper probability bounds in Sect. 1.2, two examples are presented. Section 1.3 discusses a "toy" problem where the upper bounds are estimated for the Rosenbrock function in two dimensions: a polynomial whose numerical optimization is known to be challenging. The discussion includes a statistical convergence study of U_P that underscores the danger of "under-sampling." The second application, discussed in Sect. 1.4, analyzes a more realistic engineering problem that involves a high-fidelity finite element simulation of structural vibration response. Section 1.5 concludes the discussion and offers recommendations.

1.2 Concentration-of-Measure Inequalities of Probability Bounds

This section gives a brief summary of the derivations presented in Refs. [1,6] to establish the upper bound of a probability-of-failure. The approach is based on concentration-of-measure inequalities [7]. We do not attempt to re-derive these results and focus instead on the equations needed to understand the numerical implementations and applications of Sects. 1.3 and 1.4.

1.2.1 Upper Probability Bounds Based on Concentration-of-Measure Inequalities

The proposal of Ref. [1] for uncertainty quantification and system certification is that the fluctuation of a function is more-or-less concentrated about its mean value. The extent of this concentration is controlled by a "diameter" quantity D_F that can be estimated experimentally or numerically. This concept of concentration is exploited to bound the probability that a function exceeds its expected (or mean) behavior. Here, the function represents a complex engineered system whose performance we wish to evaluate. The terminology "deviation from the mean," likewise, indicates a variation away from the nominal performance of the system. The probability of exceeding a given fluctuation represents the failure probability of the system.

The main result that is illustrated with several applications is that the probability of observing a fluctuation of a given magnitude away from the mean value is bounded according to:

$$\mathrm{Prob}\left[F(X) - \mu \geq \delta\right] \leq e^{-2\left(\frac{\delta}{D_F}\right)^2}, \tag{1.4}$$

where $F(\bullet)$ is a real-valued function that depends on N input variables $X = (X_1; X_2; \ldots X_N)$ and δ is the level of fluctuation away from the mean value, denoted by μ, of the function:

$$\mu \equiv E[F] = \int_{X \in ([-1;+1])^N} F(X) \cdot dX. \tag{1.5}$$

Equation (1.5) implies a multi-dimensional integration in the space $\Re^N$. Without loss of generality it is assumed that the variables X_k are scaled in the interval $-1 \leq X_k \leq +1$. The symbol D_F of Eq. (1.4) denotes the McDiarmid diameter of the function that controls the extent to which fluctuations are concentrated (or not) about the mean value. To derive an upper bound of probability at any level of fluctuation δ, it suffices to know the diameter D_F that is formally defined as:

$$D_F^2 = \sum_{1 \leq k \leq N} D_k^2, \tag{1.6}$$

where D_k is a partial diameter that defines the influence on function values of the kth input variable X_k:

$$D_k = \sup_{\{X, X^* \in U^{(k)}\}} \left(F(X) - F(X^*)\right). \tag{1.7}$$

Equation (1.7) seeks the maximum range of function $F(\bullet)$ over a specific space $U^{(k)}$. This space, within which the optimization is performed, is the subspace of the hyper-cube domain $([-1;+1])^N$ where the kth input variable X_k is constrained to be constant:

$$U^{(k)} = \left\{X \in ([-1;+1])^N \text{ such that } X_k = \text{constant}\right\}. \tag{1.8}$$

Note that keeping the kth input variable constant and equal to a known value leads to a partial diameter D_k that becomes conditioned on X_k. This is not what is meant in Eqs. (1.7) and (1.8). Instead, the optimization is defined in the N-dimensional space $([-1;+1])^N$ while enforcing the constraint that X_k remains constant and equal to an ***unknown*** value. This unknown value of X_k must also be optimized while solving for Eq. (1.7).

1.2.2 Numerical Implementation of the Technique

Often, an engineered system is designed such that its performance does not deviate "too much" from the nominal (or mean) target performance. "Failure" can therefore be conceptually defined as exceeding a fluctuation of response that would be detrimental to the required performance. The corresponding probability-of-failure is:

$$P_F = \text{Prob}\left[F(X) - \mu \geq \delta\right], \tag{1.9}$$

and the reliability of the engineered system is defined, as before, by Eq. (1.2).

Equations (1.4) through (1.6) define a rather simple procedure to derive an upper bound of reliability. It rests on the ability to evaluate, either experimentally or numerically, the McDiarmid diameter. Equation (1.6) indicates that D_F accumulates the partial diameters D_k in quadrature; the simplicity of this definition, however, is deceiving.

Estimating a partial diameter D_k can be viewed, in fact, as a two-level, nested optimization. The innermost optimization optimizes the kth variable while the other variables are kept constant and equal to fixed values, that is, $X_1 = \lambda_1$, $X_2 = \lambda_2, \ldots X_{k-1} = \lambda_{k-1}$, $X_{k+1} = \lambda_{k+1}, \ldots X_N = \lambda_N$. The variable X_k is optimized to search for the maximum range $F_k^{(Max)}(\lambda_{(-k)}) - F_k^{(Min)}(\lambda_{(-k)})$ shown below, where the symbol $\lambda_{(-k)}$ denotes the vector of $\Re^{N-1}$ defined as $\lambda_{(-k)} = (\lambda_1; \lambda_2; \ldots \lambda_{k-1}; \lambda_{k+1}; \ldots; \lambda_N)$. (The subscript "-k" is used to indicate that all variables are included, except the kth one.) This first step is embedded in an outer optimization that solves for the partial diameter D_k by optimizing the $(N-1)$ variables of vector $\lambda_{(-k)}$. These two steps must be repeated for each variable.

The numerical implementation of the nested optimizations (1.7) and (1.8) can be written as:

$$D_k = \max_{\{\lambda_{(-k)} \in ([-1;+1])^{N-1}\}} F_k^{(Max)}(\lambda_{(-k)}) - F_k^{(Min)}(\lambda_{(-k)}), \tag{1.10}$$

where

$$F_k^{(Min)}(\lambda_{(-k)}) = \min_{\{-1 \leq X_k \leq +1\}} F(\lambda_1; \lambda_2; \cdots \lambda_{k-1}; X_k; \lambda_{k+1}; \cdots \lambda_N), \tag{1.11}$$

and

$$F_k^{(Max)}(\lambda_{(-k)}) = \max_{\{-1 \leq X_k \leq +1\}} F(\lambda_1; \lambda_2; \cdots \lambda_{k-1}; X_k; \lambda_{k+1}; \cdots \lambda_N). \tag{1.12}$$

It can be verified that substituting the minimum (1.11) and maximum (1.12) in Eq. (1.10) is equivalent to searching for the maximum range, written as "sup(F(X) – F(X*))" in Eq. (1.7). Equations (1.10) through (1.12) reveal that the calculation of a partial diameter D_k requires the resolution of three optimization problems over different subspaces of $([-1;+1])^N$. This is a rather expensive proposition since the resolution of these optimizations must be repeated for every variable $X_1, X_2, \ldots, X_N$ of the problem.

Three strategies are implemented to estimate the McDiarmid diameter. The first two algorithms, applied in Sect. 1.3 to the two-dimensional Rosenbrock polynomial function and in Sect. 1.4 to the three-dimensional structural vibration problem, rely on the formulation of Eqs. (1.10) through (1.12).

In the first strategy, the functions $F_k^{(Min)}(\bullet)$ and $F_k^{(Max)}(\bullet)$ are explored for several values of $\lambda_{(-k)}$ by solving the minimization and maximization problems of Eqs. (1.11) and (1.12). After enough "samples" have been obtained, multi-dimensional emulators are best-fitted to the pairs $(\lambda_{(-k)}; F_k^{(Max)}(\lambda_{(-k)}))$ and $(\lambda_{(-k)}; F_k^{(Min)}(\lambda_{(-k)}))$. The emulators are defined as polynomial models in subspace $([-1;+1])^{N-1}$. Their orders are kept as low as possible, provided that the goodness-of-fit is acceptable, to avoid over-fitting the training data. These fast-running emulators are used to solve the optimization of Eq. (1.10). The procedure is then repeated for every variable X_k of the problem.

The second strategy adopts a nested implementation of Genetic Algorithms (GAs) to optimize Eq. (1.10). The GA-based implementation provided by the Global Optimization Toolbox in MATLAB™ [8] is employed, where it is noted that the default settings with respect to selection, crossover, and mutation functions are used for the analyses discussed herein. As GAs are well-accepted in the numerical optimization community, only the details relevant to the nested implementation are provided. First, the "outer" GA initiates a population of individuals. Each of these individuals is then passed to an "inner" GA that searches the space associated with the kth parameter to determine the maximum range of the function, while keeping the $\lambda_{(-k)}$ variables ***fixed***. That is, each execution of the inner GA seeks the maximum range for a fixed subset of $\lambda_{(-k)}$ variables, instead solving Eqs. (1.11) and (1.12) separately. After all individuals from the outer GA are evaluated in this manner, they are ranked, selected, etc., and the procedure repeats for each population of individuals until termination of the outer GA.

The third approach implemented for comparison with the first two consists of relying on a large number of Monte Carlo samples, then, searching for the minimum and maximum values of functions $F_k^{(Min)}(\bullet)$ and $F_k^{(Max)}(\bullet)$. The implementation of this sampling strategy is as follows. First, a predefined number of samples M, are generated from the space $([-1;+1])^N$. This produces M vectors of randomly sampled variables, denoted as the "unprimed" samples. Next, for each of the M vectors, ***only*** the kth variable is re-sampled; these new vectors are denoted as the "primed" samples. The final step is then to evaluate the unprimed and primed samples and store the absolute value of the difference between the mth unprimed and the mth primed evaluation (where $m = 1, 2, \ldots, M$). The maximum difference found among the M samples is then designated as the partial diameter D_k. In this way, the two-step optimization is being condensed into a single step, whereby the $\lambda_{(-k)}$ variables are held constant for two random (i.e. unknown) values of the kth variable. The quality of the estimation rests entirely on whether or not the two random values of the kth variable generate the minimum ***and*** maximum values of functions $F_k^{(Min)}(\bullet)$ and $F_k^{(Max)}(\bullet)$, for the "correct" values of the $\lambda_{(-k)}$ variables. Lastly, in order to assess the convergence properties of the sampling strategy, the above procedure is repeated ten times.

1.3 Application to the 2D Rosenbrock Function

The first application of the concentration-of-measure inequality-based reliability is to a two-dimensional polynomial function. It represents a "toy" problem that does not reflect the practical difficulties of real engineered applications. This problem, nevertheless, offers the advantage of simplicity. It also allows for the assessment of convergence properties, discussed in this section.

1.3.1 The Two-Dimensional Rosenbrock Function

The Rosenbrock function is a simple, analytic polynomial whose numerical evaluation is trivial. The results discussed are obtained with an implementation developed with the software MATLAB™ [8]. After estimating the McDiarmid diameter described by Eq. (1.6), the upper bound in Eq. (1.4) is calculated where "failure" is defined as the probability that the Rosenbrock function deviates from the mean value defined by Eq. (1.5) by more than a given fluctuation δ. The results are presented as curves that show the upper probability bounds as a function of δ.

The Rosenbrock function in the N-dimensional space $(X_1; X_2; \ldots X_N)$ is defined analytically as:

$$Y = \sum_{k=1\cdots(N-1)} (1 - X_k)^2 + 100 \cdot \left(X_{k+1} - X_k^2\right)^2, \tag{1.13}$$

where each variable X_k is scaled in $[-1; +1]$. In two dimensions, the function becomes:

$$Y = (1 - X_1)^2 + 100 \cdot \left(X_2 - X_1^2\right)^2. \tag{1.14}$$

The Rosenbrock function is a benchmark test commonly encountered in numerical optimization because it varies rapidly and defines "valleys" [9]. This makes it challenging to search for a global minimum using gradient-based optimization algorithms. Another reason for selecting this function is the existence of previous work documented in Refs. [10, 11]. Reference [12] presents results obtained with a variant defined in 17 dimensions.

Figure 1.2 illustrates the Rosenbrock function in two dimensions. The formation of a "valley," that makes numerical optimization difficult, can be observed to be centered at the coordinates $(X_1; X_2) = (0;0)$. The valley bifurcates into two branches: one of them progresses towards the corner $(X_1; X_2) = (+1;-1)$ while the other branch extends towards $(X_1; X_2) = (+1;+1)$. It can be verified that the global minimum of the Rosenbrock function is reached when $X_k = +1$ for all variables, whether in 2D or for an arbitrary number of dimensions, which is the only point where the function is equal to zero. In two dimensions, the maximum of the function over $([-1;+1])^2$ is equal to 404.

Another theoretical result [7] combines the minimum and maximum values to bound the McDiarmid diameter as:

$$\frac{1}{\sqrt{N}} \leq \frac{D_F}{\max\limits_{\{X\in([-1;+1])^N\}} F(X) - \min\limits_{\{X\in([-1;+1])^N\}} F(X)} \leq \sqrt{N}. \tag{1.15}$$

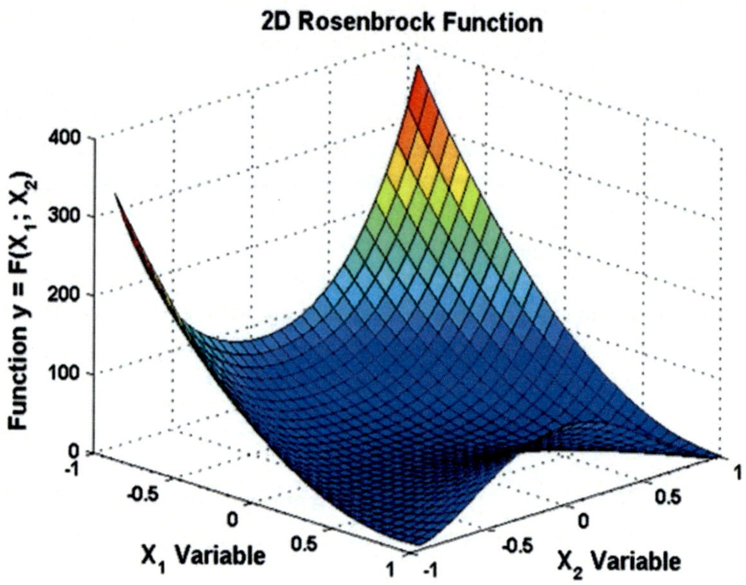

Fig. 1.2 Illustration of the Rosenbrock function in 2D space

The numerical application of Eq. (1.15) indicates that the McDiarmid diameter is bounded in the interval $285.67 \leq D_F \leq 571.34$. These bounds, although they are somewhat "loose," can be used to assess the quality of estimations obtained with different numerical implementations. The next statistics useful for the analysis are the mean and standard deviation. In 2D, the mean value (μ_Y) and standard deviation (σ_Y) are obtained analytically by integrating the function over the hyper-cube space $([-1;+1])^2$. The intermediate derivations for these statistics are omitted for brevity and instead given as:

$$\mu_Y = \frac{1}{\text{Volume}\left([-1;+1]^2\right)} \cdot \int_{-1}^{+1} \int_{-1}^{+1} (1-X_1)^2 + 100 \cdot \left(X_2 - X_1^2\right)^2 dX_1 dX_2$$
$$\approx 54.667, \tag{1.16}$$

and

$$\sigma_Y = \sqrt{\frac{1}{\text{Volume}\left([-1;+1]^2\right)} \cdot \int_{-1}^{+1} \int_{-1}^{+1} \left((1-X_1)^2 + 100 \cdot \left(X_2 - X_1^2\right)^2 - \mu_Y\right)^2 dX_1 dX_2}$$
$$\approx 65.433. \tag{1.17}$$

It is noted that the standard deviation of the 2D Rosenbrock function is large relative to its mean value, due to the presence of corners where values of the polynomial become large (see Fig. 1.2). This is a property encountered in other engineering applications but is not important to justify the method, as indicated in Ref. [12] where it does not hold in the case of the 17D Rosenbrock variant. The mean and standard deviation values, μ_Y and σ_Y, are used to guide the selection of appropriate function fluctuations δ in the analysis that follows.

1.3.2 *Estimation of the McDiarmid Diameter and Upper Probability Bounds*

The analysis begins by estimating the true-but-unknown failure probabilities using statistical sampling. Collecting this information is, here, possible because the Rosenbrock function can be evaluated numerically in a fraction of second. These highly accurate estimations of probability are used for comparison with the McDiarmid-based upper bounds of Eq. (1.4).

Figure 1.3 illustrates the empirical histogram of Rosenbrock function values. It is obtained from ten million (10 M) uniform and uncorrelated samples of variables (X_1; X_2) in hyper-cube $([-1;+1])^2$. Analyzing 10 M samples suffices to yield statistically converged estimates, an assertion that is verified by repeating the estimation ten times and observing that the statistical estimates do not change in their fourth significant digit from one trial to the next. The resulting statistics are

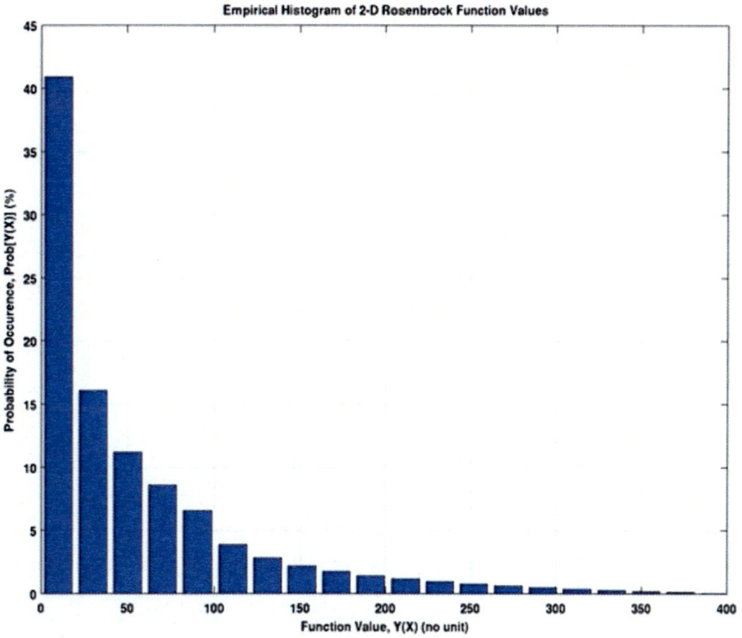

Fig. 1.3 Histogram of values sampled for the 2D Rosenbrock function

Table 1.1 Statistics of the 2D Rosenbrock function estimated with 10 M samples

Definition	Symbol	Estimate	"Exact" value
Minimum value	$Y^{(Min)}$	0.00	0.00
Maximum value	$Y^{(Max)}$	404.00	404.00
Total range	$Y^{(Max)} - Y^{(Min)}$	404.00	404.00
Mean value	μ_Y	54.67	54.67
Standard deviation value	σ_Y	65.43	65.43

listed in Table 1.1, where they are compared to the exact values presented in Sect. 1.3.1. The "exact" failure probabilities are estimated from these samples in a frequentist manner, simply by counting how many values exceed a given level of fluctuation δ relative to the mean value μ_Y.

Next, the conditional functions $F_k^{(Min)}(\bullet)$ and $F_k^{(Max)}(\bullet)$ of Eqs. (1.11) and (1.12) are estimated numerically. Because the function is defined in two dimensions, the symbol $\lambda_{(-k)}$ that denotes the combination of variables kept constant during the optimization reduces to a single variable. The conditional functions $F_1^{(Min)}(\bullet)$ and $F_1^{(Max)}(\bullet)$ represent the minimum and maximum of the Rosenbrock function, respectively, as variable X_2 is kept constant and fixed to a known value. Likewise, $F_2^{(Min)}(\bullet)$ and $F_2^{(Max)}(\bullet)$ are the minimum and maximum as variable X_1 is kept constant. A total of 50,000 equally-spaced samples are selected in the interval $[-1; +1]$ to generate these constant values "X_2 = constant" (for $F_1^{(Min)}$ and $F_1^{(Max)}$) or "X_1 = constant" (for $F_2^{(Min)}$ and $F_2^{(Max)}$).

Each time that one of these fixed values is selected for one of the variables, the other variable of the function is optimized to search for $F_k^{(Min)}$ and $F_k^{(Max)}$. It is emphasized that these are simple, one-dimensional constrained optimization problems. They must, however, be repeated twice to search for the minimum and maximum values for each "sample" considered (which here, is 50,000 evaluations). Once $F_k^{(Min)}$ and $F_k^{(Max)}$ have been evaluated numerically a sufficient number of times over the design space, the partial function range $G_k(\bullet)$ is defined as:

$$G_k(\lambda_{(-k)}) = F_k^{(Max)}(\lambda_{(-k)}) - F_k^{(Min)}(\lambda_{(-k)}). \tag{1.18}$$

Because the partial function range $G_k(\bullet)$ depends on a single variable, its functional dependency on $X_{(-k)}$ can be suggested graphically. The run numbers are sorted in ascending magnitudes of $X_{(-k)}$ and the resulting values are shown in Fig. 1.4a, b for $G_1(\bullet)$ and $G_2(\bullet)$, respectively. Figure 1.4a shows that the partial function range $G_1(X_2)$ is smoothly varying, with the exception of a sharp discontinuity at $X_2 = 0.19$. Figure 1.4b suggests a quadratic functional form for $G_2(X_1)$. Going back to Eq. (1.10), it can be inferred that the partial diameter D_k is simply the maximum of function $G_k(\bullet)$, which can be read directly from the figure. Of course, higher-dimensional problems necessitate a numerical optimization to estimate D_k from the function $G_k(\bullet)$.

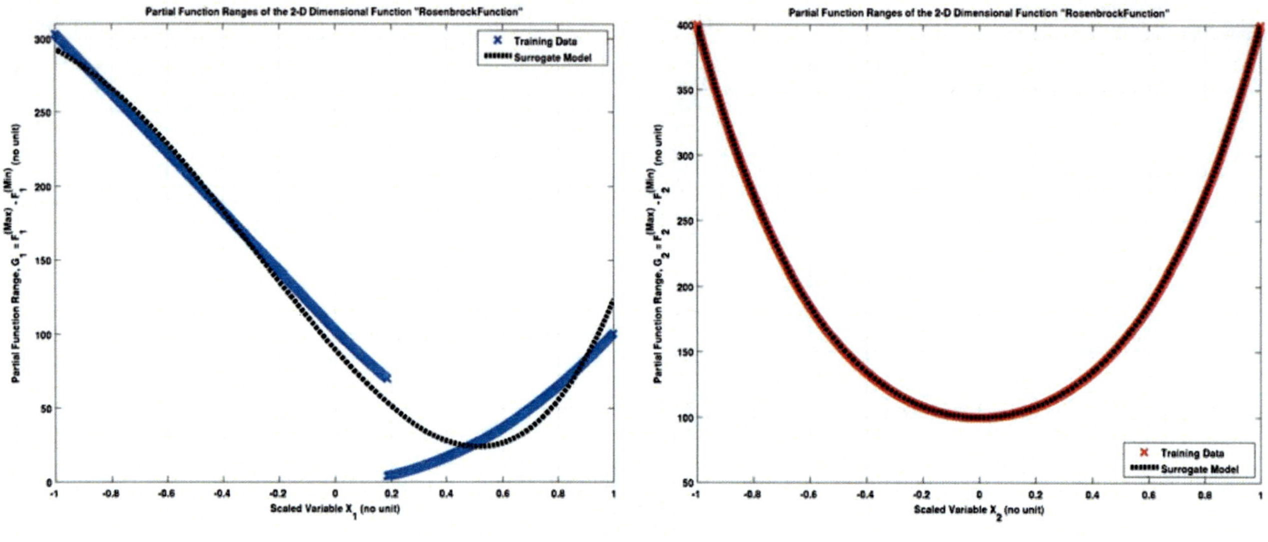

Fig. 1.4 Sorted training data and polynomial emulators. **(a)** *Left*: partial function range $G_1(X_2)$ versus X_2. **(b)** *Right*: partial function range $G_2(X_1)$ versus X_1

Table 1.2 "Exact" and approximate values of the McDiarmid partial diameters, D_k

	McDiarmid partial diameter, D_k (no unit)	
Input variable, X_k	"Exact" value of the diameter	Optimization-based estimation
D_1	303.00	123.79
D_2	400.00	400.00
D_F	501.80	418.72

The next step is to develop a low-order polynomial emulator from the training data accumulated for each partial function range $G_k(\bullet)$. This step is, here, not needed since the maximum values D_k can be read directly from Fig. 1.4a, b; the polynomial emulators are nevertheless developed for consistency when solving higher dimensional problems as discussed in [12]. Fourth-order polynomials are best-fitted to the 50,000 evaluations of functions $G_1(\bullet)$ and $G_2(\bullet)$. The goodness-of-fit obtained for $G_1(\bullet)$ is 8.60% Mean Square Error (MSE). Given that the function exhibits a discontinuity, this level of accuracy is deemed appropriate. The goodness-of-fit of the fourth-order polynomial developed to emulate $G_2(\bullet)$ is $5.29 \times 10^{-7}\%$ MSE. Figure 1.4a, b visually compare the polynomial emulators to the training data. Clearly, the discontinuity of function $G_1(\bullet)$ in Fig. 1.4a cannot be approximated by a polynomial, irrespective of its order. What matters most, however, is that the maximum of the data be represented appropriately because it is what the optimization is after.

The surrogate models shown in Fig. 1.4a, b are optimized to search for the maxima, which are the partial diameters D_k. Results of the optimization are tabulated in Table 1.2 where they are compared to the exact solutions. It is observed that the first partial diameter D_1 is severely under-estimated; the second optimization converges to the correct value for D_2. The McDiarmid diameter D_F is then calculated from Eq. (1.6) as:

$$D_F = \sqrt{D_1^2 + D_2^2} = \sqrt{(123.79)^2 + (400.00)^2} = 418.72. \tag{1.19}$$

This estimate complies with the interval $285.67 \leq D_F \leq 571.34$ of Eq. (1.15). Note that for the remainder of this discussion, the value $D_F = 418.72$ is used, as opposed to the correct answer listed in Table 1.2.

The upper bounds of failure probability defined in Eq. (1.4) are obtained for different levels of fluctuation δ using the mean statistic of Eq. (1.16) and McDiarmid diameter of Eq. (1.19). Figure 1.5 compares the upper bounds to the true-but-unknown probabilities estimated from 10 M uniform and uncorrelated Monte Carlo samples.

Levels of fluctuation ranging from 0 to 202 are used to display the results in Fig. 1.5. A fluctuation of $\delta = 202$ represents three standard deviations away from the mean value of the Rosenbrock function. It can be observed that, irrespective of the magnitude of fluctuation δ used, the upper bounds are too conservative to be useful for sensible decision-making. This is expected near the mean value, $\mu_Y = 54.67$, where the concentration-of-measure theory may not apply. But it should not be the case in the "tail" of the distribution. For example, the upper bound is $U_P = 61.6\%$ at the last level of fluctuation, $\delta = 202$, while the true probability is only $P_F = 2.2\%$.

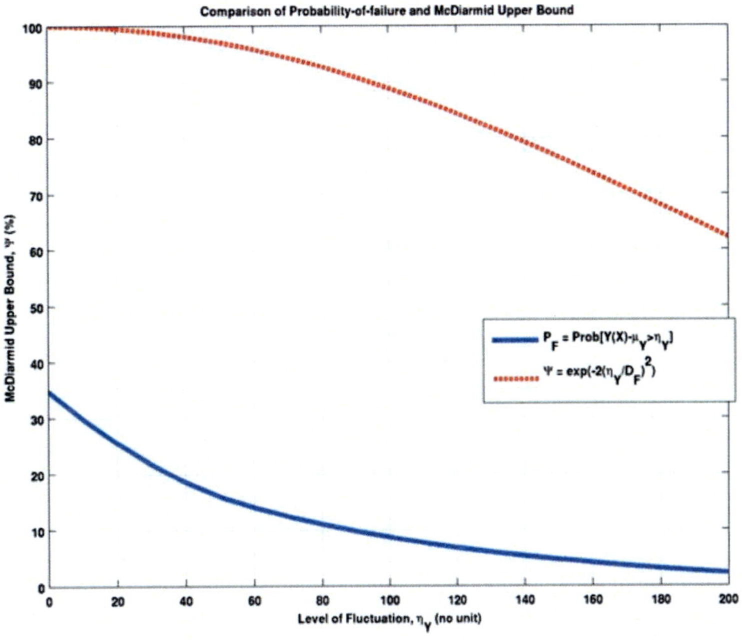

Fig. 1.5 Upper bounds U_P and estimated probabilities P_F of the 2D Rosenbrock function

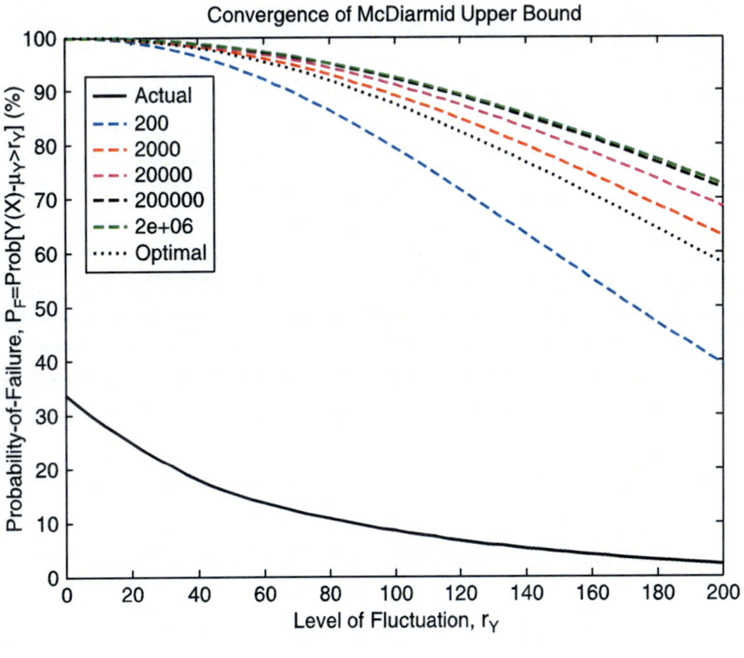

Fig. 1.6 Sampling-based upper probability bounds U_P of the 2D Rosenbrock function

1.3.3 Convergence Behavior of Sampling-Based Estimates

As mentioned at the end of Sect. 1.2.2, two other strategies are available to estimate the partial diameters D_k. The second approach is based on the nested GA optimization described previously and the third approach is based on sampling the hyper-space $([-1;+1])^N$, assuming uniform and uncorrelated probability distributions for the variables X_k, to search for the minimum and maximum functions $F_k^{(Min)}(\bullet)$ and $F_k^{(Max)}(\bullet)$. A comparison with the optimization-based results reported Fig. 1.5 is briefly discussed next.

Figure 1.6 compares the upper bounds U_P of the GA optimization-based solution (dotted line) to five sampling-based estimations obtained with different numbers of Monte Carlo runs (dashed lines). Including the replicates, the sizes of these sampling-based simulations (i.e. $10 \times M$) vary from $M = 200$ to $M = 2$ million samples, where it is noted that Fig. 1.6 reflects

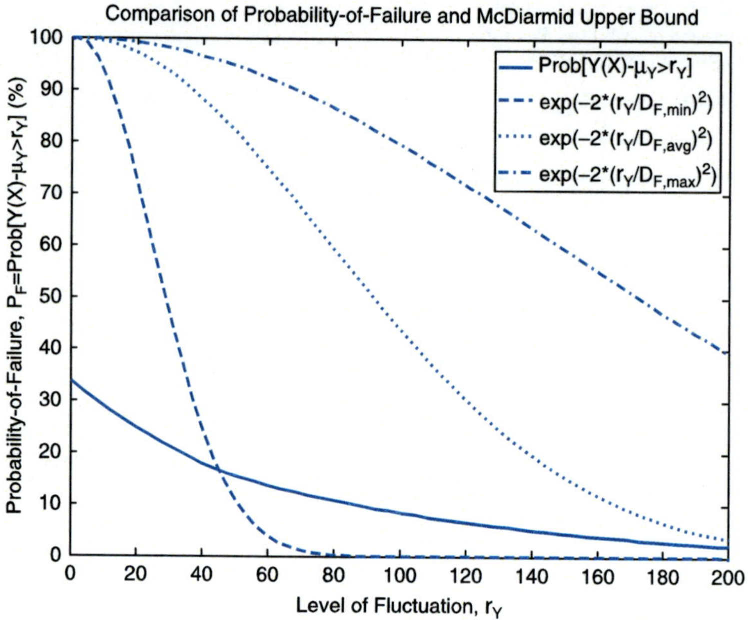

Fig. 1.7 Statistics of sampling-based upper probability bounds U_P obtained with 200 runs only

the "best" sampling-based estimation of the upper bound U_P (i.e. the maximum McDiarmid diameter) realized from the ten replicates. The "true" failure probabilities P_F, estimated via Monte Carlo sampling, are also included for reference (solid, black line).

Comparing Figs. 1.5 and 1.6, it is seen that both optimization-based implementations yield very nearly the same upper bounds. However, a somewhat counterintuitive result is also seen in that increasing the number of samples produces upper bounds that converge to a solution that does ***not*** correspond to either of the optimization-based solutions. This is almost entirely due to the underestimation of the first partial diameter D_1, which is a common feature of both optimization-based implementations.

A second observation that is somewhat more disconcerting is that the upper bounds obtained through sampling can be smaller than the values provided by the optimization-based solver. This behavior opens the door to the possibility of estimating a value U_P that would ***not*** be an upper limit, therefore violating Eq. (1.3). Figure 1.7 illustrates that this dangerous situation can manifest itself when only a 200 samples are used, where it is seen that the "worst" sampling-based estimation of the upper bound (i.e. the minimum McDiarmid diameter) falls ***below*** the estimate of true probability-of-failure. This is an unfortunate observation as it suggests that analyzing a small number of samples may not be a viable alternative to an optimization-based solution.

1.4 Application to the High-Fidelity Model of a Three-Story Frame

In this section, the upper probability bounds are applied to an engineering application. The example is the linear vibration of a three-story frame structure simulated numerically with a high-fidelity Finite Element (FE) model. The number of variables is not significantly different from the dimensionality of the Rosenbrock function discussed previously. The main difficulty of this application is to estimate the McDiarmid-based inequality, given that the FE model is computationally expensive.

1.4.1 Description of the Three-Story Frame Structure

The three-story structure of Fig. 1.8a is the system employed in this study. It consists of aluminum columns and plates assembled using bolted joints. Each floor plate measures 30.5 × 30.5 × 2.5 cm, and is connected to the floors above and/or below by four aluminum columns measuring 17.7 × 2.5 × 0.6 cm. Each connection comprises four bolts that act to sandwich the end of the column between the floor plate and a 2.5 × 2.5 × 1.3 cm end cap. The structure slides on rails that permit movement in one direction only, as shown in Fig. 1.8b, which is referred to as the "weak" bending direction of the columns.

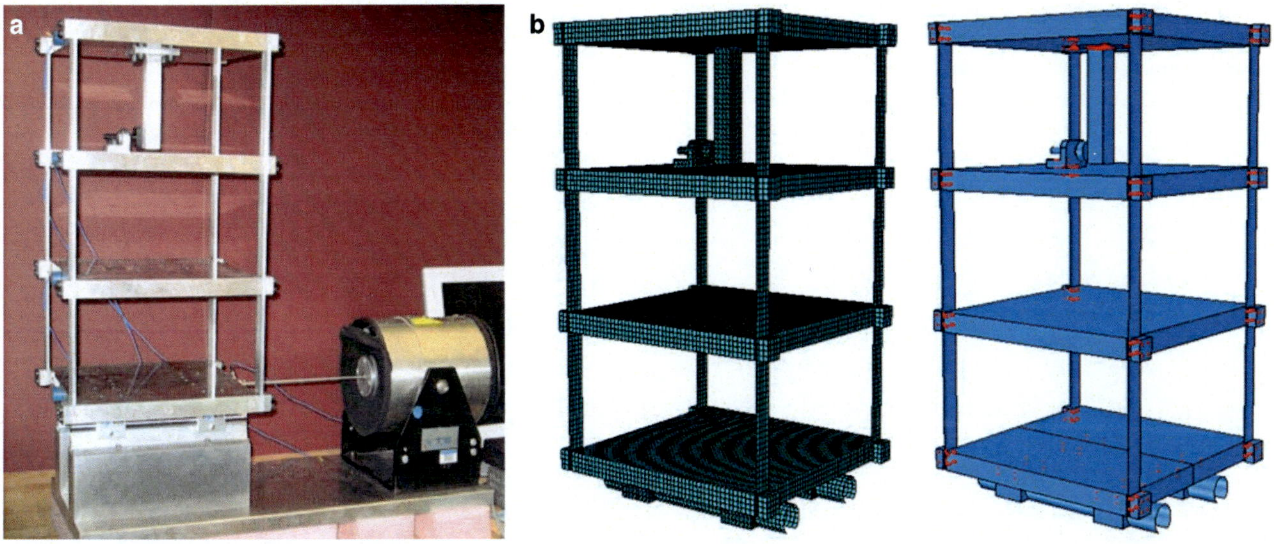

Fig. 1.8 High-fidelity finite element model of the three-story frame structure. (**a**) *Left*: frame structure. (**b**) *Right*: finite element discretization

Table 1.3 Definition of uncertainty variables of the frame structure

Variable, X_k	Definition of variable
X_1	Elastic modulus of all Aluminum-based components
X_2	Mass density of all Aluminum-based components
X_3	Bolt "radius of influence" for end cap connections

Additionally, a small column is suspended from the third floor, and a bumper mechanism is attached to the second floor. Depending upon the amplitude of the force imparted by the shaker and the gap between the suspended column and bumper, a contact nonlinearity may be introduced into the system. This capability is not exercised here.

Figure 1.8b illustrates the highly refined FE model developed using the commercially available software Abaqus 6.10-1 [13]. All floors and end caps are modeled with linear, hexagonal continuum elements (Abaqus C3D8R). The bottom supports, suspended column, and bumper assembly are modeled with modified quadratic, tetrahedral continuum elements (Abaqus C3D10M). The latter element is formulated to properly account for pressures associated with contact, an important point to consider as all contact is modeled explicitly in the model. The vertical columns spanning between the base floor and the other levels are modeled with shell elements (Abaqus S4R), due to their enhanced capability to capture bending behavior, which is the dominant action experienced by those members.

Figure 1.8b indicates the high geometrical fidelity with which the three-story frame structure is discretized. The mechanics are also modeled with high fidelity since all sources of contact and friction are represented explicitly. A mesh convergence study, omitted herein for brevity, provides a mesh seed of 3 mm, which is slightly less than one-eighth of an inch. The corresponding discretization, shown in Fig. 1.8b, yields predictions of the resonant frequencies that are "converged-enough" relative to the overall level of experimental variability. Details on the modeling and analysis of the frame structure are available from Ref. [14].

1.4.2 Upper Probability Bounds of the Three-Story Frame Structure

Table 1.3 presents the uncertain variables of the 3D FE model. The first two variables are straightforward in their meaning, whereas the third variable warrants explanation. It denotes the "radius of influence" associated with the bolts that connect the columns to the floors by way of the end caps. These radii are identified as red dots in Fig. 1.8b. The effect of this variable is somewhat analogous to bolt torque. Varying the value changes the rigidity of the column-to-floor connection. That is, at small values of the radius, the column-to-floor connections behave more like pinned connections (free to rotate) whereas for larger values, the beam-to-floor connections behave more like fixed connections. At very large values, the columns begin to act as rigid structural members.

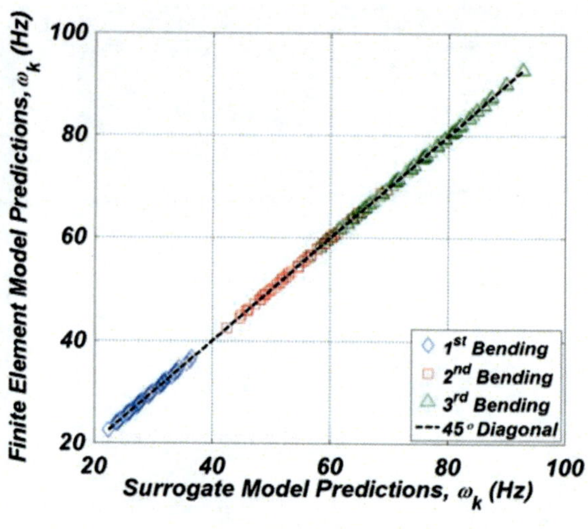

Fig. 1.9 Goodness-of-fit of bending frequencies predicted by the surrogate models

The question illustrated in this application is: given the modeling uncertainty embodied by the variables of Table 1.3, what is the probability P_F that a bending frequency of the structure deviates from the mean frequency by more than, say, three Hertz? This question can be formulated as a reliability problem, where the probability of predicting a frequency is integrated for all frequency values that exceed the mean statistic by, at least, 3 Hz. This true-but-unknown value of probability P_F is estimated, first, with Monte Carlo sampling. The question is then addressed using the McDiarmid-based inequality of Eq. (1.4). Comparing the upper bound U_P to the true-but-unknown probability P_F assesses the usefulness of this approach for decision-making.

Compared to the Rosenbrock function, the 3D FE model requires a significant computational demand to calculate the low-frequency bending modes. To address this difficulty, a design-of-experiments is populated to generate the training datasets needed to develop fast-running, polynomial emulators. These polynomial surrogates are defined as quadratic models:

$$Y = \beta_0+\beta_1X_1+\beta_2X_2+\beta_3X_3+\beta_{12}X_1X_2+\beta_{13}X_1X_3+\beta_{23}X_2X_3+\beta_{11}X_1^2+\beta_{22}X_2^2+\beta_{33}X_3^2, \tag{1.20}$$

and yield the predictions of resonant frequency presented in Fig. 1.9. The figure plots the bending frequencies predicted by the 3D FE model as a function of those predicted by the polynomial surrogates of Eq. (1.20). The fact that the points shown are aligned with the 45°-angle diagonal indicates that the polynomial equations emulate well the computationally expensive FE model. The high goodness-of-fit lends credence to employing these surrogates. The optimization-based and sampling-based evaluations of the McDiarmid diameter of Eqs. (1.6) and (1.10) through (1.12) are performed using the polynomial surrogates of Eq. (1.20).

Figure 1.10 shows the empirical distribution of first bending frequencies predicted using Monte Carlo sampling. This histogram is obtained by analyzing 10 M samples, replicated ten times. The mean and standard deviation statistics are $\mu_Y 29.38$ Hz and $\sigma_Y = 2.51$ Hz, which gives 8.54 % variability relative to the mean. The statistics of minimum and maximum function values are equal to 22.43 and 36.46 Hz, respectively. Using Eq. (1.15), the McDiarmid diameter should fall within the bounds as:

$$\frac{1}{\sqrt{3}} \leq \frac{D_F}{36.46 - 22.43} \leq \sqrt{3}, \tag{1.21}$$

or $8.10 \leq D_F \leq 24.30$. Table 1.4 indicates the partial diameters estimated from 10 M samples, replicated ten times. Combining definition (1.6) and the results listed in the table yields a McDiarmid diameter of $D_F = 9.67$ Hz, which "clears" the inequalities of Eq. (1.21).

Figure 1.11 depicts the upper probability bounds U_P as a function of frequency fluctuation, δ. Solutions are obtained with five sampling sizes, ranging from $M = 300$ to $M = 3$ million samples. The upper bound reached by relying on numerical optimization is compared to the sampling-based bounds. Also included for comparison are the true-but-unknown probabilities obtained by integrating the empirical histogram of Fig. 1.10 for each value of δ.

The first observation from Fig. 1.11 is that the sampling-based estimation of upper probability bounds converges toward a solution that does ***not*** correspond with the optimization-based solution. A similar behavior was noticed for the 2D Rosenbrock function discussed in Sect. 1.3. It is confirmed that, as in the latter case, the optimization procedure fails

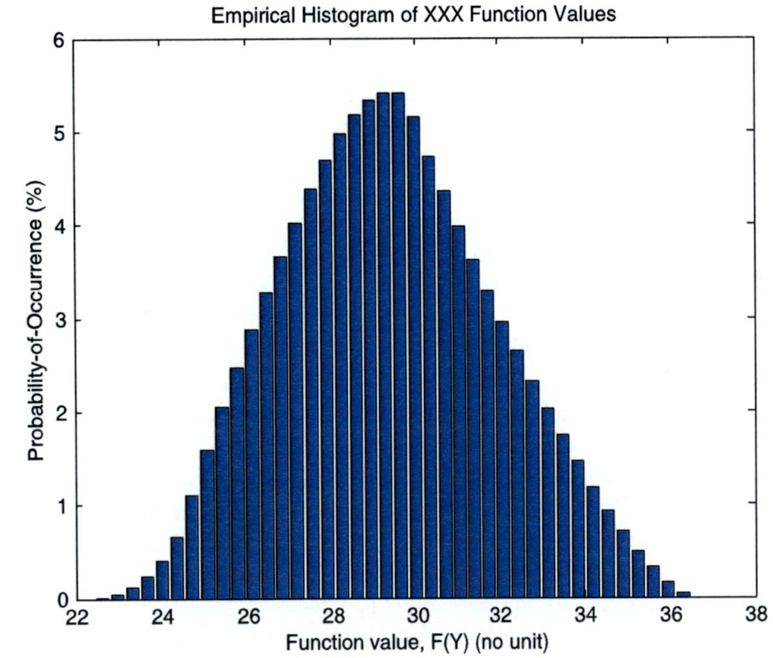

Fig. 1.10 Histogram of first bending frequencies sampled for the 3D FE model

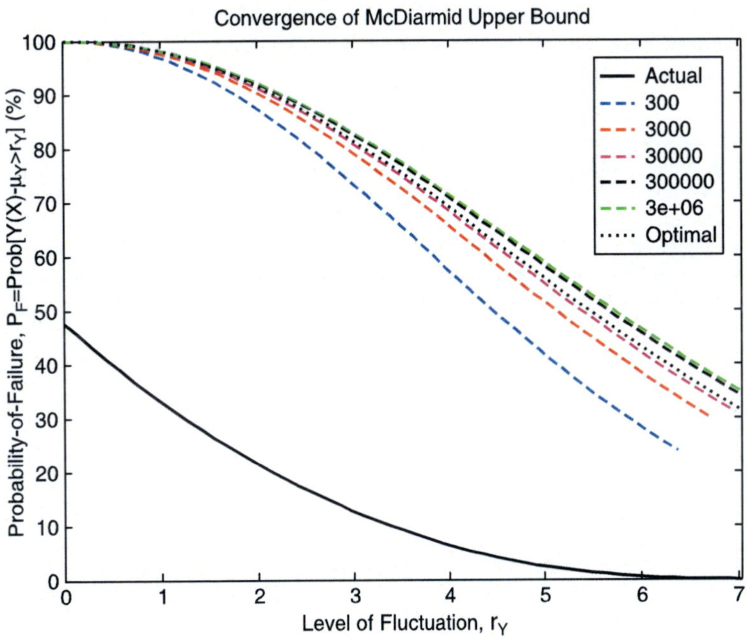

Fig. 1.11 Sampling-based upper probability bounds U_P of the 3D FE model

Table 1.4 Partial diameters D_k of the 3D FE model estimated from 10 M samples

Variable, X_k	Variable definition	McDiarmid partial diameter, D_k
X_1	Elastic modulus of Aluminum components	6.56
X_2	Mass density of Aluminum components	6.59
X_3	Bolt "radius of influence" of end connections	2.67

to produce the optimal (i.e. largest) partial diameters (as approximated in Table 1.4), thus yielding a reduced McDiarmid diameter. A second finding from Fig. 1.11, which has been consistently observed throughout this work, is the "looseness" of upper bounds U_P relative to the true-but-unknown probabilities P_F. These two observations, and those from the previous sections, make it difficult to justify using the McDiarmid inequalities to bound a failure probability.

1.5 Conclusion

A bounding theory is investigated for its ability to derive an upper limit for a probability-of-failure. The approach investigated is based on the McDiarmid concentration-of-measure theory, which postulates that fluctuations of a function are more-or-less concentrated about its mean value. This is an attractive proposition because, first, the method avoids postulating probability laws that are often unknown and, second, it substitutes efficient numerical optimization to slow-converging statistical sampling. Two examples of the application of this theory are presented. The first one involves the two-dimensional Rosenbrock function and the second example is a realistic engineering problem wherein a finite element model is employed to simulate the vibration of a three-story frame structure.

Comparisons performed between the upper bounds and highly-accurate estimates of the true-but-unknown probabilities indicate that these bounds are too "loose," or conservative, for useful decision-making. The finding applies whether fluctuations of the function are assessed near the mean value or in the tails of the statistical distributions. In addition, it is observed that estimating the McDiarmid diameter is computationally expensive, which reduces the potential benefits if the problem studied is defined in a high-dimensional space.

These observations suggest that the theory not only leads to upper bounds that are inefficient, but that can also turn out to be sub-optimal if their numerical estimation is based on too few runs of the computational model. One explanation is that the McDiarmid-based inequalities yield upper bounds that are not necessarily realized by any model of the family considered. This means that there may not be any model, included in the uncertainty space defined for analysis, whose predictions reach the upper bound U_P of the inequality "$P_F \leq U_P$."

The Center of the Predictive Science Academic Alliance Program at Caltech has recently advanced another theory to provide "tight" bounds of a probability-of-failure [15–17]. The main advantage of this novel theory over the approach investigated herein is that the upper bounds, while optimal, are guaranteed to be reached by a model within the uncertainty space defined. The recent work also comes with a numerical implementation that could yield significant computational savings. Future work will investigate this novel theory, and the accompanying implementation, to provide upper bounds of a probability-of-failure.

Acknowledgements This work is performed under the auspices of the Verification and Validation (V&V) program for Advanced Scientific Computing (ASC) at Los Alamos National Laboratory (LANL). The first author is grateful to Frederick J. Wysocki, V&V program manager at LANL, for his continuing support. Funding for the second author was sponsored by the U.S. Department of Energy, Nuclear Energy Division, Advanced Modeling and Simulation Office (NE-71), Nuclear Energy Advanced Modeling and Simulation (NEAMS) Program, Verification, Validation and Uncertainty Quantification (VU) Program Element. The second author is sincerely grateful to Brian J. Williams (CCS-6) for his continued support. LANL is operated by the Los Alamos National Security, LLC for the National Nuclear Security Administration of the U.S. Department of Energy under contract DE-AC52-06NA25396.

References

1. Lucas LJ, Owhadi H, Ortiz M (2008) Rigorous verification, validation, uncertainty quantification and certification through concentration-of-measure inequalities. J Comput Methods Appl Mech Eng 197:4591–4609
2. Hemez FM, Atamturktur SH, Unal C (2009) Prediction with quantified uncertainty of temperature and rate dependent material behavior. In: 11th AIAA non-deterministic approaches conference, Palm Springs, May 4–7, 2009
3. Maupin R, Hylok J, Rutherford A, Anderson M (2005) Validation of a threaded assembly, part I: overview. In: 6th European conference on structural dynamics, Paris, Sept 5–7, 2005
4. Hylok J, Rutherford A, Maupin R, Anderson M, Groethe M (2005) Validation of a threaded assembly, part II: experiments. In: 6th European conference on structural dynamics, Paris, Sept 5–7, 2005
5. Rutherford A, Maupin R, Hylok J, Anderson M (2005) Validation of a threaded assembly, part III: validation. In: 6th European conference on structural dynamics, Paris, Sept 5–7, 2005
6. Owhadi H, Sullivan TJ, McKerns M, Ortiz M, Scovel C (2010) Optimal uncertainty quantification. Technical report 2010-03, Applied and Computational Mathematics, California Institute of Technology, Pasadena
7. McDiarmid C (1989) On the method of bounded differences. In: Siemons J (ed) Surveys in combinatorics. Volume 141 of the london mathematical society lecture. Cambridge University Press, Cambridge, pp 148–188
8. The MathWorks, Inc. (2011) MATLAB R2011a product help, Natick
9. Rosenbrock HH (1960) An automatic method for finding the greatest or least value of a function. Comput J 3:175–184
10. Hemez FM, Atamturktur SH (2011) The dangers of sparse sampling for the quantification of margin and uncertainty. Reliab Eng Syst Saf 96:1220–1231
11. Hemez FM (2010) Performance of the Morris one-at-a-time sensitivity analysis to explore large-dimensional functions. Technical report LA-UR-10-0069, Los Alamos National Laboratory, Los Alamos
12. Stull CJ, Hemez FM (2012) Optimal inequalities to bound a performance probability using McDiarmid concentration of measure theory. Technical report LA-UR-12-24769, Los Alamos National Laboratory, Los Alamos

13. Dassault Systèmes (2010) Abaqus 6.10 online documentation, Providence
14. Stull CJ, Hemez FM (2012) "On the use of info-gap decision theory to select from among models of varying complexity. Technical report LA-UR-12-0379, Los Alamos National Laboratory, Los Alamos
15. Sullivan TJ, Owhadi H (2012) Distances and diameters in concentration inequalities: from geometry to optimal assignment of sampling resources. Int J Uncertain Quantif 2(1):21–38
16. Owhadi H, Scovel C, Sullivan TJ, McKerns M, Ortiz M (2012) Optimal uncertainty quantification. SIAM Rev (submitted, Published at arXiv:1009.0679)
17. Sullivan TJ, McKerns M, Meyer D, Theil F, Owhadi H, Ortiz M (2012) Optimal uncertainty quantification for legacy data observations of Lipschitz functions. Math Model Numer Anal (submitted, Published at arXiv:1202.1928)

Chapter 2
Remaining Fatigue Life Predictions Considering Load and Model Parameters Uncertainty

Maurizio Gobbato, Joel P. Conte, and John B. Kosmatka

Abstract Fatigue-driven damage propagation is one of the most unpredictable failure mechanisms for a large variety of mechanical and structural systems subjected to cyclic and/or random operational loads during their service life. Therefore, monitoring the critical components of these systems, assessing their structural integrity, recursively predicting their remaining fatigue life (RFL), and providing a cost-efficient reliability-based inspection and maintenance (RBIM) plan are crucial tasks. In contribution to these objectives, the authors developed a comprehensive reliability-based fatigue damage prognosis methodology for recursively predicting and updating the RFL of critical structural systems and/or sub-assemblies. An overview of the proposed framework is provided in the first part of the paper. Subsequently, a set of experimental fatigue test data is used to validate the proposed methodology at the reliability component level. The proposed application example analyzes the fatigue-driven crack propagation process in a center-cracked 2024-T3 aluminum plate subjected to a sinusoidal load with random amplitude. Four probabilistic models of increasing load amplitude uncertainty together with damage evolution model parameter uncertainty and measurement uncertainty are considered in this study. The results obtained demonstrate the efficiency of the proposed framework in recursively updating and improving the RFL estimations and the benefits provided by a nearly continuous monitoring system.

Keywords Damage prognosis • Probabilistic damage detection and quantification • Bayesian inference • Load uncertainty • Remaining fatigue life

2.1 Introduction

Integrated structural health monitoring and damage prognosis (SHM-DP) methodologies, coupled with sensor-based nondestructive evaluation (NDE) techniques, are becoming fundamental engineering tools for (i) assessing the current structural integrity of several mechanical and structural systems, (ii) predicting their remaining service life, and (iii) providing cost-efficient reliability-based inspection and maintenance (RBIM) programs. As discussed earlier by the authors [1, 2], in NDE-based SHM-DP methodologies, these objectives are achieved through (i) periodic or nearly continuous NDE inspections, (ii) a rigorous probabilistic treatment of the NDE inspection results as well as all pertinent sources of measurement uncertainty, (iii) mechanics-based predictive modeling techniques, and (iv) well-established component and system reliability analysis methods. In order to obtain accurate and meaningful damage prognosis results, the NDE-based SHM-DP system must be capable of accurately detecting all damage locations and, for each of them, correctly identifying all damage mechanisms progressing in time. In this respect, fatigue-driven damage propagation is one of the most unpredictable failure mechanisms for a large variety of mechanical and structural systems subjected to cyclic and/or random operational loads during their service life [1, 2]. Therefore, monitoring the most critical sub-components of these systems, assessing their structural integrity, recursively predicting the remaining fatigue life (RFL) of the entire system, and determining a cost-efficient RBIM program are fundamental tasks. To this end, the authors developed a comprehensive reliability-based fatigue

M. Gobbato (✉)
Risk Management Solutions, Inc., 7575 Gateway Blvd., Newark, CA 94560, USA
e-mail: maurizio.gobbato@rms.com

J.P. Conte • J.B. Kosmatka
Department of Structural Engineering, University of California, San Diego, 9500 Gilman Dr., La Jolla, CA 92093-0085, USA
e-mail: jpconte@ucsd.edu; jkosmatka@ucsd.edu

T. Simmermacher et al. (eds.), *Topics in Model Validation and Uncertainty Quantification, Volume 5*,
Conference Proceedings of the Society for Experimental Mechanics Series 41,
DOI 10.1007/978-1-4614-6564-5_2, © The Society for Experimental Mechanics, Inc. 2013

damage prognosis methodology for recursively predicting and updating the RFL of monitored structures and/or structural sub-assemblies [1, 2]. The developed methodology is briefly overviewed herein in Sect. 2.2

In the second part of the paper, a set of experimental fatigue test data, available in the literature [3], is used to provide an experimental verification and validation of the proposed framework at the local reliability component level [1]. The experimental dataset consists of 68 crack propagation trajectories obtained from a series of fatigue tests performed on 68 nominally identical center-cracked 2024-T3 aluminum plates. The results obtained from this numerical application emphasize (i) the importance and the benefits of a periodic NDE monitoring system, (ii) the efficiency of the proposed recursive Bayesian updating scheme in making use of new NDE inspection results, (iii) the accuracy in predicting (after each NDE inspection) the RFL of the monitored component, and (iv) the robustness of the proposed framework against increasing levels of load uncertainty.

2.2 General Overview and Description of Proposed Framework for Remaining Fatigue Life Predictions

As can be inferred from the flowchart in Fig. 2.1, the proposed reliability-based damage prognosis framework consists of five analytical steps: Bayesian inference analysis, probabilistic load hazard analysis, probabilistic damage evolution analysis, probabilistic global performance analysis, and damage prognosis analysis. Through these five analysis steps, it is possible to estimate the RFL of a monitored structural or mechanical system every time a new set of NDE inspection results becomes available. These inspection results, at time t_p ($p = 0, 1, 2, \ldots$), consist of the measured damage size vector, $\mathbf{A}_{\mathrm{m}}^p = \mathbf{a}_{\mathrm{m}}^p$, at the inspected damage locations [1, 2] In the first step of the proposed methodology, *Bayesian inference analysis*, $\mathbf{a}_{\mathrm{m}}^p$ is used to compute the posterior joint PDF, $f''_{\mathbf{A}_{\mathrm{a}}^p|\boldsymbol{\Theta}_{\mathrm{mat}},\boldsymbol{\Theta}_{\mathrm{dam}}^p}(\mathbf{a}_{\mathrm{a}}^p|\boldsymbol{\theta}_{\mathrm{mat}}, \boldsymbol{\theta}_{\mathrm{dam}}^p)$, of the actual damage size vector, $\mathbf{A}_{\mathrm{a}}^p$, conditional on the material ($\boldsymbol{\Theta}_{\mathrm{mat}}$) and damage evolution model ($\boldsymbol{\Theta}_{\mathrm{dam}}^p$) parameters, as well as all the NDE measurement results up to time t_p. Simultaneously, $\mathbf{a}_{\mathrm{m}}^p$ is also used to provide the posterior joint PDF, $f''_{\boldsymbol{\Theta}_{\mathrm{dam}}^p}(\boldsymbol{\theta}_{\mathrm{dam}}^p)$, of the damage evolution model parameters ($\boldsymbol{\Theta}_{\mathrm{dam}}^p$). These posterior PDFs are constructed as described in [1,2] by making use of (i) a probability of detection (*POD*) curve [4,5] and (ii) a damage size measurement model [1,6,7] which respectively characterize, in probabilistic terms, the detection capability and the measurement accuracy of the NDE technique.

The second step of the methodology, *probabilistic load hazard analysis*, defines the joint PDF of a vector of load intensity measures, **IM**. In this study, the vector **IM** is defined as $\mathbf{IM} = \{\mathbf{IM}_{\mathrm{m}}, \mathbf{IM}_{\mathrm{a}}\}$. The vector $\mathbf{IM}_{\mathrm{m}}$ characterizes the average (mean) load intensity acting on the structure, while $\mathbf{IM}_{\mathrm{a}}$ defines the intensity/amplitude of the superimposed random load

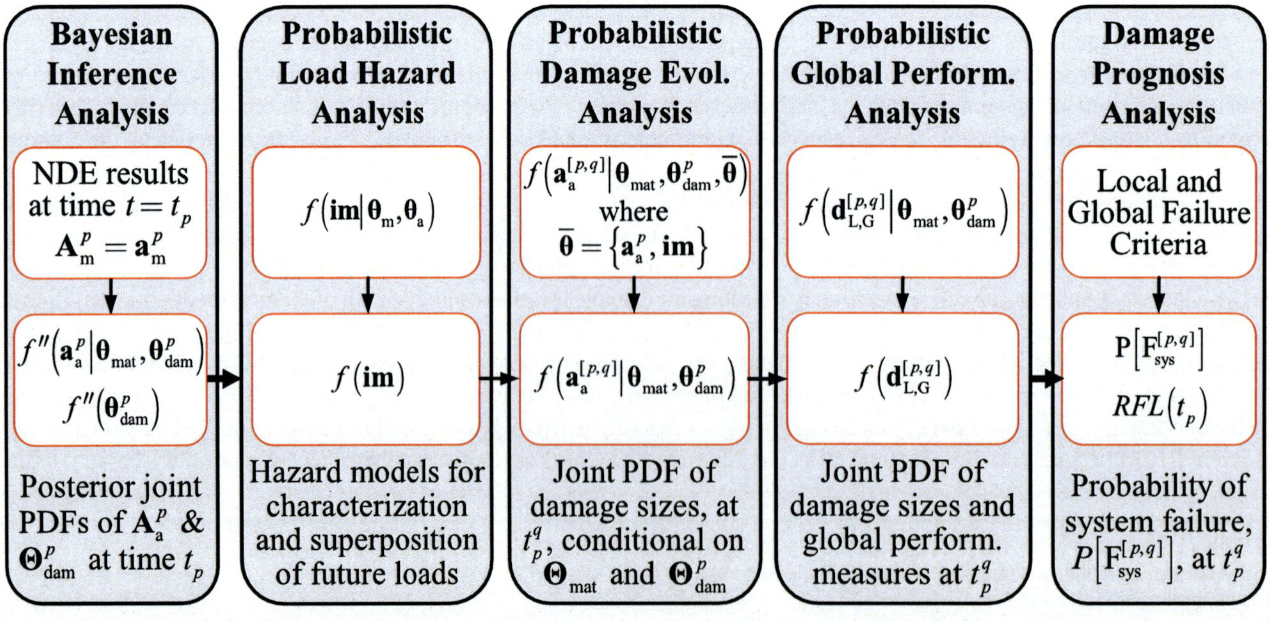

Fig. 2.1 Schematic illustration of proposed damage prognosis framework

(or load fluctuations) about the mean load [2, 8]. The joint PDF of **IM** is herein denoted as $f_{\mathbf{IM}}(\mathbf{im})$ and is computed by unconditioning the conditional joint PDF $f_{\mathbf{IM}|\boldsymbol{\Theta}_m,\boldsymbol{\Theta}_a}(\mathbf{im}\,|\theta_m,\theta_a)$ with respect to (w.r.t.) the distribution parameter vectors $\boldsymbol{\Theta}_m$ and $\boldsymbol{\Theta}_a$ [2, 8].

The third step of the proposed methodology, *probabilistic damage evolution analysis*, aims at computing the conditional joint PDF of the damage size vector $\mathbf{A}_a^{[p,q]}$ at future times $t_p^q = t_p + q \cdot \Delta\tau$, with t_p being the current time, $q \in \{1, 2, \ldots\}$, and $\Delta\tau$ a suitable fixed time interval related to the time-scale of the damage propagation process of interest [1, 2, 8]. This conditional joint PDF, denoted as $f_{\mathbf{A}_a^{[p,q]}|\boldsymbol{\Theta}_{mat},\boldsymbol{\Theta}_{dam}^p}\left(\mathbf{a}_a^{[p,q]}\,|\theta_{mat},\theta_{dam}^p\right)$, is computed by unconditioning the conditional joint PDF $f_{\mathbf{A}_a^{[p,q]}|\boldsymbol{\Theta}_{mat},\boldsymbol{\Theta}_{dam}^p,\mathbf{A}_a^p,\mathbf{IM}}\left(\mathbf{a}_a^{[p,q]}|\theta_{mat},\theta_{dam}^p,\mathbf{a}_a^p,\mathbf{im}\right)$ w.r.t. $\mathbf{A}_a^p$ and **IM** as described in detail in [1, 2, 8]. Since this study focuses on the analysis of a mechanical component subjected to harmonic load, all time related parameters ($t_p, t_p^q, \Delta\tau$) are express in terms of number of load cycles (N). Current time, t_p, is replaced by the current number of load cycles, N_p; the time interval between two consecutive damage propagation/prognosis evaluations, $\Delta\tau$, is substituted with the number of load cycles, ΔN; the future time t_p^q is replaced by the number of load cycles $N_p^q = N_p + q \cdot \Delta N$ with $q \in \{1, 2, \ldots\}$.

The fourth step shown in Fig. 2.1, *probabilistic global performance analysis*, estimates the joint PDF of $\mathbf{A}_a^{[p,q]}$ and a set of properly selected global performance metrics/measures (collected in $\boldsymbol{\Psi}_g^{[p,q]}$) at time $t_p^q = t_p + q \cdot \Delta\tau$; i.e., the joint PDF $f_{\mathbf{A}_a^{[p,q]},\boldsymbol{\Psi}_g^{[p,q]}}\left(\mathbf{a}_a^{[p,q]}, \boldsymbol{\psi}_g^{[p,q]}\right)$. As an illustration, the random vector $\boldsymbol{\Psi}_g^{[p,q]}$ can collect a given set of natural frequencies of the monitored structure, the reduced (due to damage) flutter velocity of an aircraft wing, etc. The joint PDF $f_{\mathbf{D}_{L,G}^{[p,q]}}\left(\mathbf{d}_{L,G}^{[p,q]}\right) = f_{\mathbf{A}_a^{[p,q]},\boldsymbol{\Psi}_g^{[p,q]}}\left(\mathbf{a}_a^{[p,q]}, \boldsymbol{\Psi}_g^{[p,q]}\right)$, with $\mathbf{D}_{L,G}^{[p,q]} = \left\{\mathbf{A}_a^{[p,q]}, \boldsymbol{\Psi}_g^{[p,q]}\right\}$, thus contains both local (through $\mathbf{A}_a^{[p,q]}$) and global (through $\boldsymbol{\Psi}_g^{[p,q]}$) structural integrity information. The practical computation of this joint PDF is described in detail in [1, 8].

Once the joint PDF $f_{\mathbf{D}_{L,G}^{[p,q]}}\left(\mathbf{d}_{L,G}^{[p,q]}\right)$ is determined, well established component and system reliability analysis approaches are used (in the *damage prognosis analysis* step) to estimate the probability of failure, $P\left[\mathrm{F}_{sys}^{[p,q]}\right]$, of the entire system at future times or, in equivalent terms, its RFL [1, 2]. The RFL, estimated at current time t_p, is defined herein as $RFL(t_p) = t_F(\bar{p}_F, t_p) - t_p$, where $t_F(\bar{p}_F, t_p)$ is the predicted time at which the probability of *system failure*, $P\left[\mathrm{F}_{sys}^{[p,q]}\right]$, reaches the critical threshold $\bar{p}_F$. For structures subjected to harmonic loads or other cyclic loading conditions, the RFL can be expressed as $RFL(N_p) = N_F(\bar{p}_F, N_p) - N_p$ where $N_F(\bar{p}_F, N_p)$ is the predicted number of load cycles at which the probability of system failure up-crosses the critical value $\bar{p}_F$.

It should be emphasized that, even though NDE measurement uncertainty is taken into account in the proposed methodology, the probabilistic characterizations of all the quantities mentioned above are conditional on the particular realization/scenario of NDE inspection results. The application of the proposed framework with a different realization (i.e., a different set) of NDE inspection results would lead to different outcomes. Therefore, in order to provide an unconditional estimate of all damage prognosis results (i.e., Bayesian updating results, RFL, time-varying probability of failure), it is necessary to perform multiple analyses across an ensemble of NDE measurement results. This approach, which will be considered in future studies, provides a robust statistical analysis for assessing and evaluating the overall performance and benefits of the proposed NDE-based SHM-DP methodology.

2.3 Application Example

A set of experimental fatigue test data, published in the literature [3], is now used to verify and validate the proposed framework at the local reliability component level (i.e., single damage mechanism at a single damage location) [1, 2] with special emphasis on the recursive Bayesian updating results and the RFL predictions. The dataset used consists of 68 crack propagation trajectories obtained from fatigue tests performed on center-cracked 2024-T3 aluminum plates. All the specimens tested had the same nominal geometry: length $L = 558.80$ mm, width $w = 152.40$ mm, thickness $t = 2.54$ mm, and a center crack of initial length $2a_a^0 = 18.00$ mm (see Fig. 2.2). The tests were performed under load control using a sinusoidal input at 20 Hz which was producing a constant stress range of $\Delta\sigma = \sigma_{max} - \sigma_{min} = 48.28$ MPa with a stress ratio equal to $R_\sigma = \sigma_{min}/\sigma_{max} = 0.2$. Consequently, the load intensity measure vector $\mathbf{IM} = \{\mathbf{IM}_m, \mathbf{IM}_a\}$ is defined as $\mathbf{IM}_m = (\sigma_{min} + \sigma_{max})/2 = 36.21$ MPa and $\mathbf{IM}_a = \Delta\sigma/2 = 24.14$ MPa. However, in this study, the amplitude of the sinusoidal load, $\mathbf{IM}_a$, is considered unknown and therefore treated as a random variable. Four probabilistic models of increasing load amplitude uncertainty are defined and used to validate the proposed framework

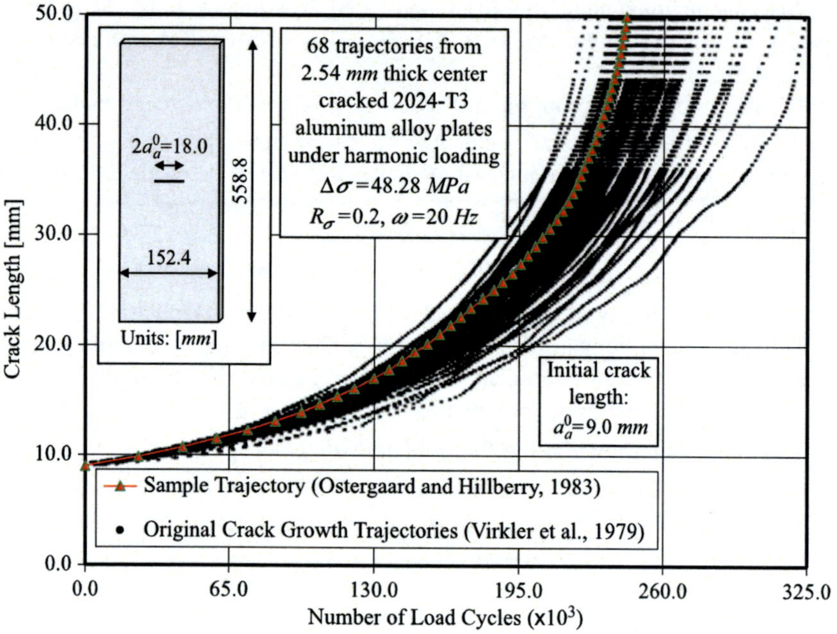

Fig. 2.2 Experimental dataset consisting of 68 crack propagation trajectories

at the local reliability component level. The first probabilistic model, namely LU0, represents the deterministic case where $\mathbf{IM}_{\mathrm{a}}(\mathrm{LU0}) = \Delta\sigma/2 = 24.14\,\mathrm{MPa}$ (i.e., the real experimental scenario) [2]. The second model, denoted as $\mathrm{LU1}_{\mathcal{U},sp}$, assumes the load amplitude to be uniformly distributed as $\mathbf{IM}_{\mathrm{a}}(\mathrm{LU1}_{\mathcal{U},sp}) \sim \mathcal{U}[21.002, 27.037]\mathrm{MPa}$. The third one, namely $\mathrm{LU2}_{\mathcal{U},sp}$, considers $\mathbf{IM}_{\mathrm{a}}(\mathrm{LU2}_{\mathcal{U},sp}) \sim \mathcal{U}[17.381, 29.451]\mathrm{MPa}$. Finally, the fourth and most severe probabilistic model, $\mathrm{LU3}_{\mathcal{U},sp}$, takes $\mathbf{IM}_{\mathrm{a}}(\mathrm{LU3}_{\mathcal{U},sp}) \sim \mathcal{U}[9.656, 33.796]\mathrm{MPa}$. These probabilistic models were chosen such that $\Delta\mathbf{IM}_{\mathrm{a}}(\mathrm{LU1}_{\mathcal{U},sp}) = 27.037 - 21.002 = 0.25 \times \mathbf{IM}_{\mathrm{a}}(\mathrm{LU0})$, $\Delta\mathbf{IM}_{\mathrm{a}}(\mathrm{LU2}_{\mathcal{U},sp}) = 29.451 - 17.381 = 0.50 \times \mathbf{IM}_{\mathrm{a}}(\mathrm{LU0})$, and $\Delta\mathbf{IM}_{\mathrm{a}}(\mathrm{LU3}_{\mathcal{U},sp}) = 33.796 - 9.656 = 1.00 \times \mathbf{IM}_{\mathrm{a}}(\mathrm{LU0})$.

Following previous research work on the same experimental dataset [9], each of the 68 crack propagation trajectories, shown in, is curve-fitted with the well-known Paris-Erdogan law [10]

$$\frac{da_a}{dN} = C\,(\Delta K)^m \tag{2.1}$$

where da_a/dN, expressed in *mm/cycle*, represents the rate of crack propagation of the actual crack length (a_a). C and m are the two damage evolution model parameters to be determined by curve fitting each experimental crack propagation trajectory. Lastly, the quantity ΔK, denotes the range of the stress intensity factor at the crack-tip within a given load cycle [11]. The recursive Bayesian inference scheme, used in this study, requires the initial prior PDFs of the crack length, A_a^0, and the damage evolution model parameters, Θ_{dam}^0, at time $N_0 = 0$ cycles [1, 8]. These two initial prior PDFs are denoted as $f'_{A_a^0}(a_a^0)$ and $f'_{\boldsymbol{\Theta}_{\mathrm{dam}}^0}(\theta_{\mathrm{dam}}^0)$, respectively. The random variable A_a^0 is assumed to be statistically independent of $\boldsymbol{\Theta}_{\mathrm{dam}}^0$ and its prior PDF is taken as the scaled Beta distribution defined in [8] while the initial prior PDF of $\boldsymbol{\Theta}_{\mathrm{dam}}^0$ is derived from the statistical analysis of the experimental dataset as described in detail in [11].

Among the 68 experimental trajectories, one sample trajectory [12] is selected and highlighted in Fig. 2.2. This sample trajectory (unknown in a real-world application) is used as underlying true crack propagation path in the verification and validation of the proposed damage prognosis Framework [2, 10]. This trajectory is used to randomly generate NDE measurements, $\mathbf{A}_{\mathrm{m}}^p$, at evenly spaced intervals. These simulated measurements are consistent with the *POD* curve and the damage measurement model described and used in [2, 10]. The values of the *POD* curve model parameters used in this specific application example are taken as $\hat{\alpha}_0 = 11.0$ and $\hat{\alpha}_1 = 5.0$ (see Ref. [11]). On the other hand, the values of the sizing model parameters used in the sequel are: $\hat{\beta}_0 = -1.5\,\mathrm{mm}$, $\hat{\beta}_1 = 1.2\,\mathrm{mm}$, and $\hat{\sigma}_\varepsilon = 2.0\,\mathrm{mm}$ (see Refs. [1,2,11]). In this specific application example, the selected sample trajectory is monitored over the time window [0 200,000] load cycles and the interval of time between two subsequent NDE inspections is taken as $\Delta N = 20{,}000$ load cycles. At each NDE inspection, five statistically independent NDE measurements (i.e., $n_{\mathrm{MS}}^p = 5\;\forall p \geq 0$) are taken and used in the proposed Bayesian inference scheme [1, 2] to recursively compute the posterior conditional PDF $f''_{A_{\mathrm{a}}^p|\boldsymbol{\Theta}_{\mathrm{dam}}^p}(a_{\mathrm{a}}^p\,|\,\theta_{\mathrm{dam}}^p)$, the posterior unconditional PDF

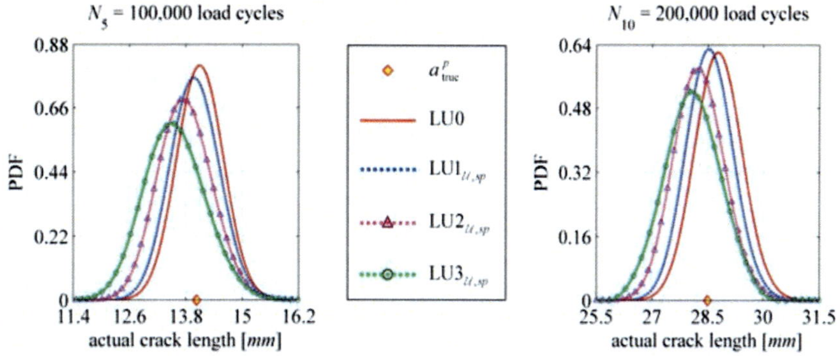

Fig. 2.3 Posterior PDFs of the actual crack length, $f''_{A^p_a}(a^p_a)$, estimated through the proposed recursive Bayesian inference scheme at $N_5 = 100,000$ (i.e., $p = 5$) and $N_{10} = 200,000$ load cycles (i.e., $p = 10$)

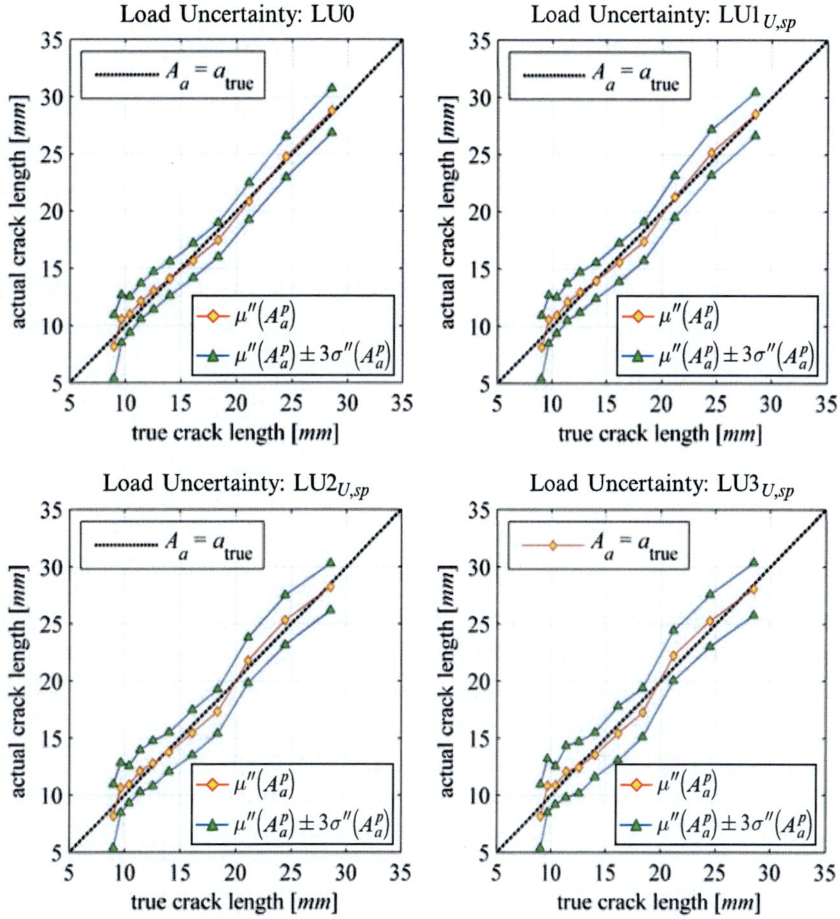

Fig. 2.4 Comparisons between the posterior mean, $\mu''(A^p_a)$, of the actual crack length and the corresponding underlying true crack length immediately after each NDE inspection and for each one of the four increasingly severe levels of load uncertainty considered in this study (i.e., $LU0, LU1_{U,sp}, LU2_{U,sp}, LU3_{U,sp}$)

$f''_{A^p_a}(a^p_a)$, and the posterior joint PDF $f''_{\Theta^p_{dam}}(\theta^p_{dam}) = f''_{\ln C^p, m^p}(\ln C^p, m^p)$. For the sake of conciseness, only a selected set of the Bayesian updating results for $f''_{A^p_a}(a^p_a)$ is presented herein. More specifically, Fig. 2.3 shows the posterior PDFs of the crack length at times $N_5 = 100{,}000$ load cycles (i.e., immediately after the sixth NDE inspection) and $N_{10} = 200{,}000$ load cycles (i.e., immediately after the 11th and last NDE inspection). As can be noticed, the higher the uncertainty in the load the larger the variance of the posterior PDF and, in general, the larger the bias. However, the increase in the variance is kept within very acceptable limits. These conclusions can be extended to any inspection time over the entire time window [0 200,000] load cycles as illustrated by the four sub-plots in Fig. 2.4. The results provided in this latter figure demonstrate the overall efficiency and robustness of the proposed recursive Bayesian inference scheme in tracking the true underlying crack propagation path.

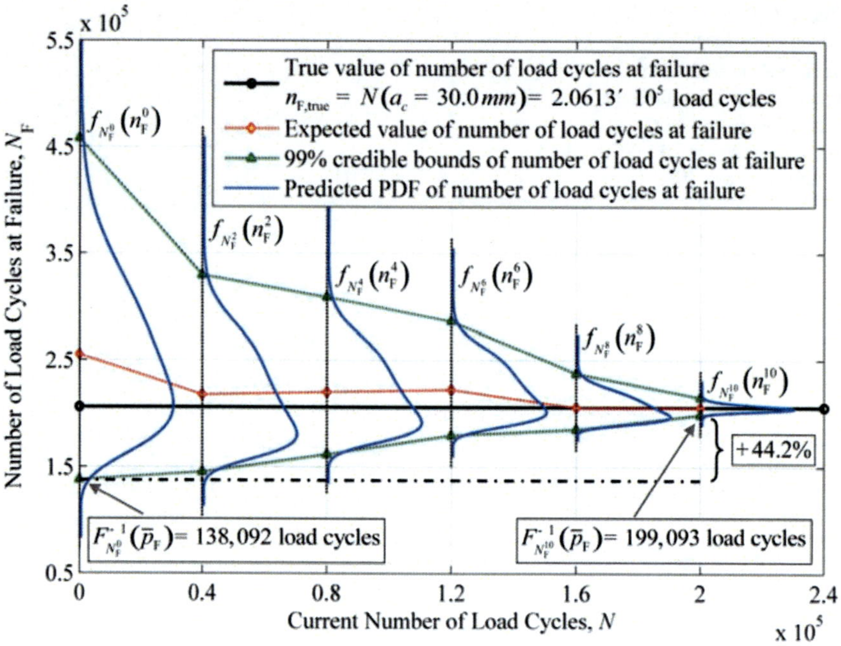

Fig. 2.5 Remaining fatigue life predictions obtained using $\mathrm{LU1}_{\mathcal{U},sp}$, $\Delta N = 20{,}000$ load cycles, $n^p_{\mathrm{MS}} = 5\,(\forall p \geq 0)$, $\hat{\sigma}_\varepsilon = 2.0\,\mathrm{mm}$, and $\bar{p}_{\mathrm{F}} = 0.5\,\%$ All PDFs are scaled to the same height for better visualization

After each NDE inspection at time N_p (with $p \geq 0$), the crack propagation path is modeled and simulated using the Paris-Erdogan law, as detailed in [2, 10], in order to estimate the PDF of the actual crack length, $A_{\mathrm{a}}^{[p,q]}$, at future load cycle $N_p^q = N_p + q \cdot \Delta N$ (with $q = 1, 2, \ldots, \bar{q}$). Once this information becomes available, it is then possible to compute the time-varying probability of failure ($p_{\mathrm{F}}^{[p,q]}$), at the local reliability component level as [1, 2, 11]

$$p_{\mathrm{F}}^{[p,q]} = P\left[A_{\mathrm{a}}^{[p,q]} \geq a_c\right] = \int_{a_c}^{+\infty} f_{A_{\mathrm{a}}^{[p,q]}}\left(a_{\mathrm{a}}^{[p,q]}\right)\, da_{\mathrm{a}}^{[p,q]} \tag{2.2}$$

where a_c denotes the pre-defined and deterministic critical crack length (taken as $a_c = 30.0$ mm in this specific application example). Equivalently, it is also possible to derive (see Refs. [2, 10]) the PDF of the number of cycles at failure (N_{F}^p) denoted as $f_{N_{\mathrm{F}}^p}\left(n_{\mathrm{F}}^p\right)$; i.e., the PDF of the number of cycles at which the pre-defined critical crack length is reached. With this information in hand, the estimate (at time N_p) for the RFL of the monitored mechanical component can be expressed as $RFL(N_p) = N_{\mathrm{F}}(\bar{p}_{\mathrm{F}}, N_p) - N_p = N_{\mathrm{F},\bar{p}_{\mathrm{F}}}^p - N_p = F_{N_{\mathrm{F}}^p}^{-1}(\bar{p}_{\mathrm{F}}) - N_p$, where the term $F_{N_{\mathrm{F}}^p}^{-1}(\cdot)$ denotes the inverse cumulative distribution function (CDF) of the random variable N_{F}^p. Figure 2.5 shows the RFL results associated with the NDE monitoring of the selected sample trajectory when the random sinusoidal amplitude load is assumed to follow the PDF model defined earlier as $\mathrm{LU1}_{\mathcal{U},sp}$. For the sake of clarity, only the results from the damage prognosis analyses performed at $N_p = p \cdot \Delta N$ (with $p = 0$, 2, 4, 6, 8, 10) are shown in Fig. 2.5. It is observed that the RFL predictions become progressively more and more accurate (i.e., less bias) as new NDE data become available and are assimilated by the proposed framework. For the particular case shown below, the initial coefficient of variation (COV) of the random variable N_{F}^0 at $N_0 = 0$, — equal to $COV(N_{\mathrm{F}}^0) = 26.3\%$ — is reduced to $COV(N_{\mathrm{F}}^{10}) = 1.5\%$ at $N_{10} = 200{,}000$ load cycles (i.e., more than 17 times smaller). Additionally, the value of the number of cycles at failure with a probability of not being exceeded of $\bar{p}_{\mathrm{F}} = 0.5\ \%$ — i.e., $N_{\mathrm{F}}(\bar{p}_{\mathrm{F}}, N_p) = N_{\mathrm{F},\bar{p}_{\mathrm{F}}}^p = F_{N_{\mathrm{F}}^p}^{-1}(\bar{p}_{\mathrm{F}})$ — is also extended by 44.2 % from time $N_0 = 0$ to time $N_{10} = 200{,}000$ load cycles. In other words, the initial estimate for the fatigue life $N_{\mathrm{F}}(\bar{p}_{\mathrm{F}}, N_0) = F_{N_{\mathrm{F}}^0}^{-1}(\bar{p}_{\mathrm{F}}) = 138{,}092$ load cycles is extended to $N_{\mathrm{F}}(\bar{p}_{\mathrm{F}}, N_{10}) = F_{N_{\mathrm{F}}^{10}}^{-1}(\bar{p}_{\mathrm{F}}) = 199{,}093$ load cycles. More formally, this relative gain in the predicted number of cycles at failure at time N_p can be defined as

$$\delta N_{\mathrm{F}}^p = \frac{N_{\mathrm{F}}(\bar{p}_{\mathrm{F}}, N_p) - N_{\mathrm{F}}(\bar{p}_{\mathrm{F}}, N_0)}{N_{\mathrm{F}}(\bar{p}_{\mathrm{F}}, N_0)}, \quad p \geq 1 \tag{2.3}$$

and its corresponding value can be computed after each NDE inspection at time N_p (with $p \geq 1$). Figure 2.6 shows the relative gain in the predicted number of cycles at failure for the four increasingly severe levels of load uncertainty considered in this

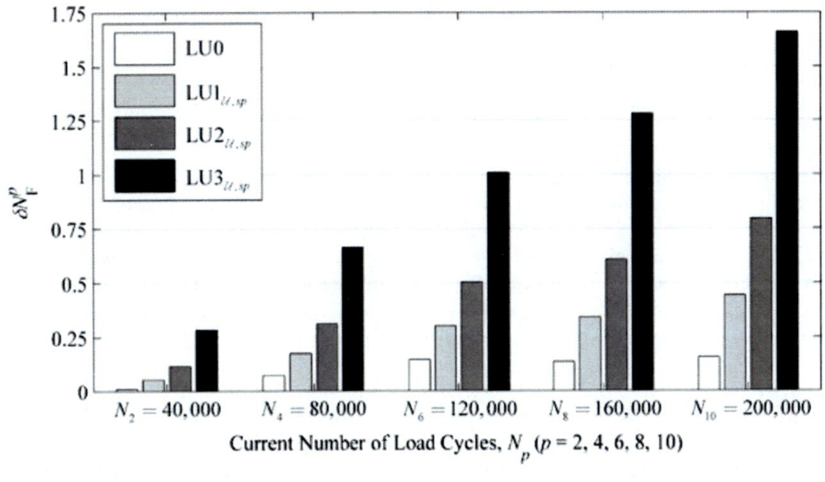

Fig. 2.6 Relative gain in the predicted number of load cycles at failure after the NDE inspections performed at times $p \cdot \Delta N$ load cycles (with $\Delta N = 20{,}000$ load cycles and $p = 2, 4, 6, 8, 10$)

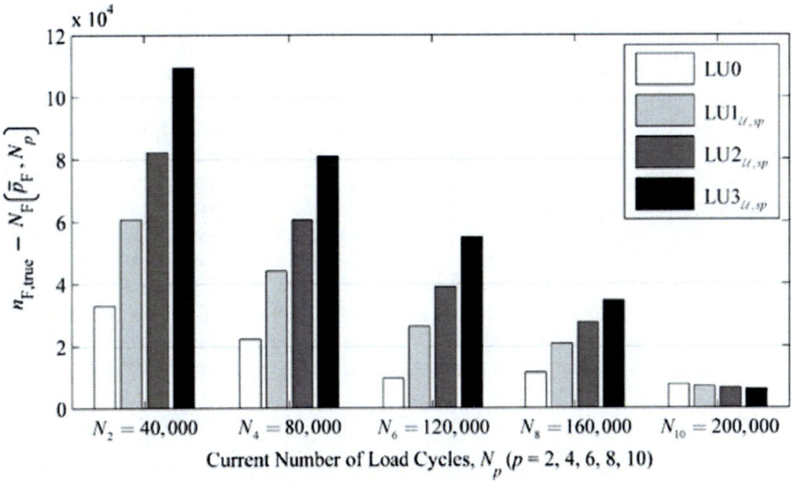

Fig. 2.7 Difference between true and predicted number of load cycles at failure after the NDE inspections performed at times $p \cdot \Delta N$ load cycles (with $\Delta N = 20{,}000$ load cycles and $p = 2, 4, 6, 8, 10$)

study (i.e., $\mathrm{LU0}, \mathrm{LU1}_{\mathcal{U},sp}, \mathrm{LU2}_{\mathcal{U},sp}, \mathrm{LU3}_{\mathcal{U},sp}$) and demonstrates how the proposed framework is a fundamental and powerful analysis tool towards an optimal life cycle cost management of a given monitored system. Simultaneously, as shown in Fig. 2.7, the difference between the true ($n_{\mathrm{F,true}}$) and the predicted number of cycles at failure progressively decreases as new NDE results become available and are assimilated by the proposed framework. In particular, it should be noticed that the predicted quantity $N_{\mathrm{F}}(\bar{p}_{\mathrm{F}}, N_p)$ (with $p = 2, 4, 6, 8, 10$) is never larger than the underlying truth (i.e., $n_{\mathrm{F,true}}$). This study thus demonstrates that the proposed methodology can lead to an extent of the initially predicted RFL with a consequent economical gain without compromising the minimum safety requirements; however, in other practical scenarios, it could also lead to an increase of safety by detecting a fault earlier than expected and therefore avoiding a very costly catastrophic failure.

2.4 Conclusions

A reliability-based damage prognosis framework for predicting the remaining fatigue life (RFL) of monitored mechanical/structural components and/or systems was presented in the first part of this paper. The methodology relies on periodic NDE inspections results to (i) assess and update the current state of damage of the monitored component/system, and (ii) predict the probability distribution of the actual damage extent at future times. This information is then used to estimate the probability of failure at future times or, equivalently, the RFL. The proposed framework is formulated to explicitly account for the uncertainties related to: NDE detection capability and measurement accuracy, material model parameters, damage evolution model parameters, and future operational loads. In the second part of the paper, a set of experimental fatigue test data was used to provide a numerical verification and a first experimental validation of the proposed damage prognosis framework at the local reliability component level. Load and damage evolution model parameters uncertainty were explicitly considered in this study. The results presented in this paper demonstrate how the proposed framework is capable of recursively predicting and updating the RFL of a given monitored component It was also shown how a nearly continuous NDE monitoring coupled with the proposed framework can lead to an extension of the initially predicted RFL with consequent economical benefits over the service life of the monitored component/structure.

Acknowledgements The work presented in this paper stems from a research project funded by the Educational Collaboration between the Los Alamos National Laboratory (LANL) and the University of California, San Diego, (UCSD) on "A Damage Prognosis System for Unmanned Aerial Vehicles", contract number: 72232-001-03. Partial support of this work was also provided by the UCSD Academic Senate Research Grant RJ086G-CONTE. The authors wish to thank Prof. Yongming Liu of the Department of Civil and Environmental Engineering at Clarkson University and Dr. Xuefei Guan (Research Scientist at Siemens Corporate Research) for providing the experimental dataset used in this study.

References

1. Gobbato M, Conte JP, Kosmatka JB, Farrar CR (2012) A reliability-based framework for fatigue damage prognosis of composite aircraft structures. Prob Eng Mech 29:176–188
2. Gobbato M, Kosmatka JB, Conte JP A recursive bayesian approach for fatigue damage prognosis: an experimental validation at the reliability component level. Mech Syst Signal Process (Under review)
3. Virkler DA, Hillberry BM, Goel PK (1979) The statistical nature of fatigue crack propagation. Trans ASME 101(2):148–153
4. Berens AP (1989) NDE reliability analysis, metals handbook, vol 17, 9th edn. ASM International, New York, pp 689–701
5. Staat M (1993) Sensitivity of and influences on the reliability of an HTR-module primary circuit pressure boundary. In: 12th international conference on structural mechanics in reactor technology (SMiRT), Amsterdam
6. Zhang R, Mahadevan S (2001) Fatigue reliability analysis using non-destructive inspection. J Struct Eng 127(8):957–965
7. Zheng R, Ellingwood BR (1998) Role of non-destructive evaluation in time-dependent reliability analysis. Struct Saf 20(4):325–339
8. Gobbato M (2011) Reliability-based framework for fatigue damage prognosis of bonded structural elements in aerospace composite structures. Ph.D. thesis, Department of Structural Engineering, University of California, San Diego
9. Kotulski ZA (1998) On efficiency of identification of a stochastic crack propagation model based on Virkler experimental data. Arch Mech 50(5):829–847
10. Paris PC, Erdogan FA (1963) Critical analysis of crack propagation laws. J Basic Eng Trans ASME 85(Series D):528–534
11. Gobbato M, Kosmatka JB, Conte JP (2012) A recursive approach for remaining fatigue life predictions of monitored structural systems. In: Proceedings of the 53rd AIAA/ASME/ASCE/AHS/ASC structures, structural dynamics, and materials conference, Honolulu, 23–26 Apr 2012
12. Ostergaard DF, Hillberry BM (1983) Characterization of the variability in fatigue crack propagation data. In: Bloom JM, Ekvall JC (eds) Probabilistic fracture mechanics and fatigue methods: applications for structural design and maintenance, ASTM STP 798. ASTM, Philadelphia, pp 97–115

Chapter 3
Fast Computing Techniques for Bayesian Uncertainty Quantification in Structural Dynamics

Costas Papadimitriou and Dimitra-Christina Papadioti

Abstract A Bayesian probabilistic framework for uncertainty quantification and propagation in structural dynamics is reviewed. Fast computing techniques are integrated with the Bayesian framework to efficiently handle large-order models of hundreds of thousands or millions degrees of freedom and localized nonlinear actions activated during system operation. Fast and accurate component mode synthesis (CMS) techniques are proposed, consistent with the finite element (FE) model parameterization, to achieve drastic reductions in computational effort when performing a system analysis. Additional substantial computational savings are also obtained by adopting surrogate models to drastically reduce the number of full system re-analyses and parallel computing algorithms to efficiently distribute the computations in available multi-core CPUs. The computational efficiency of the proposed approach is demonstrated by updating a high-fidelity finite element model of a bridge involving hundreds of thousands of degrees of freedom.

Keywords Bayesian inference • Structural dynamics • Component mode synthesis • Surrogate models • HPC

3.1 Introduction

In structural dynamics, Bayesian inference [1–3] is used for quantifying and calibrating uncertainty models based on vibration measurements, as well as propagating these modeling uncertainties in system simulations to obtain updated robust predictions of system performance, reliability and safety [4]. The Bayesian tools for identifying system and uncertainty models as well as performing robust prediction analyses are Laplace methods of asymptotic approximation and more accurate stochastic simulation algorithms, such as MCMC [5] and Transitional MCMC [6]. These tools involve solving optimization problems, generating samples for tracing and then populating the important uncertainty region in the parameter space, as well as evaluating integrals over high-dimensional spaces of the uncertain model parameters. A moderate to very large number of repeated system analyses are required to be performed over the space of uncertain parameters. Consequently, the computational demands depend highly on the number of system analyses and the time required for performing a system analysis.

To reliably update models, high fidelity FE model classes, often involving a large number of DOFs, should be introduced to simulate structural behavior. For such large-order finite element models the computational demands in implementing asymptotic approximations as well as stochastic simulation techniques may be excessive. The present work proposes methods for drastically reducing the computational demands at the system, algorithm and computer hardware levels involved in the implementation of Bayesian tools. At the system level, CMS techniques [7] are integrated with Bayesian techniques to efficiently handle large-order models of hundreds of thousands or millions degrees of freedom and localized nonlinear actions activated during system operation. Fast and accurate CMS techniques are obtained [8], consistent with the FE model parameterization, to achieve drastic reductions in computational effort. The CMS allows the repeated computations to be carried out in a significantly reduced space of generalized coordinates. At the level of the TMCMC algorithm, surrogate models are adopted to drastically reduce the number of computationally expensive full model runs. At the computer hardware level, parallel computing algorithms are used to efficiently distribute the computations in available multi-core CPUs [9].

C. Papadimitriou (✉) • D.-C. Papadioti
Department of Mechanical Engineering, University of Thessaly, Volos, Greece
e-mail: costasp@uth.gr; dxpapadioti@uth.gr

T. Simmermacher et al. (eds.), *Topics in Model Validation and Uncertainty Quantification, Volume 5*,
Conference Proceedings of the Society for Experimental Mechanics Series 41,
DOI 10.1007/978-1-4614-6564-5_3, © The Society for Experimental Mechanics, Inc. 2013

3.2 Bayesian Uncertainty Quantification and Propagation Framework

Consider a class M_m of structural dynamics models used to predict various output quantities of interest $\underline{f}(\underline{\theta}_m|\mathrm{M}_m)$ of a system, where $\underline{\theta}_m$ is a set of parameters in this model class that need to be estimated using experimental data $D \equiv \{\underline{\hat{y}}\}$. Following a Bayesian formulation [2, 3] and assuming that the observation data and the model predictions satisfy the prediction error equation

$$\underline{\hat{y}} = \underline{f}(\underline{\theta}_m|\mathrm{M}_m) + \underline{e} \tag{3.1}$$

where the error term $\underline{e} \sim N(\underline{0}, \Sigma)$ is a zero-mean Gaussian vector with covariance $\Sigma \equiv \Sigma(\underline{\theta}_e)$ depending on the parameters θ_e of the prediction error model class M_e, the updated distribution $p(\underline{\theta}|D,\mathrm{M})$ of the augmented parameter set $\underline{\theta} = (\underline{\theta}_m, \underline{\theta}_e)$, given the data D and the combined model class $\mathrm{M} = (\mathrm{M}_m, \mathrm{M}_e)$, results from the application of the Bayes theorem as follows

$$p(\underline{\theta}|D,\mathrm{M}) = \frac{p(D|\underline{\theta},\mathrm{M})\pi(\underline{\theta}|\mathrm{M})}{p(D|\mathrm{M})} \tag{3.2}$$

where

$$p(D|\underline{\theta},\mathrm{M}) = \frac{|\Sigma(\underline{\theta}_e)|^{-1/2}}{(2\pi)^{N/2}} \exp\left[-\frac{1}{2}J(\underline{\theta};\mathrm{M})\right] \tag{3.3}$$

is the likelihood of observing the data from the model class,

$$J(\underline{\theta};\mathrm{M}) = [\underline{\hat{y}} - \underline{f}(\underline{\theta}_m|\mathrm{M})]^T \Sigma^{-1}(\underline{\theta}_e)[\underline{\hat{y}} - \underline{f}(\underline{\theta}_m|\mathrm{M})] \tag{3.4}$$

is the measure of fit between the experimental and model predicted properties, $\pi(\underline{\theta})$ is the prior probability distribution of the model parameters based on previous knowledge and/or user experience, and $p(D|\mathrm{M})$ is the evidence of the model class.

The Bayesian probabilistic framework can also be used to compare two or more competing model classes and select the optimal model class based on the available data. Consider a family M_i, $i = 1,\cdots,\kappa$, of κ alternative, competing, parameterized FE and prediction error model classes, and let $\underline{\theta}_i \in R^{N_{\theta_i}}$ be the free parameters of the model class M_i. The posterior probabilities $P(\mathrm{M}_i|D)$ of the various model classes given the data D is [10, 11]

$$P(\mathrm{M}_i|D) = \frac{p(D|\mathrm{M}_i)P(\mathrm{M}_i)}{p(D|\mathrm{M}_1,\cdots,\mathrm{M}_\kappa)} \tag{3.5}$$

where $P(\mathrm{M}_i)$ is the prior probability and $p(D|\mathrm{M}_i)$ is the evidence of the model class M_i. The optimal model class M_{best} is selected as the one that maximizes $P(\mathrm{M}_i|D)$ given by (3.5).

For large enough number of measured data, the posterior distribution of the model parameters in (3.2) can be asymptotically approximated by a Gaussian distribution [3] centered at the most probable value $\hat{\theta}$ of the model parameters with covariance equal to the inverse of the Hessian $h(\theta)$ of the function

$$g(\theta;\mathrm{M}) = -\ln p(D|\underline{\theta},\mathrm{M}) = \frac{1}{2}J(\underline{\theta}_m;\mathrm{M}) + \frac{1}{2}|\Sigma(\underline{\theta}_e)| - \ln \pi(\underline{\theta}|\mathrm{M}) \tag{3.6}$$

evaluated at the most probable value $\underline{\hat{\theta}}$ of the model parameters. The most probable value $\underline{\hat{\theta}}$ maximizes the posterior probability distribution $p(\underline{\theta}|D,\mathrm{M})$ or, equivalently, minimizes the function $g(\underline{\theta};\mathrm{M})$. For model selection, an asymptotic approximation based on Laplace's method is also used to give an estimate of the integral involved in the estimation of the evidence $p(D|\mathrm{M})$ in (3.2) and (3.5) [11]. For the case for which analytical expressions for the gradient of $J(\underline{\theta};\mathrm{M})$ with respect to the structural model parameters θ_m are available, computationally efficient gradient-based optimization algorithms can be used to obtain the optimal value of the model parameters by minimizing the function $g(\underline{\theta};\mathrm{M})$. Specifically, for linear structural dynamics models and experimental data consisting of modal frequencies and mode shapes, such analytical expressions are available and can be computed using efficient adjoint techniques (e.g. [12]). However, there are certain classes of nonlinear systems where such analytical expressions or adjoint techniques are not applicable or it is inconvenient to introduce within commercially available structural dynamics solvers. In such cases, non-gradient-based optimization algorithms can be used to obtain the most probable value of the model parameters. Once the most probable value has been computed, the Hessian required in the asymptotic approximation can be estimated using either higher-order adjoint techniques [12], if applicable, or using numerical differentiation techniques.

It should be noted that the asymptotic expression is approximate. Moreover, even for large number of experimental data, it may fail to give a good representation of the posterior probability distribution in the case of multimodal distributions. In addition, the asymptotic approximation fails to provide acceptable estimates for un-identifiable cases manifested for relatively large number of model parameters in relation to the information contained in the data. For more accurate estimates, one should use stochastic simulation algorithms (e.g. MCMC [5], Transitional MCMC – TMCMC [6], Delayed Rejection Adaptive Metropolis – DRAM [13]) to generate samples that populate the posterior probability distribution function in (3.2) and then evaluate the integrals involved in propagating uncertainties into robust predictions. Among the stochastic simulation algorithms available, the transitional MCMC algorithm [6] is one of the most promising algorithms for selecting the most probable model as well as finding and populating with samples the importance region of interest of the posterior probability distribution, even in the unidentifiable cases and multi-modal posterior distributions. In addition, the TMCMC method yields an estimate of the evidence $p(D|\mathrm{M})$ of a model class M, required for model class selection [10, 14], based on the samples generated by the algorithm.

The samples generated at the final stage of the TMCMC algorithm can further be used for estimating the probability integrals encountered when interested in robust predictions of various performance quantities of interest. Specifically, consider an output quantity q of interest in structural dynamics simulations. Posterior robust predictions of q are obtained by taking into account the updated uncertainties in the model parameters given the measurements D. Let $p(q|\underline{\theta},\mathrm{M})$ be the conditional probability distribution of q given the values of the parameters. Using the total probability theorem, the posterior robust probability distribution $p(q|D,\mathrm{M})$ of q, taking into account the model M and the data D, is given by Papadimitriou et al. [4]

$$p(q|D,\mathrm{M}) = \int p(q|\underline{\theta},\mathrm{M})p(\underline{\theta}|D,\mathrm{M})d\underline{\theta} \tag{3.7}$$

as an average of the conditional probability distribution $p(q|\underline{\theta},\mathrm{M})$ weighting by the posterior probability distribution $p(\underline{\theta}|D,\mathrm{M})$ of the model parameters. Let also $G(q)$ be a function of the output quantity of interest q. A posterior robust performance measure of the system given the data D is

$$E[G(q)|D,\mathrm{M})] = \int G(q)p(\underline{\theta}|D,\mathrm{M})d\underline{\theta} \tag{3.8}$$

The evaluation of the multi-dimensional integrals in (3.7) and (3.8) cannot be performed analytically. Asymptotic approximations are given in [4]. Alternatively, stochastic simulation methods can be conveniently used to estimate the integral from the samples $\underline{\theta}^{(i)}$, $i = 1,\cdots,N$, generated from the posterior probability distribution $p(\underline{\theta}|D,\mathrm{M})$. In this case, the integrals (3.7) and (3.8) can be approximated by

$$f(q|D,\mathrm{M}) \approx \frac{1}{N}\sum_{i=1}^{N} f(q|\underline{\theta}^{(i)},\mathrm{M}) \tag{3.9}$$

and

$$E[G(q)|D,\mathrm{M})] \approx \frac{1}{N}\sum_{i=1}^{N} G(q^{(i)}) \tag{3.10}$$

respectively, where $q^{(i)} \equiv q(\underline{\theta}^{(i)})$

3.3 Fast Computing Techniques for Large Order Finite Element Models

3.3.1 Component Model Synthesis for Parameter Estimation in Structural Dynamics

At the system level, dynamic reduction techniques such as CMS can be implemented with Bayesian uncertainty quantification and propagation framework in order to alleviate the computational burden associated with each model run in the re-analyses required in the optimization and stochastic simulation methods. CMS techniques have been successfully employed for model reduction in optimization and stochastic simulation algorithms involved in model updating [15, 16]. CMS techniques [7] divide the structure into components with mass and stiffness matrices that are reduced using fixed-interface and constrained modes. Dividing the structure into components and reducing the number of physical coordinates to a much smaller number of generalized coordinates certainly alleviates part of the computational effort. However, at each iteration or TMCMC sampling point one needs to re-compute the eigen-problem and the interface constrained modes for each component. This procedure is usually a very time consuming operation and computationally more expensive that solving directly the original matrices for the eigenvalues and the eigenvectors. It was recently shown [8] that for certain parameterization schemes for which the mass and stiffness matrices of a component depend linearly on only one of the free model parameters to be updated, often encountered in finite element model updating formulations, the full re-analyses of the component eigen-problems are avoided. The eigenproperties and the interface constrained modes as a function of the model parameters can be computed inexpensively from the eigenproperties and the interface constrained modes that correspond to a nominal value of the model parameters.

Specifically let Δ_j be the set of structural components that depend on the j-th parameter θ_j. Consider the case for which the stiffness matrix of a component $s \in \Delta_j$ depends linearly θ_j and the mass matrix is independent of θ_j, i.e. $K_{(s)} = \bar{K}_{(s)}\theta_j$ and $M_{(s)} = M_0^{(s)}$. It can be readily derived that the stiffness and mass matrices of the Craig-Bampton reduced system admits the representation

$$\hat{K}_{CB} = \hat{K}_0^{CB} + \sum_{j=1}^{N_\theta} \hat{K}_{,j}^{CB}\theta_j \quad \text{and} \quad \hat{M}_{CB} = \hat{M}_0^{CB} \tag{3.11}$$

where the coefficient matrices $\hat{K}_0^{CB}$, $\hat{K}_{,j}^{CB}$ and $\hat{M}_0^{CB}$ in the expansion (3.11) are assembled from the component stiffness and mass matrices. It is important to note that the assembled matrices $\hat{K}_0^{CB}$, $\hat{K}_{,j}^{CB}$ and $\hat{M}_0^{CB}$ of the Craig-Bampton reduced system in the expansion (3.11) are independent of the values of $\underline{\theta}$. In order to save computational time, these constant matrices are computed and assembled once and, therefore, there is no need this computation to be repeated during the iterations involved in optimization or TMCMC sampling algorithms for model updating due to the changes in the values of the parameter vector $\underline{\theta}$. This is an important result which saves substantial computational effort since it avoids (a) re-computing the fixed-interface and constrained modes for each component, and (b) assembling the reduced matrices from these components. The formulation guarantees that the reduced system is based on the exact component modes for all values of the model parameters.

The aforementioned formulation can readily be extended to treat the more general case in which the stiffness and mass matrices of a component depends nonlinearly on a single parameter, that is $K_{(s)} = \bar{K}_{(s)}f(\theta_j)$ and $M_{(s)} = \bar{M}_{(s)}\rho(\theta_j)$, where $f(\theta_j)$ and $\rho(\theta_j)$ are scalar nonlinear functions of a parameter. In this case the linear representation (3.11) is no longer applicable for such components. However, the reduced stiffness and mass matrices for each component as a function of the model parameters can be readily obtained from the eigen-properties and constrained interface modes obtained from a single analysis for a nominal value of the component parameter θ_j. It turns out that substantial computational savings arise from the fact that the re-analyses of the fixed-interface and constrained modes for each component required at each iteration or TMCMC samping point is completely avoided. In this general case, the reduced stiffness and mass component matrices have to be re-assembled in order to derive the Craig-Bampton reduced system matrices. Similar to the linear case [8], the computational demands in FE model updating formulations are again substantially reduced without compromising the solution accuracy.

3.3.2 Surrogate Models

At the level of the TMCMC algorithm, surrogate models can be used to reduce the computational time by avoiding the full model runs at a large number of sampling points in the parameters space. This is done by exploiting the function evaluations that are available at the neighbour points from previous full model runs in order to generate an estimate at a new sampling point in the parameter space. Surrogate models are well-suited to be used with MCMC algorithms, including the TMCMC algorithm [6]. The kriging technique [17] is used to approximate the function evaluation at a sampling point using the function evaluations at neighbor points in the parameter space. To ensure a high quality approximation, certain conditions are imposed in order a surrogate estimate be accepted. Specifically, the estimate is accepted based on a minimum number of neighbour design points that depend on the dimension of the uncertain parameter space. The surrogate point has to belong to the convex hull of the design points so that an interpolation is performed, while extrapolations are prohibited. The neighbour design points are selected as the ones closest to the surrogate estimate and also within the hyper- ellipse of the TMCMC proposal covariance matrix scaled to include the minimum number of design points. The estimate is also accepted based on local optimality conditions for the selected surrogate scheme, guaranteeing that the error in the surrogate estimate provided by the kriging technique is smaller than a user-defined value. Details of the integration of the kriging technique within the TMCMC algorithm can be found in [9]. An order of magnitude reduction in the number of full model runs involved in TMCMC algorithm has been reported which results in additional computational savings.

3.3.3 *Parallel Computing Algorithms*

At the computer hardware level, high performance computing (HPC) techniques can be used to reduce the computational time. Most MCMC algorithms involve a single Markov chain and are thus not parallelizable. In contrast, the TMCMC algorithm involves a large number of independent Markov chains that can run in parallel. Thus, the TMCMC algorithm is very-well suited for parallel implementation in a computer cluster [9]. Specifically, parallelization is activated at every stage of the TMCMC algorithm exploiting the large number of short, variable length, chains that need to be generated starting from the leader samples determined from the TMCMC algorithm at the particular stage. Static and dynamic scheduling schemes can be conveniently used to optimally distribute these chains in a multi-host configuration of complete heterogeneous computer workers. The static scheduling scheme distributes the chains in the workers using a weighted round-robin algorithm so that the number of likelihood evaluations is arranged to be the same for each computer worker. The static scheduling scheme is computational efficient when the computational time for a likelihood evaluation is the same independently of the location of sample in the parameter space as well as when surrogate estimates are not activated. The dynamic scheduling scheme is more general, ensuring a more efficient balancing of the loads per computer worker in the case of variable run time of likelihood function evaluations and unknown number of surrogates activated during estimation. Specifically, each worker is periodically interrogated at regular time intervals by the master computer about its availability and samples from TMCMC chains are submitted to the workers on a first come first serve basis to perform the likelihood function evaluations so that the idle time of the multiple workers is minimised. Details of the parallel implementation of the TMCMC algorithm are given in [9].

3.4 Application on Finite Element Model Updating of a Bridge

The efficiency of the proposed fast computing tools in the Bayesian framework is demonstrated by updating a FE model of the Metsovo bridge (Fig. 3.1a) using simulated modal data. A detailed FE model of the bridge is created using 3-dimensional tetrahedron quadratic Lagrange FEs. An extra coarse mesh, chosen to predict the lowest 20 modal frequencies and mode shapes of the bridge, results in a minimum 97,636 FEs and 562,101 DOF. The size of the elements in the extra coarse mesh is the maximum possible one that can be considered, with typical element length of the order of the thickness of the deck cross-section.

Let ω_c be the cut-off frequency which represents the highest modal frequency that is of interest in FE model updating. Herein, the cut-off frequency is selected to be equal to the 20th modal frequency of the nominal model. That is $\omega_c = 4.55$ Hz. For demonstration purposes, the bridge is divided into nine physical components with eight interfaces between components as shown in Fig. 3.1b. For each component it is selected to retain all modes that have frequency less than $\omega_{\max} = \rho\,\omega_c$, where the ρ values affect computational efficiency and accuracy of the CMS technique. The total number of internal DOFs before the model reduction is applied and the number of modes retained for various ρ values are given in Table 3.1. For the case $\rho = 8$, a total of 286 internal modes out of the 558,801 are retained for all 9 components. The total number of DOFs of the reduced model is 3,586 which consist of 286 fixed interface generalized coordinates and 3,300 constraint interface DOFs for all components. It is clear that a two orders of magnitude reduction in the number of DOFs is achieved using CMS.

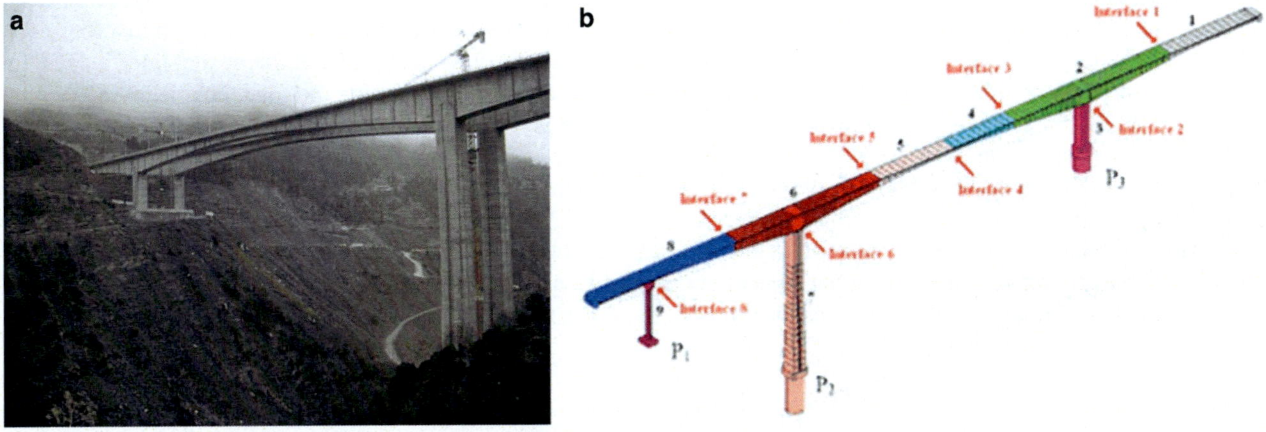

Fig. 3.1 (**a**) Metsovo bridge, (**b**) components of FE model of the bridge

Table 3.1 Number of DOF and percentage modal frequency error for the full (unreduced) and reduced models

	Full model	Reduced model (Retained modes) $\rho = 8$	$\rho = 5$	$\rho = 2$	$\rho = 8$ $\nu = 200$	$\rho = 5$ $\nu = 200$	$\rho = 2$ $\nu = 200$
Internal DOF	558,801	286	100	31	286	100	31
Interface DOF	3,300	3,300	3,300	3,300	306	306	306
Total DOF	562,101	3,586	3,400	3,331	592	406	337
Highest percentage error (%)	0.00	0.02	0.17	1.10	0.20	0.30	1.20

Table 3.1 also shows the fractional error between the modal frequencies computed using the complete FE model and the ones computed using the CMS technique for $\rho = 2, 5$ and 8. It is seen that the error fall below 0.02 % for $\rho = 8$, 0.17 % for $\rho = 5$ and 1.10% for $\rho = 2$. A very good accuracy is achieved for the case of $\rho = 5$.

For the specific application, a large number of generalized coordinates for the reduced system arises from the interface DOFs. A further reduction in the number of generalized coordinates for the reduced system can be achieved by retaining only a fraction of the constrained interface modes [18]. For each interface, it is selected to retain all modes that have frequency less than $\omega_{max} = \nu\omega_c$, where ν is user and problem dependent. Results are computed for $\nu = 200$. The number of interface modes retained is given in Table 3.1. It can also be seen that the fractional error for the lowest 20 modes of the structure fall below 1.20% for $\nu = 200$. In particular, the value of $\nu = 200$ and $\rho = 5$ gives sufficiently accurate results and the number of retained interfaces modes for all interfaces is 306. The reduced system has 406 DOFs from which 100 generalized coordinates are fixed-interface modes for all components and the rest 306 generalized coordinates are constrained interface modes. Using CMS, the number of generalized coordinates is drastically reduced.

For demonstration purposes, the FE model is parameterized using five parameters associated with the modulus of elasticity of one or more structural components shown in Fig. 3.1b. Specifically, the first two parameters θ_1 and θ_2 account respectively for the modulus of elasticity of the pier components 3 and 7 of the bridge. The parameter θ_3 accounts for the modulus of elasticity of the components 1 and 2 of the deck, the parameter θ_4 accounts for the components 4 and 5, while the parameter θ_5 accounts for the components 6 and 8. The component 9 is not parameterized.

The estimation of the parameter values and their uncertainties of the FE model is based on modal frequencies and mode shapes. Simulated, noise contaminated, measured modal frequencies and mode shapes are generated by adding a 1 and 3% Gaussian noise to the modal frequencies and modeshape components, predicted by the nominal non-reduced FE models. The added Gaussian noise reflects the differences observed in real applications between the predictions from a model of a structure and the actual (measured) behavior of the structure. 38 sensors are placed on the bridge to monitor vertical and transverse accelerations. The measured data contain the values of the ten lowest modal frequencies and modeshapes. The model parameters are introduced to scale the nominal values of the properties that they model so that the value of the parameters equal to one corresponds to the nominal value of the FE model.

The model updating is performed using the stochastic simulation algorithm TMCMC with the following settings of the TMCMC parameters: tolCov = 1.0, $\beta = 0.2$ and 1,000 samples per TMCMC stage [6]. The number of FE model runs for the five-parameter model class depends on the number of TMCMC stages which was estimated to be 19. The resulting number of FE model re-analyses are 19,000. The parallelization features of TMCMC [9] were also exploited, taking advantage of the available four-core multi-threaded computer unit to simultaneously run eight TMCMC samples in parallel. For comparison purposes, the computational effort for solving the eigenvalue problem of the original unreduced FE model is approximately 139 s. Multiplying this by the number of 19,000 TMCMC samples and considering parallel implementation in a four-core multi-threaded computer unit, the total computational effort for the model class is expected to be of the order 7 days. In contrast, for the reduced-order models for $\rho = 8$, the computational demands for running the model class are reduced to approximately 13 h (759 min), while for the reduced-order models for $\rho = 8$ and $\nu = 200$ these computational demands are drastically reduced to 14 min. It is thus evident that a drastic reduction in computational effort for performing the structural identification based on a set of monitoring data is achieved from approximately 7 days for the unreduced model class to 14 min for the reduced model classes corresponding to $\rho = 8$ and $\nu = 200$, without compromising the predictive capabilities of the proposed parameter estimation methodology. This results in a factor of over 500 reduction in computational effort. It should be noted that kriging technique further reduces the computational effort by approximately one order of magnitude so that the updating of the 562,101 DOF finite element model, requiring 19,000 model runs, can be performed in 2 min which is a remarkable reduction in computational effort.

3.5 Conclusions

Asymptotic approximations and stochastic simulation algorithms (e.g. the TMCMC algorithm) used in Bayesian model uncertainty quantification, calibration and propagation requires a large number of FE model simulation runs. For large size FE models with hundred of thousands or even million DOFs and localized nonlinearities, the computational demands involved in the optimization or TMCMC sampling algorithms may be excessive. Drastic reductions can be achieved at the system, algorithm and computer equipment level. At the system level, CMS techniques that exploit certain schemes often encountered in FE model parameterization are shown to be effective in completely avoiding the large number of eigen-problem re-analyses within the components or interfaces, required during the application of the optimization or TMCMC sampling algorithms. Thus parameterization consistent CMS techniques result in drastic reduction of the computational effort. At the level of the algorithm, surrogate models are well adapted to the TMCMC algorithm for significantly reducing the number of full model runs required, without sacrificing the accuracy in the surrogate estimates. At the computer hardware level, parallel computing algorithms are also very well suited to be used with TMCMC algorithm to efficiently distribute the computations in available multi-core CPUs. Application of the framework to uncertainty calibration of a structural model using vibration measurements was emphasized in this work. The method has also been successfully applied to structural health monitoring for identifying the location and severity of damage [8]. The fast computing techniques implemented within the Bayesian framework can also be used for updating robust model-based predictions and reliability given monitoring data.

Acknowledgements This research has been co-financed by the European Union (European Social Fund – ESF) and Greek national funds through the Operational Program "Education and Lifelong Learning" of the National Strategic Reference Framework (NSRF) – Research Funding Program: Aristeia.

References

1. Beck JL, Katafygiotis LS (1998) Updating models and their uncertainties. I: Bayesian statistical framework. ASCE J Eng Mech 124(4): 455–460
2. Beck JL (2010) Bayesian system identification based on probability logic. Struct Control Health Monit 17(7):825–847
3. Yuen KV (2010) Bayesian methods for structural dynamics and civil engineering. Wiley, Singapore
4. Papadimitriou C, Beck JL, Katafygiotis LS (2001) Updating robust reliability using structural test data. Probab Eng Mech 16(2):103–113
5. Metropolis N, Rosenbluth AW, Rosenbluth MN, Teller AH, Teller E (1953) Equation of state calculations by fast computing machines. J Chem Phys 21130(62):1087–1092
6. Ching J, Chen YC (2007) Transitional Markov Chain Monte Carlo method for Bayesian updating, model class selection, and model averaging ASCE J Eng Mech 133:816–832
7. Craig RR Jr, Bampton MCC (1965) Coupling of substructures for dynamic analysis. AIAA J 6(7):678–685
8. Papadimitriou C, Papadioti DC (2012) Component mode synthesis techniques for finite element model updating. Comput Struct. doi:10.1016/j.compstruc.2012.10.018
9. Angelikopoulos P, Papadimitriou C, Koumoutsakos P (2012) Bayesian uncertainty quantification and propagation in molecular dynamics simulations: a high performance computing framework. J Chem Phys 137(14). doi:10.1063/1.4757266
10. Beck JL, Yuen KV (2004) Model selection using response measurements: Bayesian probabilistic approach. ASCE J Eng Mech 130(2):192–203
11. Papadimitriou C, Katafygiotis LS (2004) Bayesian modeling and updating. In: Nikolaidis N, Ghiocel DM, Singhal S (eds) Engineering design reliability handbook. CRC, Boca Raton, FL
12. Ntotsios E, Papadimitriou C (2008) Multi-objective optimization algorithms for finite element model updating. In: ISMA2008 international conference on noise and vibration engineering, Leuven, pp 1895–1909
13. Haario H, Laine M, Mira A, Saksman E (2006) DRAM: efficient adaptive MCMC. Stat Comput 16:339–354
14. Muto M, Beck JL (2008) Bayesian updating and model class selection using stochastic simulation J Vib Control 14:7–34
15. Goller B (2011) Stochastic model validation of structural systems. Ph.D. Dissertation, Department of Engineering Mechanics, University of Innsbruck
16. Goller B, Broggi M, Calvi A, Schueller GI (2011) A stochastic model updating technique for complex aerospace structures. Finite Elem Anal Des 47(7):739–752
17. Lophaven SN, Nielsen HB, Sondergaard J (2002) Dace, a matlab kriging toolbox, technical report, IMM-TR-2002-12, DTU, DK-2800 Kgs, Lyngby
18. Castanier MP, Tan Y-C, Pierre C (2001) Characteristic constraint modes for component mode synthesis. AIAA J 39(6):1182–1187

Chapter 4
Bayesian Uncertainty Quantification and Propagation in Nonlinear Structural Dynamics

Dimitrios Giagopoulos, Dimitra-Christina Papadioti, Costas Papadimitriou, and Sotirios Natsiavas

Abstract A Bayesian uncertainty quantification and propagation (UQ&P) framework is presented for identifying nonlinear models of dynamic systems using vibration measurements of their components. The measurements are taken to be either response time histories or frequency response functions of linear and nonlinear components of the system. For such nonlinear models, stochastic simulation algorithms are suitable Bayesian tools to be used for identifying system and uncertainty models as well as perform robust prediction analyses. The UQ&P framework is applied to a small scale experimental model of a vehicle with nonlinear wheel and suspension components. Uncertainty models of the nonlinear wheel and suspension components are identified using the experimentally obtained response spectra for each of the components tested separately. These uncertainties, integrated with uncertainties in the body of the experimental vehicle, are propagated to estimate the uncertainties of output quantities of interest for the combined wheel-suspension-frame system. The computational challenges are outlined and the effectiveness of the Bayesian UQ&P framework on the specific example structure is demonstrated.

Keywords System identification • Uncertainty identification • Bayesian inference • Nonlinear dynamics • Substructuring

4.1 Introduction

Structural model updating methods (e.g. [1]) are used to reconcile linear and nonlinear mathematical models of a mechanical system with available experimental data from component or system tests. For complex structural dynamics models it is often the case that these models are usually discretized linear finite element (FE) models for a large part of the structure with localized nonlinearities in isolated structural parts. Examples include vehicle models that consist of linear structural components, such as the body of the vehicle, and nonlinear structural components, such as the suspension and the wheel components, which may exhibit strongly nonlinear behaviour. To build high fidelity models for such complex structures with localized nonlinearities, one should reconcile linear and nonlinear structural models with experimental data available at both the component and system level. This work presents the challenges of Bayesian uncertainty quantification and propagation of complex nonlinear structural dynamics models and results of identification of nonlinear models of a small scale experimental vehicle consisting of linear and nonlinear components. The goal is to build high fidelity models of the components to simulate the behaviour of the combined system.

Bayesian techniques [2, 3] have been proposed to quantify the uncertainty in the parameters of a structural model, select the best model class from a family of competitive model classes [4,5], as well as propagate uncertainties for robust response

D. Giagopoulos
Department of Mechanical Engineering, University of Western Macedonia, Kozani, Greece
e-mail: dgiagopoulos@uowm.gr

D.-C. Papadioti · C. Papadimitriou
Department of Mechanical Engineering, University of Thessaly, Volos, Greece
e-mail: costasp@uth.gr; dxpapadioti@uth.gr

S. Natsiavas (✉)
Department of Mechanical Engineering, Aristotle University, Thessaloniki, Greece
e-mail: natsiava@auth.gr

T. Simmermacher et al. (eds.), *Topics in Model Validation and Uncertainty Quantification, Volume 5*,
Conference Proceedings of the Society for Experimental Mechanics Series 41,
DOI 10.1007/978-1-4614-6564-5_4, © The Society for Experimental Mechanics, Inc. 2013

and reliability predictions [6]. Posterior probability density functions (PDFs) are derived that quantify the uncertainty in the model parameters based on the data. For nonlinear structural models, the measurements are taken to be either response time histories or frequency response functions of nonlinear systems. Computationally intensive stochastic simulation algorithms (e.g., Transitional MCMC [7]) are suitable tools for identifying system and uncertainty models as well as for performing robust prediction analyses. These algorithms require a large number of system analyses to be performed over the space of uncertain parameters. However, for relatively large order FE models involving hundreds of thousands or even million degrees of freedom and localized nonlinear actions activated during system operation, such re-analyses may require excessive computational time. Methods for drastically reducing the computational demands at the system, algorithm and hardware levels involved in the implementation of Bayesian framework have recently been developed. At the system level, efficient computing techniques can be integrated with the Bayesian framework to handle large order models and localized nonlinear action. Specifically, component mode synthesis and multilevel substructuring techniques can achieve substantial reductions in computational effort.

At the system level, efficient computing techniques are integrated with Bayesian techniques to efficiently handle large order models of hundreds of thousands or millions degrees of freedom (DOF) and localized nonlinear actions activated during system operation. Specifically, fast and accurate component mode synthesis (CMS) techniques have recently been proposed [8], consistent with the FE model parameterization, to achieve drastic reductions in computational effort. In addition, automated multilevel substructuring techniques [9] are used to achieve substantial reductions in computational effort in the re-analysis of linear substructures. At the level of the Transitional MCMC (TMCMC) algorithm, surrogate models are adopted to drastically reduce the number of computationally expensive full model runs [10]. At the computer hardware level, parallel computing algorithms are proposed to efficiently distribute the computations in available multi-core CPUs [10].

In this work the Bayesian UQ&P framework is applied to identify models of linear and nonlinear components of a small scale experimental model of a vehicle. The identification of the uncertainty models of the nonlinear wheel and suspension components is investigated using the experimentally obtained response spectra. The uncertainty models for the vehicle frame are also obtained using experimental data. The uncertainty is propagated to output quantities of interest for the combined wheel-suspension-frame system. The computational challenges and efficiency of the Bayesian UQ&P framework are outlined. The effectiveness of the framework on the specific example structure is discussed.

4.2 Review of Bayesian Formulation for Parameter Estimation and Model Class Selection

Consider a parameterized FE model class M of a nonlinear structure and let $\underline{\theta}_m \in R^N_{\underline{\theta}}$ be the structural model parameters to be estimated using a set of measured response quantities. In nonlinear structural dynamics, the measured quantities may consist of full response time histories $D = \{\underline{\hat{y}}_k \in R^{N_0}, k = 1, \cdots, N\}$ at N_0 DOF and at different time instants $t = k\Delta t$, where k is the time index and N is the number of sampled data with sampling period Δt, or response spectra $D = \{\underline{\hat{y}}_k \in R^{N_0}, k = 1, \cdots, N\}$ at different frequencies ω_k, where $k$is a frequency domain index. In addition, let $\{\underline{y}_k(\underline{\theta}_m) \in R^{N_0}, k = 1, \cdots, N\}$ be the model response predictions (response time histories or response spectra), corresponding to the DOFs where measurements are available, given the model class M and the parameter set $\underline{\theta}_m \in R^{N_\theta}$. It is assumed that the observation data and the model predictions satisfy the prediction error equation

$$\underline{\hat{y}}_k = \underline{y}_k(\underline{\theta}_m|\mathrm{M}_m) + \underline{e}_k \tag{4.1}$$

where the error term $\underline{e}_k \sim N(\underline{\mu}, \Sigma(\underline{\theta}_e))$ is a Gaussian vector with mean zero and covariance $\Sigma(\underline{\theta}_e)$. It is assumed that the error terms $\underline{e}_k$, $k = 1, \cdots, N$ are independent. This assumption may be reasonable for the case where the measured quantities are the response spectra. However, for measured response time histories this assumption is expected to be violated for small sampling periods. The effect of correlation in the prediction error models is not considered in this study. The notation $\Sigma(\underline{\theta}_e)$ is used to denote that a model is postulated for the prediction error covariance matrix that depends on the parameter set $\underline{\theta}_e$.

Bayesian methods are used to quantify the uncertainty in the model parameters as well as select the most probable FE model class among a family of competitive model classes based on the measured data. The structural model class M is augmented to include the prediction error model class that postulates zero-mean Gaussian models. As a result, the parameter set is augmented to include the prediction error parameters $\underline{\theta}_e$. Using PDFs to quantify uncertainty and following the Bayesian formulation (e.g. [2, 3, 11]), the posterior PDF $p(\underline{\theta}|D,\mathrm{M})$ of the structural model and the prediction error parameters $\underline{\theta} = (\underline{\theta}_m, \underline{\theta}_e)$ given the data D and the model class M can be obtained in the form

$$p(\underline{\theta}|D,\mathrm{M}) = \frac{[p(D|\mathrm{M})]^{-1}}{(2\pi \det \Sigma(\underline{\theta}_e))^{NN_0/2}} \exp\left[-\frac{1}{2}J(\underline{\theta})\right] \pi(\underline{\theta}|\mathrm{M}) \tag{4.2}$$

where

$$J(\theta) = \sum_{r=1}^{m} [\underline{y}(\underline{\theta}_m) - \underline{\hat{y}}]^T \Sigma^{-1}(\underline{\theta}_e)[\underline{y}(\underline{\theta}_m) - \underline{\hat{y}}]^T \tag{4.3}$$

is the weighted measure of fit between the measured and model predicted quantities, $\pi(\underline{\theta}|\mathrm{M})$ is the prior PDF of the model parameters $\underline{\theta}$ and $p(D|\mathrm{M})$ is the evidence of the model class M.

For a large enough number of experimental data, and assuming for simplicity a single dominant most probable model, the posterior distribution of the model parameters can be asymptotically approximated by the multi-dimensional Gaussian distribution [2, 11] centered at the most probable value $\hat{\underline{\theta}}$ of the model parameters that minimizes the function $g(\underline{\theta};\mathrm{M}) = -\ln p(\underline{\theta}|D,\mathrm{M})$ with covariance equal to the inverse of the Hessian $h(\underline{\theta})$ of the function $g(\underline{\theta};\mathrm{M})$ evaluated at the most probable value. For a uniform prior distribution, the most probable value of the FE model parameters $\underline{\theta}$ coincides with the estimate obtained by minimizing the weighted residuals in (4.3). An asymptotic approximation based on Laplace's method is also available to give an estimate of the model evidence $p(D|\mathrm{M})$ [11]. The estimate is also based on the most probable value of the model parameters and the value of the Hessian $h(\underline{\theta})$ evaluated at the most probable value.

The Bayesian probabilistic framework is also used to compare two or more competing model classes and select the optimal model class based on the available data. Consider a family $\mathrm{M} = \{\mathrm{M}_i,\ i = 1, \cdots, \mu\}$, of μ alternative, competing, parameterized FE and prediction error model classes and let $\underline{\theta}_i \in R^{N_{\underline{\theta}_i}}$ be the free parameters of the model class M_i. The posterior probabilities $P(\mathrm{M}_i|D)$ of the various model classes given the data D is [4]

$$P(\mathrm{M}_i|D) = \frac{p(D|\mathrm{M}_i)P(\mathrm{M}_i)}{p(D|\mathrm{M}_{Fam})} \tag{4.4}$$

where $P(\mathrm{M}_i)$ is the prior probability and $p(D|\mathrm{M}_i)$ is the evidence of the model class M_i. The optimal model class M_{best} is selected as the one that maximizes $P(M_i|D)$ given by (4.4). For the case where no prior information is available, the prior probabilities are assumed to be $P(M_i) = 1/\mu$, so the model class selection is based solely on the evidence values.

The asymptotic approximations may fail to give a good representation of the posterior PDF in the case of multimodal distributions or for unidentifiable cases manifested for relatively large number of model parameters in relation to the information contained in the data. For more accurate estimates, one should use SSA to generate samples that populate the posterior PDF in (4.2). Among the SSA available, the TMCMC algorithm [7] is one of the most promising algorithms for selecting the most probable model class among competitive ones, as well as finding and populating with samples the importance region of interest of the posterior PDF, even in the unidentifiable cases and multi-modal posterior probability distributions. In addition, the TMCMC samples $\underline{\theta}^{(i)}$, $i = 1, \cdots, N_s$ drawn from the posterior distribution can be used to yield an estimate of the evidence $p(D|\mathrm{M}_i)$ required for model class selection [7, 12, 13]. The TMCMC samples can further be used for estimating the probability integrals encountered in robust prediction of various performance quantities of interest [6]. In particular, if $q(\underline{\theta})$ is an output quantity of interest conditional on the value of the parameter set $\underline{\theta}$, the posterior robust measure of q given the data and taking into account the uncertainty in $\underline{\theta}$ is obtained from the sample estimate

$$E(q|D,\mathrm{M}) = \frac{1}{N_s} \sum_{i=1}^{N} \mu_q(\underline{\theta}^{(i)};\mathrm{M}) \tag{4.5}$$

where $\mu_q(\underline{\theta}^{(i)};\mathrm{M})$ is the conditional mean value of $q(\underline{\theta})$ given the model class.

4.3 Application to a Small Scale Laboratory Vehicle Model

In order to simulate the response of a ground vehicle an experimental device was selected and set up [14]. More specifically, the selected frame structure comprises a frame substructure with predominantly linear response and high modal density plus four supporting substructures with strongly nonlinear action. First, Fig. 4.1a shows a picture with an overview of the experimental set up. In particular, the mechanical system tested consists of a frame substructure (parts with red, gray and black color), simulating the frame of a vehicle, supported on four identical substructures. These supporting substructures consist of a lower set of discrete spring and damper units, connected to a concentrated (yellow color) mass, simulating the

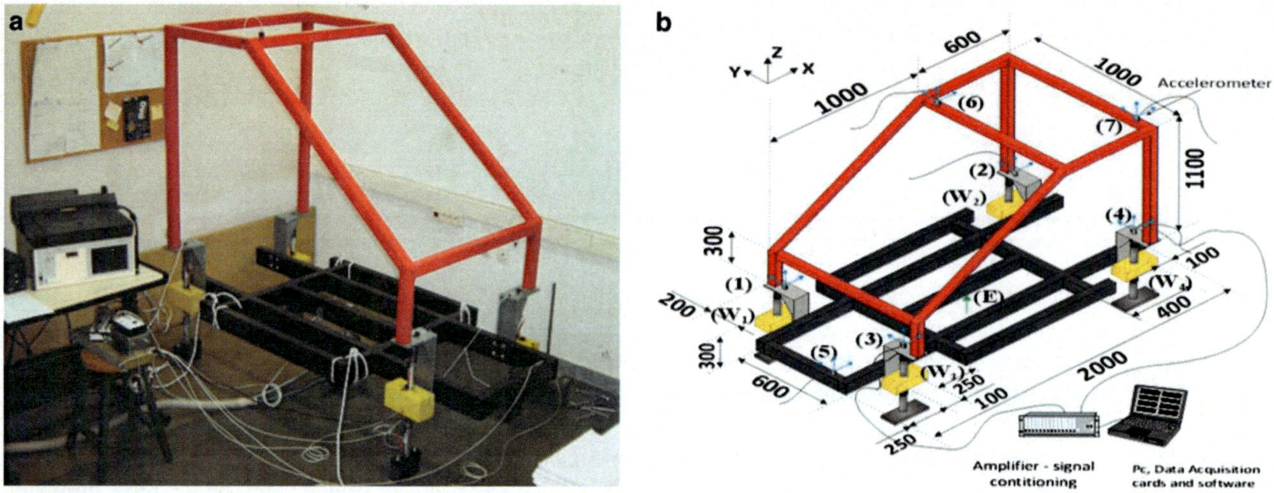

Fig. 4.1 (**a**) Experimental set up of the structure tested, (**b**) dimensions of the frame substructure and measurement points

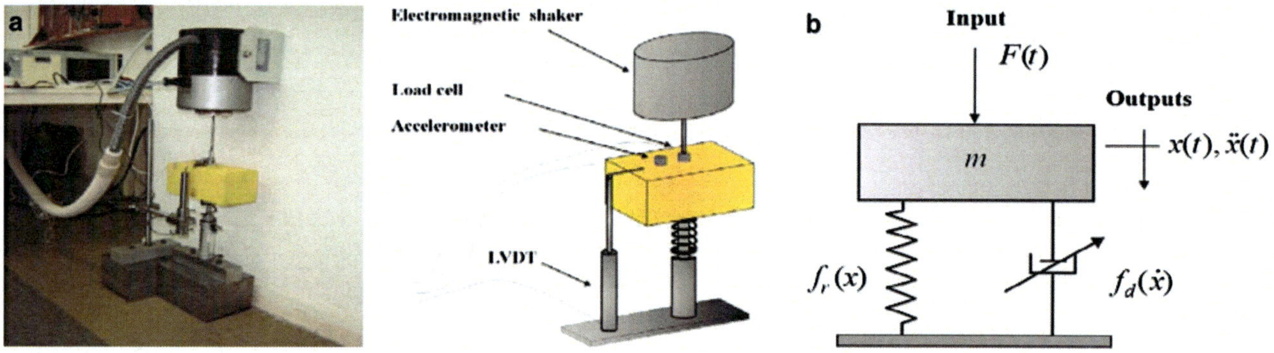

Fig. 4.2 (**a**) Experimental set up for measuring the support stiffness and damping parameters, (**b**) equivalent model

wheel subsystems, as well as of an upper set of a discrete spring and damper units connected to the frame and simulating the action of the vehicle suspension. Also, Fig. 4.1b presents more details and the geometrical dimensions of the frame subsystem. Moreover, the measurement points indicated by 1–4 correspond to connection points between the frame and its supporting structures, while the other measurement points shown coincide with characteristic points of the frame. Finally, point E denotes the point where the electromagnetic shaker is applied.

In order to identify the parameters of the four supporting subsystems, which exhibit strongly nonlinear characteristics, a series of tests was performed. To investigate this further, the elements of the supporting units were disassembled and tested separately. First, Fig. 4.2a shows a picture of the experimental setup and presents graphically the necessary details of the experimental device that was set up for measuring the stiffness and damping properties of the supports, while Fig. 4.2b shows the equivalent mechanical model.

The experimental process was applied separately to both the lower and the upper spring and damper units of the supporting substructures and can be briefly described as follows. First, the system shown in Fig. 4.2 is excited by harmonic forcing through the electromagnetic shaker up until it reaches a periodic steady state response. When this happens, both the history of the acceleration and the forcing signals are recorded at each forcing frequency. Some characteristic results obtained in this manner are presented in the following sequence of graphs. Next, Fig. 4.3a presents the transmissibility function of the system tested, obtained experimentally for three different forcing levels. Specifically, this function is defined as the ratio of the root mean square value of the acceleration to the root mean square value of the forcing signal measured at each forcing frequency. The continuous, dashed and dotted lines correspond to the smallest, intermediate and largest forcing amplitude, respectively. Clearly, the deviations observed between the forcing levels indicate that the system examined possesses nonlinear properties. Moreover, neither the applied forcing is harmonic, especially within the frequency range below $\omega = 10\,\text{Hz}$. To illustrate this, Fig. 4.3b shows two periods of the actual excitation force applied for the same three excitation levels in obtaining the results of Fig. 4.3a, which were recorder at a fundamental forcing frequency of $\omega = 4\,\text{Hz}$.

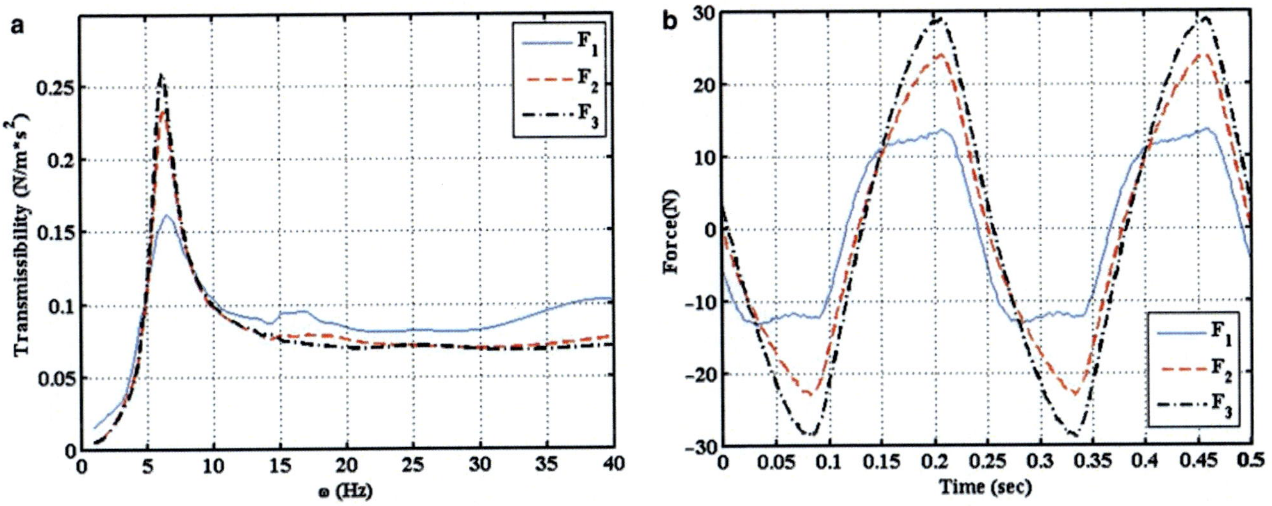

Fig. 4.3 (**a**) Transmissibility function of the support system, for three different forcing levels, (**b**) history of the external force applied with a fundamental harmonic frequency $\omega = 4\,\text{Hz}$

A number of models of the restoring and damping forces, say f_r and f_d, respectively, were tried for modeling the action of the supports and compared with the experimental results. The classic linear dependence of the restoring force on the displacement and of the damping forces on the velocity of the support unit was first assumed. However, critical comparison with the experimental results using the Bayesian model selection framework demonstrated that the outcome was unacceptable in terms of accuracy. Eventually it was found that an acceptable form of the restoring forces is the one where they remain virtually in a linear relation with the extension of the spring, namely

$$f_r(x) = kx \tag{4.6}$$

while the damping force was best approximated by the following formula

$$f_d(\dot{x}) = c_1\dot{x} + \frac{c_2\dot{x}}{c_3 + |\dot{x}|} \tag{4.7}$$

As usual, the linear term in the last expression is related to internal friction at the support, while the nonlinear part is related to the existence and activation of dry friction. More specifically, in the limit $c_3 \to 0$, the second term in the right hand side of Eq. (4.7) represents energy dissipation action corresponding to dry friction. On the other side, in the limit $c_3 \to \infty$, this term represents classical viscous action and can actually be absorbed in the first term.

4.4 Results

The value of the parameters appearing in the assumed models of the restoring and damping forces of the supports, like the coefficients k, c_1, c_2 and c_3 in Eqs. (4.6) and (4.7), are determined by applying the Bayesian uncertainty quantification and calibration methodology. Results are obtained based on experimental response spectra values for both the displacement and acceleration of either the wheel or the suspension component. It is assumed that the prediction errors in the Bayesian formulation are uncorrelated with prediction error variance $\Sigma = diag(\Sigma_1, \Sigma_2) = diag(\sigma_1^2 I, \sigma_2^2 I)$, where $\Sigma_1 = \sigma_1^2 I$ and $\Sigma_2 = \sigma_2^2 I$ are the covariance matrices for the prediction errors corresponding to the displacements and accelerations, respectively. The parameter space is six dimensional and includes $\theta = (k, c_1, c_2, c_3, \sigma_1, \sigma_2)$. Parameter estimation results are obtained using the parallelized and surrogate-based version [10] of the TMCMC algorithm [7] with 500 samples per stage. Eight computer workers were used to perform in parallel the computations involved in the TMCMC algorithm. The computational time required to run all 5,500 samples for the 11 TMCMC stages, without surrogate approximation, for the SDOF model is approximately 7 h. Surrogate modeling [10] reduces further this time by approximately one order of magnitude. For illustration purposes, results for the TMCMC samples projected in the two-dimensional parameter spaces (k_1, c_1) and (c_1, c_2)

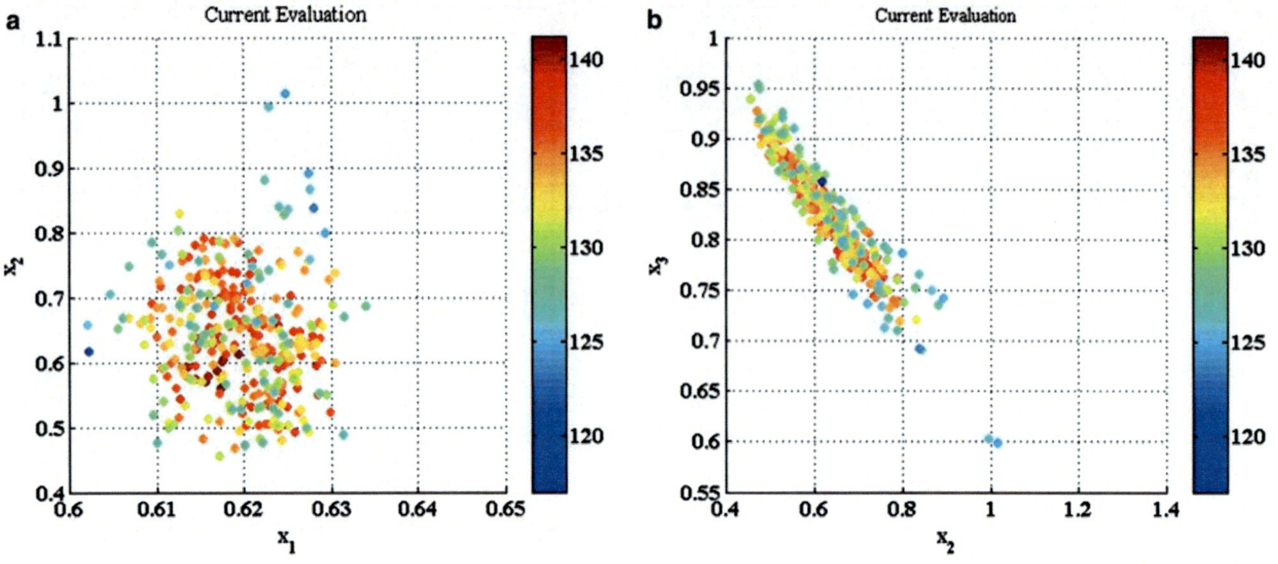

Fig. 4.4 Model parameter uncertainty: projection of TMCMC samples in the two dimensional parameter spaces

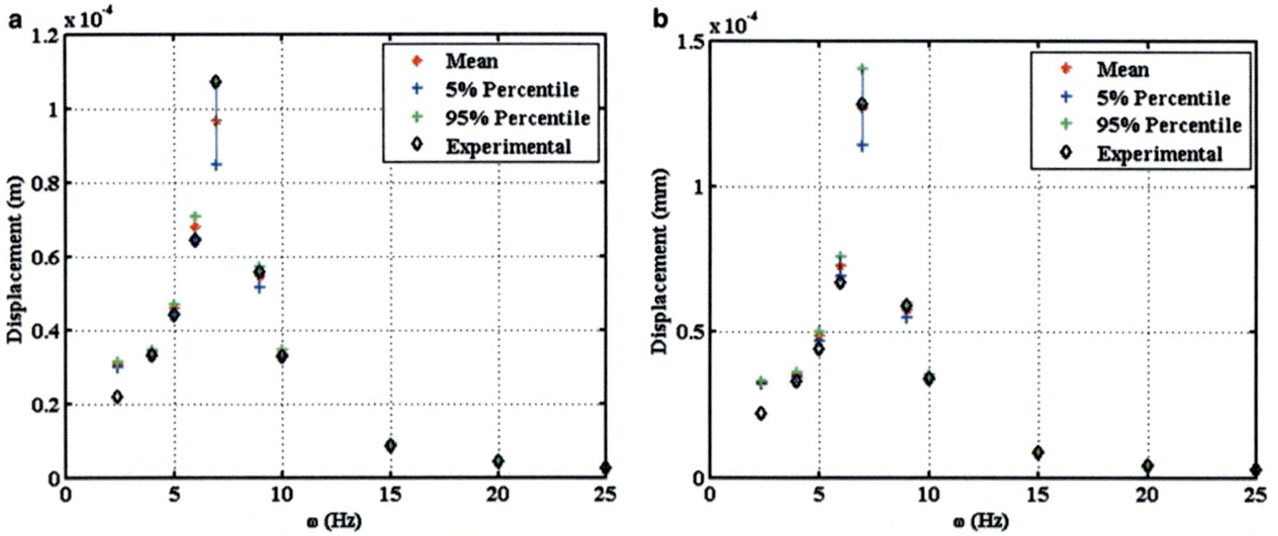

Fig. 4.5 Uncertainty propagation: displacement response spectra uncertainty along with comparisons with the experimental data for the suspension component. (**a**) Moderate excitation level, (**b**) strong excitation level

are shown in Fig. 4.4 for the SDOF system, shown in Fig. 4.2b, corresponding to the suspension component. It is clear that the uncertainties in the damping parameters c_1 and c_2 are relatively high and c_1 and c_2 are highly correlated along certain directions in the parameter space.

The parameter uncertainties are propagated through the SDOF model to estimate the uncertainties in the displacement and acceleration response spectra. The results are shown in Figs. 4.5 and 4.6 for the displacement and acceleration response spectra, respectively and are compared to the experimental values of the response spectra. An adequate fit is observed. Discrepancies between the model predictions and the experimental measurements are mainly due to the model errors related to the selection of the particular forms of the restoring force curves in (4.6) and (4.7). The Bayesian model selection strategy based on Eq. (4.5) can be used to select among alternative restoring force models in an effort to improve the observed fit.

The above procedure has been repeated for the wheel component to identify the uncertainties in the linear stiffness and nonlinear damping model. In addition, the uncertainties in nine stiffness-related parameters of the frame component were also estimated using the Bayesian methodology and the experimental values the first ten modal frequencies and the mode shape components at 72 locations of the frame [15]. The linear finite element model has 45,564 DOFs. Due to excessive computational cost arising in stochastic simulation algorithms, the model was reduced using a recently developed CMS

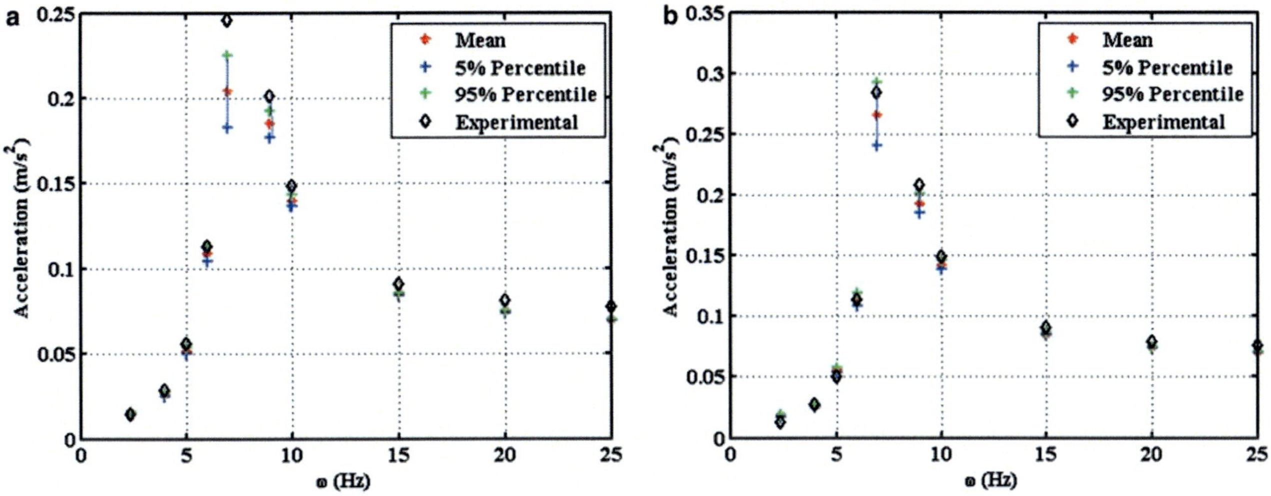

Fig. 4.6 Uncertainty propagation: acceleration response spectra uncertainty along with comparisons with the experimental data for the suspension component. (**a**) Moderate excitation level, (**b**) strong excitation level

method for FE model updating [8]. The reduced model has 30 DOFs, resulting in substantial computational savings of more than two orders of magnitude. Due to space limitations, results of the parameter estimation are not shown here.

The estimates of the model parameter values and their uncertainties for each component are used to build the model for the combined wheel-suspension-frame structure. The number of DOFs of the nonlinear model of the combined structure is 45,568. The parametric uncertainties are then propagated to uncertainties in the response of the combined structure. The CMS was again used to reduce the number of DOFs to 34 and thus drastically reduce the computational effort that arises from the re-analyses due to the large number of TMCMC samples and the nonlinearity of the combined system. Selected uncertainty propagation results are next presented. Specifically, the parameters of the wheel model and the model of the frame structure are kept to their mean values and only the uncertainties in the model parameters of the suspension components are considered. Such uncertainties are propagated to uncertainties for the acceleration transmissibility function at a point on the wheel, the connection of the wheel with the frame and an internal point on the frame as shown in Fig. 4.7. It is observed that a large uncertainty in the response spectra is obtained at the resonance region close to 3.4 Hz, which is dominated by local wheel body deflections. The response in the resonance regions close to 58 and 68 Hz is mainly dominated by deflection of the frame structure. It is observed that the uncertainties in the suspension parameters do not significantly affect the response spectra at the resonance regions. As a result, response spectra obtained experimentally in these resonance regions for the complete vehicle model are not expected to be adequate to identify uncertainties in the parameters of the suspension model.

4.5 Conclusions

A Bayesian UQ&P framework was presented for identifying nonlinear models of dynamic systems using vibration measurements of their components. The use of Bayesian tools, such as stochastic simulation algorithms (e.g., TMCMC algorithm), may often result in excessive computational demands. Drastic reduction in computational effort to manageable levels is achieved using component mode synthesis, surrogate models and parallel computing algorithms. The framework was demonstrated by identifying the linear and nonlinear components of a small-scale laboratory vehicle model using experimental response spectra available separately for each component. Such model uncertainty analyses for each component resulted in building a high fidelity model for the combined system to be used for performing reliable robust response predictions that properly take into account model uncertainties. The theoretical and computational developments in this work can be used to identify and propagate uncertainties in large order nonlinear dynamic systems that consist of a number of linear and nonlinear components.

Acknowledgements This research has been co-financed by the European Union (European Social Fund – ESF) and Greek national funds through the Operational Program "Education and Lifelong Learning" of the National Strategic Reference Framework (NSRF) – Research Funding Program: Heraclitus II. Investing in knowledge society through the European Social Fund.

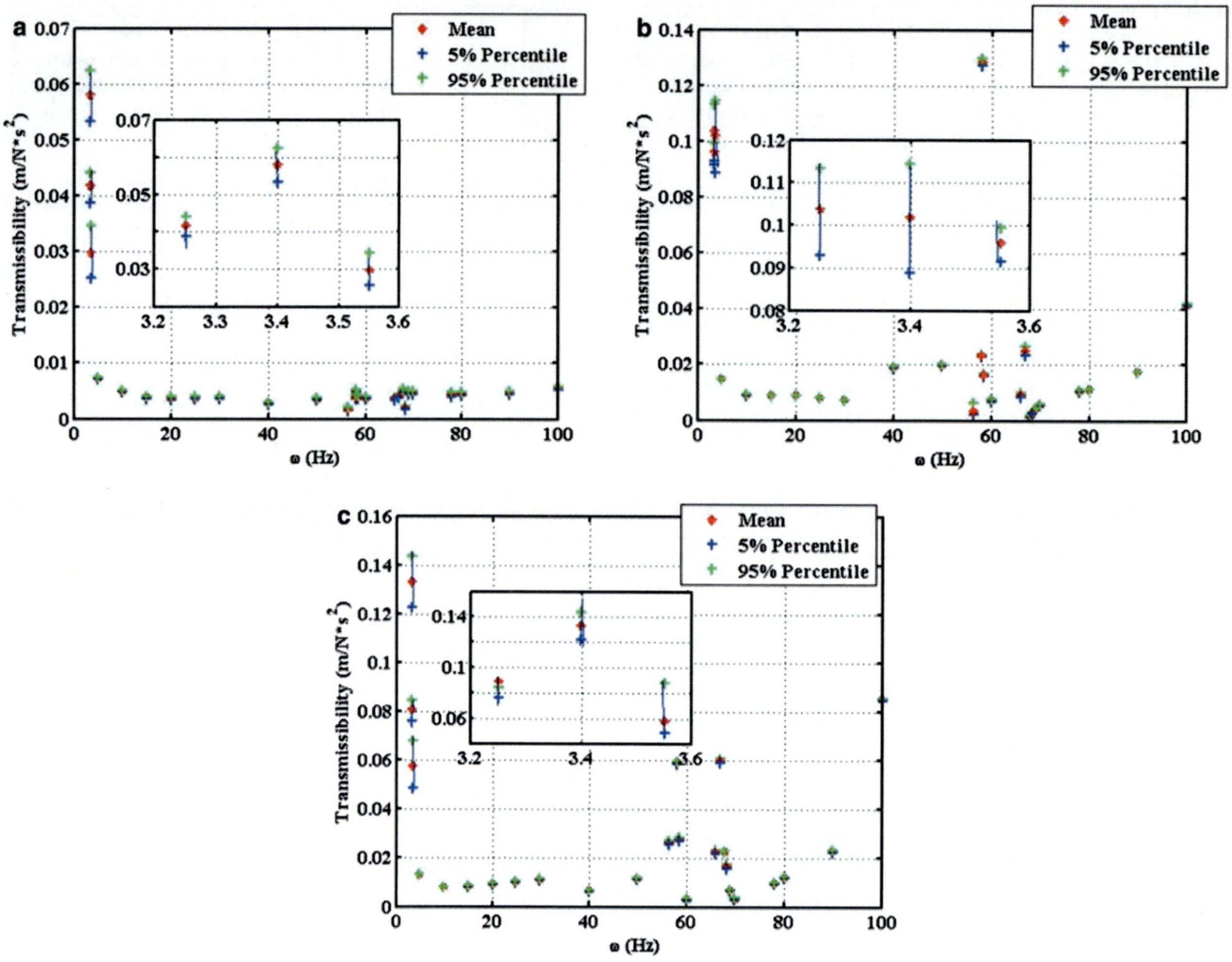

Fig. 4.7 Uncertainty propagation: acceleration transmissibility function uncertainty for combined system. (**a**) Wheel DOF, (**b**) DOF at connection between suspension and frame, (**c**) frame DOF

References

1. Yuen KV, Kuok SC (2011) Bayesian methods for updating dynamic models. Appl Mech Rev 64(1):010802
2. Beck JL, Katafygiotis LS (1998) Updating models and their uncertainties- I: Bayesian statistical framework. ASCE J Eng Mech 124(4):455–461
3. Yuen KV (2010) Bayesian methods for structural dynamics and civil engineering. Wiley, Singapore/Hoboken
4. Beck JL, Yuen KV (2004) Model selection using response measurements: Bayesian probabilistic approach. ASCE J Eng Mech 130(2):192–203
5. Yuen KV (2010) Recent developments of Bayesian model class selection and applications in civil engineering. Struct Saf 32(5):338–346
6. Papadimitriou C, Beck JL, Katafygiotis LS (2001) Updating robust reliability using structural test data. Probab Eng Mech 16:103–113
7. Ching J, Chen YC (2007) Transitional Markov Chain Monte Carlo method for Bayesian updating, model class selection, and model averaging. ASCE J Eng Mech 133:816–832
8. Papadimitriou C, Papadioti DC (2012) Component mode synthesis techniques for finite element model updating. Comput Struct. doi:10.1016/j.compstruc.2012.10.018
9. Papalukopoulos C, Natsiavas S (2007) Dynamics of large scale mechanical models using multi-level substructuring. ASME J Comput Nonlinear Dyn 2:40–51
10. Angelikopoulos P, Papadimitriou C, Koumoutsakos P (2012) Bayesian uncertainty quantification and propagation in molecular dynamics simulations: a high performance computing framework. J Chem Phys 137(14). doi:10.1063/1.4757266
11. Christodoulou K, Papadimitriou C (2007) Structural identification based on optimally weighted modal residuals. Mech Syst Signal Process 21:4–23
12. Muto M, Beck JL (2008) Bayesian updating and model class selection using stochastic simulation J Vib Control 14:7–34

13. Beck JL, Au SK (2002) Bayesian updating of structural models and reliability using Markov chain Monte Carlo simulation. ASCE J Eng Mech 128(4):380–391
14. Giagopoulos D, Natsiavas S (2007) Hybrid (numerical-experimental) modeling of complex structures with linear and nonlinear components. Nonlinear Dyn 47:193–217
15. Papadimitriou C, Ntotsios E, Giagopoulos D, Natsiavas S (2011) Variability of updated finite element models and their predictions consistent with vibration measurements. Struct Control Health Monit. doi:10.1002/stc.453

Chapter 5
Probabilistic Damage Identification of the Dowling Hall Footbridge Using Bayesian FE Model Updating

Iman Behmanesh and Babak Moaveni

Abstract This paper presents a probabilistic damage identification study on a full-scale structure, the Dowling Hall Footbridge, through Bayesian finite element (FE) model updating. The footbridge is located at Tufts University campus and is equipped with a continuous monitoring system that measures the ambient acceleration response of the bridge. A set of data is recorded once an hour or when triggered by large vibrations. The modal parameters of the footbridge are extracted based on each set of measured ambient vibration data and are used for model updating. In this study, effects of physical damage are simulated by loading a small segment of footbridge's deck with concrete blocks. The footbridge deck is divided into five segments and the added mass on each segment is considered as an updating parameter. Overall, 72 sets of data are collected during the loading period (i.e., damaged state of the bridge) and different subsets of these data are used to find the location and extent of the damage (added mass). Adaptive Metropolis Hasting algorithm with adaption on the proposal probability density function is successfully used to generate Markov Chains for sampling the posterior probability distributions of the five updating parameters. Effect of the number of data sets used in the identification process is investigated on the posterior probability distributions of the updating parameters.

Keywords Bayesian FE model updating • Adaptive metropolises hasting algorithm • Damage identification • Uncertainty analysis • Dowling hall footbridge

5.1 Introduction

Although the deterministic FE model updating methods have been used for damage identification of several real-world, large-scale structures [1–5], there are few applications of Bayesian FE model updating methods to full-scale complex structures [6–8]. In this paper, the authors investigate the challenges of implementing a Bayesian FE model updating framework for damage identification of a full-scale structure, the Dowling Hall footbridge. Information about deterministic structural identification of this footbridge is available at [9–11]. In this study, damage on the footbridge is simulated by addition of 2.29 metric tons of concrete blocks on a small segment of the bridge deck. The effect of added mass will be similar to a loss of stiffness (commonly used as damage indication) at segments of the bridge. The extracted modal parameters from acceleration time histories of the damaged structure (loaded structure) are used to find the location and extent of damage (added mass) through a Bayesian FE model updating scheme.

The footbridge is divided into five segments and the added mass of each segment is considered as a model parameter to be calibrated in the model updating process. Model updating is performed to minimize the misfit between the measured modal parameters and those from the FE model. An adaptive Metropolis-Hasting algorithm [12–15] is used to sample the posterior probability distributions of the updating parameters given the data and model class. Impact of the amount of data used in the updating process is investigated on the accuracy of probabilistic damage identification results. To do this, six different set of damage identification results are obtained based on 1, 2, 5, 10, 36, and all 72 sets of identified modal parameters.

This paper is organized in the following order. In Sect. 5.2, the Dowling Hall Footbridge and its continuous monitoring system are introduced. Section 5.3 explains how damage is simulated on the footbridge through addition of mass on a segment

I. Behmanesh • B. Moaveni (✉)
Department of Civil and Environmental Engineering, Tufts University, Medford, MA, USA
e-mail: iman.behmanesh@tufts.edu; babak.moaveni@tufts.edu

T. Simmermacher et al. (eds.), *Topics in Model Validation and Uncertainty Quantification, Volume 5*,
Conference Proceedings of the Society for Experimental Mechanics Series 41,
DOI 10.1007/978-1-4614-6564-5_5,

of bridge deck. A brief review of Bayesian FE model updating and the sampling techniques used is provided in Sect. 5.4.1 while the model updating results are reported in Sect. 5.4.2. Finally, the concluding remarks are presented in Sect. 5.5.

5.2 The Dowling Hall Footbridge and Its Continuous Monitoring System

The Dowling Hall Footbridge is located on the Medford campus of Tufts University. Figure 5.1 shows the south view of the footbridge. The bridge consists of two 22 m spans and it is 3.9 m wide. It connects Dowling Hall on its eastern end to Tufts main campus on its western end. The footbridge is composed of a steel frame with a reinforced concrete deck. More details about the structural details of the Dowling Hall Footbridge can be found in [16].

A continuous monitoring system was designed and deployed on the Dowling Hall Footbridge in the fall of 2009 and has been providing continuous data since January 2010. The monitoring system consists of eight accelerometers, which are connected to a data acquisition device and a communication system that transfers the measured data wirelessly to a host computer in the Department of Civil and Environmental Engineering at Tufts University. The monitoring program continuously samples the acceleration channels at a 2,048 Hz sampling rate. A 5-min data sample is recorded once each hour, beginning at the top of the hour or when the 1-s root-mean square (RMS) value of each acceleration channel exceeds 0.03 g. The layout of the accelerometers is shown in Fig. 5.2. More information about design and deployment of this continuous monitoring system can be found in [10]. Figure 5.3 shows the identified modal parameters of the first six most excited vibration modes based on a preliminary test data performed in April 2009. In this plot, mode shapes are interpolated between the sensor locations (indicated by empty circles) using a cubic spline. The data-driven stochastic subspace identification (SSI-Data) method [17] is applied to the cleaned ambient vibration data for modal identification of the footbridge.

Fig. 5.1 South view of Dowling Hall Footbridge

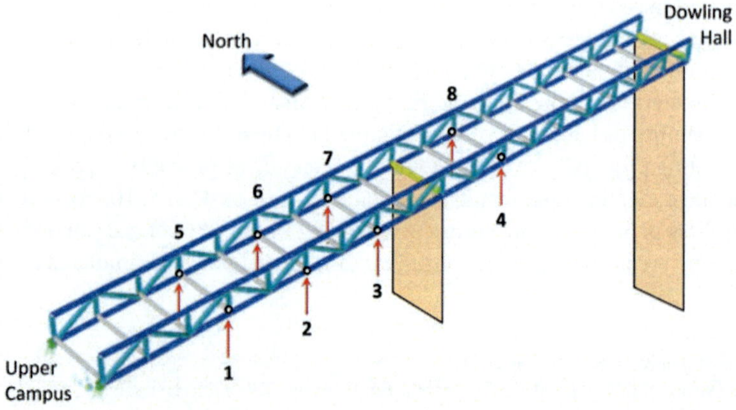

Fig. 5.2 Layout of accelerometers on the bridge

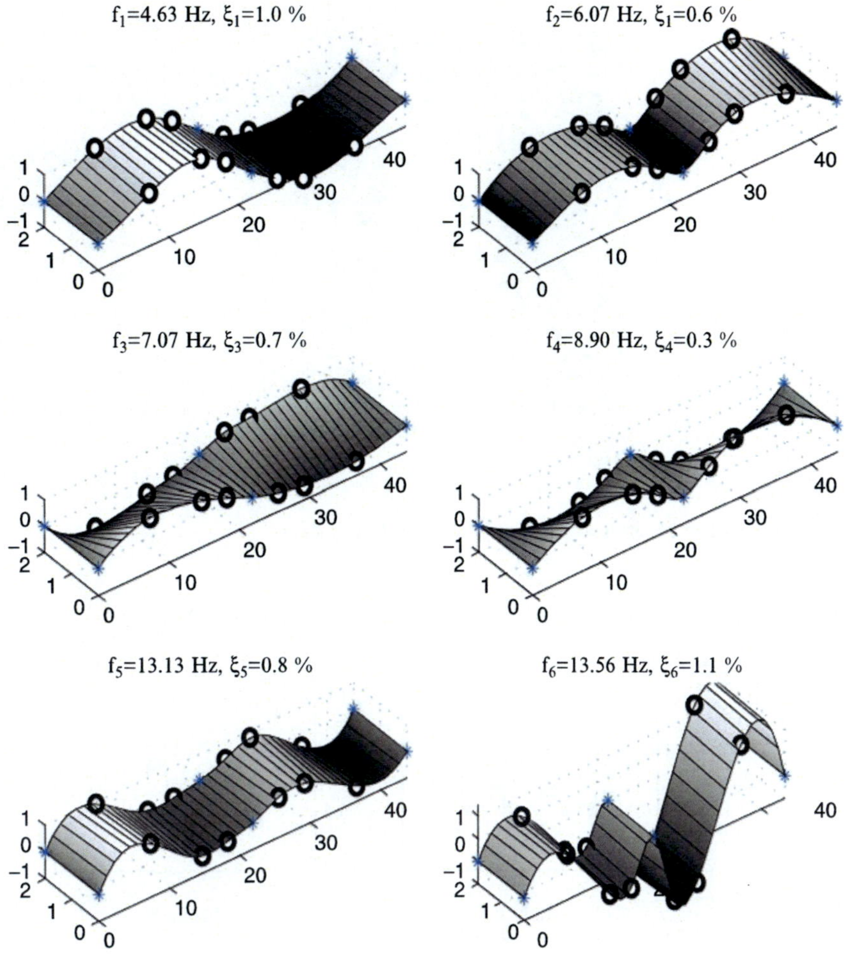

Fig. 5.3 Identified modal parameters from preliminary test data

5.3 Simulation of Structural Damage on the Footbridge

To simulate the effects of damage, a small segment of the footbridge was loaded with 2.29 metric tons (2,290 kg) of concrete blocks for 72 h. Figure 5.4 shows the blocks on the bridge deck and safety measures that were considered for this test. Seventy two sets of measurements (once every hour) were collected and their corresponding modal parameters were extracted that represents the model parameters of the damaged structure. Figure 5.5 shows the effects of added mass on identified natural frequencies of modes 1–6. The plot also shows the identified natural frequencies 24 h before and 24 h after the loading. The grey dots correspond to the undamaged condition while the black dots refer to the damaged (loaded) condition of the bridge. Significant drops of natural frequencies are observed for mode 1, 2, 3, and 4; yet natural frequencies of modes 5 and 6 remained almost unchanged. Note that all the natural frequencies that are identified from hourly recorded vibration data are plotted in this figure. In the model updating process, some of these modal parameters are considered as outliers and are excluded from the data used in updating. Detailed information about statistical information of modal parameters and the rate of data loss due to different sources during loading period are provided in Table 5.1. The most reliably identified modes are modes 3 and 4, and the most unreliable modes are modes 5 and 6. It should be noted that there are two closely spaced vibration modes around the second natural frequency (at approximately 5.9 and 6.1 Hz); however, the simplified FE model predicts only one mode at this frequency range. Therefore, a larger number of data for the second mode are excluded from the identification process.

Fig. 5.4 Concrete blocks loaded on footbridge's deck for 3 days

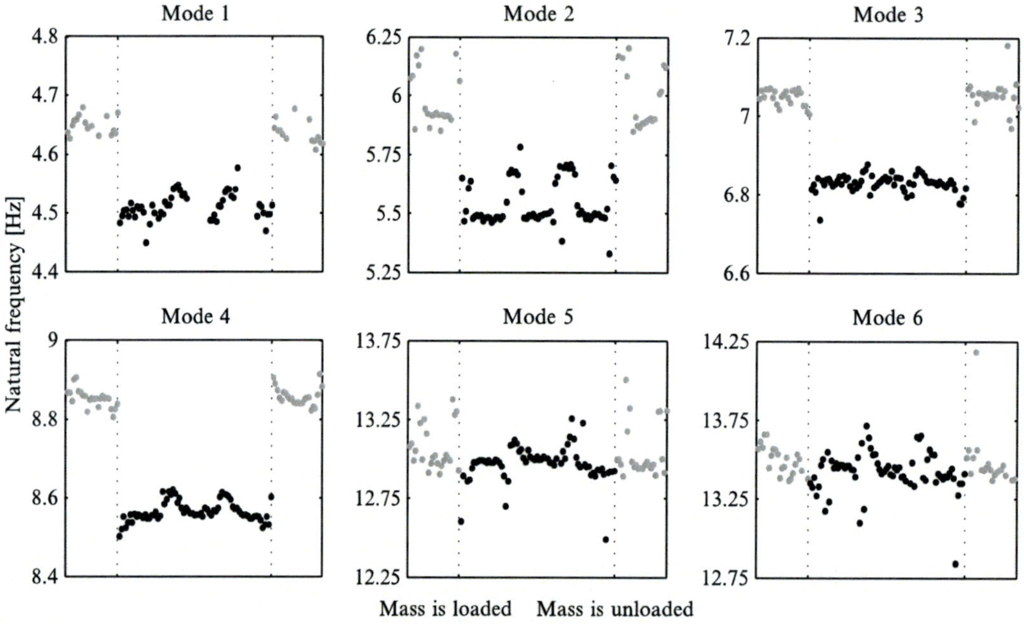

Fig. 5.5 Variation of identified natural frequencies before, during, and after loading

5.4 Bayesian FE Model Updating

The first step in the updating process consists of calibrating an initial FE model of the structure, created based on design information, to a reference FE model that corresponds to "as built" properties of the structure in its undamaged/baseline state. An initial FE model of the footbridge is created based on the design drawings and visual inspection of the footbridge, using the MATLAB-based structural analysis software FEDEASLab [18]. The modal parameters extracted from the data recorded at 4 a.m. of July 29, 2011 are selected as the reference modal parameters. The initial FE model is calibrated so that its modal parameters best match to reference modal parameters using a deterministic sensitivity-based FE model updating. The calibrated model, referred to as reference FE model, is assumed to represent the true undamaged condition of the footbridge. In the Bayesian FE model updating performed in this study, the footbridge is divided to five segments and the added mass of each segment is considered as an updating parameter. These five segments are shown in Fig. 5.6.

Table 5.1 Statistics of identified modal parameters during loading period

Mode	N_i^a	N_o^b	N_a^c	μ_f^d	σ_f^e	cov_f^f	σ_Φ^g	cov_Φ^h
1	52	3	49	4.51	2.10	0.47	2.07	6.93
2	71	51	20	5.67	2.86	0.51	3.10	11.19
3	71	1	70	6.83	1.91	0.28	1.70	5.36
4	72	0	72	6.57	2.51	0.29	0.79	2.54
5	70	12	58	12.97	8.78	0.68	6.26	21.33
6	71	2	69	13.44	10.41	0.77	3.38	10.49

[a] Total number of identified modal parameters
[b] Number of outliers
[c] Total number of available modal parameters
[d] Mean of available natural frequencies (Hz)
[e] Standard deviation of available natural frequencies (%)
[f] Coefficient of variations of available natural frequencies (%)
[g] Standard deviation of available mode shapes (%), defined as the average of the STD of each mode shape component
[h] Coefficient of variations of available mode shapes (%), defined as the average of the COV of each mode shape component

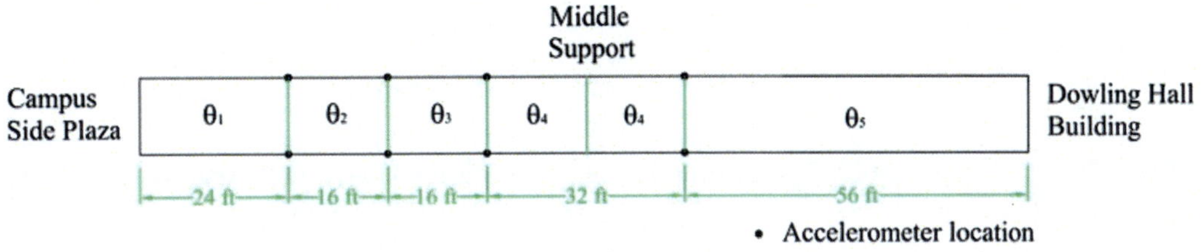

Fig. 5.6 Five segments along the footbridge deck corresponding to the five updating parameters

5.4.1 Bayesian Formulation and Sampling

This section presents a summary of the Bayesian FE model updating formulation used in this study. More detailed formulation of the Bayesian model updating process can be found in seminal publications on this topic [6, 7, 19–23]. According to the Bayes theorem, posterior probability distributions of updating parameters θgiven the measured data $\mathbf{d}$ (modal parameters in this study) and the model class M can be obtained from the likelihood function $p(\mathbf{d}|\theta,M)$, which is the probability of measured data given the updating model parameters, and the conditional prior probability distribution of model parameters $p(\theta|M)$:

$$p(\theta|\mathbf{d},M) = \mathrm{c}p(\mathbf{d}|\theta,M)p(\theta|M) \tag{5.1}$$

The normalization constant c is to ensure that the posterior probability density function (PDF) integrates to one. Only one model class is considered in this study, thus conditioning on the model class Mwill be dropped hereafter.

Data vector $\mathbf{d}$ contains the identified modal parameters extracted from one set of measurements. Modal parameters are assumed to be independently distributed from mode to mode and from natural frequencies to mode shapes. The likelihood function can be written as:

$$p(\mathbf{d}|\theta) = \prod_{\mathrm{m}=1}^{\mathrm{N_m}} p(\tilde{\omega}_\mathrm{m}|\theta)p(\tilde{\Phi}_\mathrm{m}|\theta) \tag{5.2}$$

where N_m is the number of identified modes,$\tilde{\omega}_\mathrm{m} = 2\pi\tilde{f}_\mathrm{m}$ is identified natural frequency and $\tilde{\Phi}_\mathrm{m}$ is the identified mode shape of mode m. The eigen-frequency and mode shape errors are defined as:

$$e_{\omega_m} = \tilde{\omega}_m^2 - \omega_m^2(\theta) \tag{5.3}$$

$$\mathbf{e}_{\Phi_m} = \frac{\tilde{\Phi}_m}{\|\tilde{\Phi}_m\|} - \frac{\Gamma\Phi_m(\theta)}{\|\Gamma\Phi_m(\theta)\|} \tag{5.4}$$

The matrix Γ picks the observed degrees of freedom from model-calculated mode shape $\Phi_m(\theta)$. Assuming the errors are zero-mean Gaussian distributed random variables, PDFs of the measured natural frequencies and mode shapes become:

$$p(\tilde{\omega}_{\mathrm{m}}|\theta) \propto \exp\left(-\frac{1}{2}\frac{\left(\tilde{\omega}_m^2 - \omega_m^2(\theta)\right)^2}{\sigma_{\omega_m^2}^2}\right) \tag{5.5}$$

$$p(\tilde{\Phi}_m|\theta) \propto \exp\left[-\frac{1}{2\sigma_{\Phi_m}^2 \|\tilde{\Phi}_m\|^2}\left(\frac{\tilde{\Phi}_m}{\|\tilde{\Phi}_m\|} - \frac{\Gamma\Phi_m(\theta)}{\|\Gamma\Phi_m(\theta)\|}\right)^T\left(\frac{\tilde{\Phi}_m}{\|\tilde{\Phi}_m\|} - \frac{\Gamma\Phi_m(\theta)}{\|\Gamma\Phi_m(\theta)\|}\right)\right] \tag{5.6}$$

Inserting Eqs. (5.5) and (5.6) into (5.2) yields the following likelihood function:

$$p(\mathbf{d}|\theta) \propto \exp(-\frac{1}{2}J(\theta,\mathbf{d})) \tag{5.7}$$

$$J(\theta,d) = \sum_{\mathrm{m}=1}^{N_m}\frac{\left(\tilde{\omega}_m^2 - \omega_m^2(\theta)\right)^2}{\sigma_{\omega_m^2}^2} + \sum_{m=1}^{N_m}\frac{1}{\sigma_{\Phi_m}^2\|\tilde{\Phi}_m\|^2}\left(\frac{\tilde{\Phi}_m}{\|\tilde{\Phi}_m\|} - \frac{\Gamma\Phi_m(\theta)}{\|\Gamma\Phi_m(\theta)\|}\right)^T\left(\frac{\tilde{\Phi}_m}{\|\tilde{\Phi}_m\|} - \frac{\Gamma\Phi_m(\theta)}{\|\Gamma\Phi_m(\theta)\|}\right) \tag{5.8}$$

In the case of having N_d independent sets of measured modal parameters, the probability $p(\mathbf{d}_1 : \mathbf{d}_{N_d}|\theta)$ can be states as:

$$p(\mathbf{d}_1 : \mathbf{d}_{N_d}|\theta) = \hat{c}\prod_{n=1}^{N_d}\exp\left(-\frac{1}{2}J(\theta,\mathbf{d}_n)\right) \tag{5.9}$$

The posterior PDF of Eq. (5.1) is a joint PDF, and therefore a multi-dimension integration is required to obtain the marginal distribution of each model parameter:

$$p(\theta_i|\mathbf{d}) = \int_{\Theta_{-i}} p(\theta_i,\theta_{-i}|\mathbf{d})\,d\theta_{-i} \tag{5.10}$$

These probability distributions often cannot be calculated analytically and are usually sampled numerically. Markov Chain Monte Carlo methods are used to sample complex multidimensional distributions. Among them Metropolis Hastings (MH) algorithm is the most common method in the application of FE model updating [19, 23, 24]. The algorithm is based on generating samples from a prescribed proposal probability density function and accepting the samples with a probability of move. However, direct application of standard MH algorithm is not suited for applications where parameters with highest probabilities are concentrated in small regions of the parameter space or in the case of sharply peaked or multi-modal posterior PDFs.

To avoid these difficulties, adaptive Metropolis Hasting (AMH) algorithms have been proposed [23,24]. The adaption can be done on proposal PDF or the target PDF. The adaption on target PDF such as AMH algorithm [23] or TMCMC algorithm [24] incorporates a number of intermediate target PDFs that converge to the posterior PDF of Eq. (5.1). The advantage of using these methods is their ability to sample the whole space and exploring all the peaks. However, the main disadvantage of these methods is the high computational cost required in case of complex structural models where the numbers of updating parameters increase. Another approach is to adapt the proposal PDF. Advantage of this method is that it is fast for identifiable cases where the posterior distributions of updating parameters have unique clear peaks, thus samples quickly converge to the highest probability region even when the region is sharply peaked. In this study, adaption on proposal PDF is selected instead of target PDF. For efficiency, the adaption is performed once for each block of samples, with specific standard deviation for its proposal PDF. The initial standard deviation is chosen to be relatively large, to let the samples take big jumps and be able to approach the vicinity of the high probability regions quickly. Covariance matrix of the proposal PDF for the next block of samples is tuned so that the samples acceptance ratio reaches the desired acceptance ratio value. Each block uses the standard MH to generate samples.

5.4.2 Model Updating Results

Six cases of model updating are performed using different subsets (1, 2, 5, 10, 36, 72 sets) of the available identified modal parameters. In all the cases, the prior distributions of segments' masses are considered as independent uniform distributions

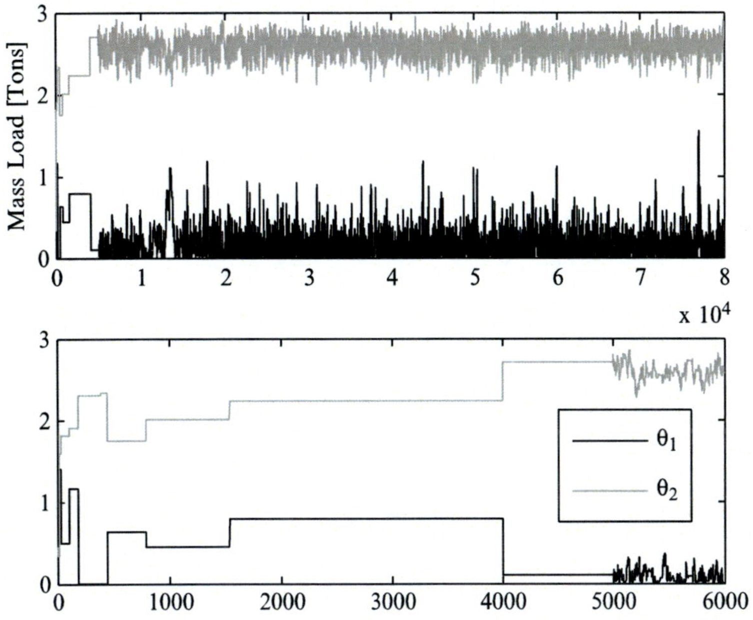

Fig. 5.7 Distribution of samples for segment 1 and 2: (**a**) all samples, (**b**) first 6,000 samples

with the lower bound of 0 and upper bound of 6.8 tons. The block size in the AMH algorithm is 5,000 samples and the desired acceptance ratio is 30%. Total number of samples is set to be 80,000 and the number of burn-in samples is 10,000. Values of standard deviation for natural frequencies in Eq. (5.8) are considered as:

$$\sigma_{\omega_m^2} = \mathbf{COV}_{\tilde{\omega}_m}^2 \tilde{\omega}_m^2 \tag{5.13}$$

where $\mathrm{COV}_{\tilde{\omega}_m}$ is taken as the average of calculated coefficients-of-variation (COVs) in Table 5.1 for all six modes. The COV of modes 2, 5, and 6 are then divided by 10 to account for their higher uncertainties. This factor is selected based on prior knowledge of system identification studies of this footbridge [10, 11]. Standard deviations of mode shapes are computed from:

$$\sigma_{\Phi_m} = \sqrt{N_s}\mathrm{COV}_{\tilde{\omega}_m} \tag{5.14}$$

where it is assumed that the COV of each mode shape component is equal to the COV of the corresponding natural frequency.

Figure 5.7a shows the distribution of samples generated using the AMH algorithm for updating parameters 1 and 2 for the case of two data sets. It can be seen that after the first block of samples, parameter space is sampled around almost zero for segment 1 and 2.6 tons for segment 2. The added mass does not exceed 1.6 tons for segment 1 and remains between 2.1 and 3.0 tons for segment 2. Note that this graph only shows the accepted samples. Figure 5.7b zooms in the first 6,000 samples. It is observed that the number of accepted samples is very low in the first 5,000 samples (first block). The proposal PDF is wide for the first block and yields to a high rejection rate of the samples. Note that the posterior PDF in this case is not sharply peaked and therefore getting close to the peak neighborhood is possible even with smaller block sizes.

Bayesian model updating is performed using the first, the first 2, 5, 10, 36, and 72 identified modal parameters. Figure 5.8 shows the posterior PDF of segment 2 identified for each of these six cases. It can be seen that using more data mitigate the variations of updating parameters. This is a phenomenon that cannot be seen in deterministic model updating, where uncertainty of model parameter does not reduce necessarily by adding more data. Smaller uncertainty on updated model parameter would yield to more accurate prediction and detection of potential damage. It is worth noting that the identified posterior PDFs show improvements in both bias and variation by adding more data especially when the total number available data is small. The mean, mode, and standard deviation (STD) of samples generated for each updating case are provided in Table 5.2. Substantial reduction can be observed on STD of samples by adding data especially in the first few cases. However, the reduction of STDs becomes less significant as the number of data sets exceeds 36. Therefore, no drastic change in estimated mass probability distributions would be expected if more data sets were available. Note that generating the Markov Chains for the case of 1 data set is much faster than for the case of 72 data sets.

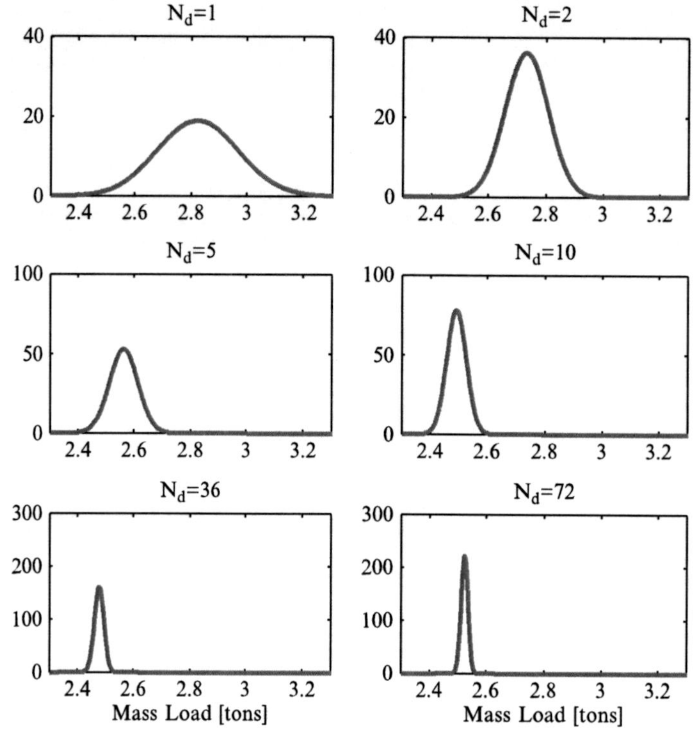

Fig. 5.8 Effect of number of data sets on posterior probability distribution of segment 2 mass

Table 5.2 Statistics (mean [ton], mode or maximum a posteriori [ton], and standard deviation [ton]) of the five updating parameters for the six considered cases

No. of	θ_1			θ_2			θ_3			θ_4			θ_5		
data sets	Mean	Mode	STD	Mean	Mode	STD	Mean	Mode	STD	Mean	Mode	STD	Mean	Mode	STD
1	0.71	0.07	0.845	2.38	2.77	0.388	0.32	0.00	0.318	0.39	0.00	0.352	0.08	0.01	0.073
2	0.16	0.01	0.169	2.60	2.74	0.115	0.09	0.01	0.093	0.21	0.03	0.200	0.07	0.01	0.064
5	0.05	0.00	0.047	2.50	2.55	0.059	0.03	0.00	0.029	0.12	0.01	0.122	0.08	0.04	0.062
10	0.02	0.00	0.019	2.46	2.49	0.040	0.01	0.00	0.012	0.06	0.00	0.065	0.06	0.03	0.046
36	0.01	0.00	0.005	2.47	2.48	0.018	0.00	0.00	0.004	0.02	0.00	0.016	0.01	0.00	0.006
72	0.00	0.00	0.003	2.52	2.53	0.013	0.00	0.00	0.002	0.01	0.00	0.006	0.00	0.00	0.003

5.5 Conclusions

A Bayesian FE model updating approach is implemented for damage identification of the Dowling Hall Footbridge. Damage is simulated physically by loading a small segment of the footbridge deck with 2.29 metric tons of concrete block. The location and extent of the added mass is determined probabilistically using Bayesian model updating. Adaptive Metropolis algorithm is used to sample the posterior probability distributions of the five updating model parameters. The adaption is performed on the proposal probability density function and the performance of the sampling technique is satisfactory.

Effect of the number of data sets used in the identification process is investigated by using different subsets (1, 2, 5, 10, 36, and 72 data sets) of available data. The bias and variation of model parameters are reduced by including more data set. The reduction diminishes with increasing the number of data set. This is a phenomenon that cannot be seen in deterministic model updating, where uncertainty of model parameter does not reduce necessarily by adding more data sets to the updating process. Thus, model parameter updated using a probabilistic approach based on several sets of measurements would yield to more accurate prediction and detection of potential damage.

Acknowledgements The authors would like to acknowledge partial support of this project by the National Science Foundation Grant No. 1125624 which was awarded under the Broadening Participation Research Initiation Grants in Engineering (BRIGE) program. The authors also acknowledge Ms. Alyssa Kody for design and performance of the bridge load test and Ms. Rachele Pesenti for her help in programming the real-time module of the data acquisition system. The opinions, findings, and conclusions expressed in the paper are those of the authors and do not necessarily reflect the views of the individuals and organizations involved in this project.

References

1. Huth O, Feltrin G, Maeck J, Kilic N, Motavalli M (2005) Damage identification using modal data: experiences on prestressed concrete bridge. J Struct Eng 131(12):1898–1910
2. Teughles A, De Roeck G (2004) Structural damage identification of the highway bridge Z24 by FE model updating. J Sound Vib 278(3):589–610
3. Reynders E, De Roeck D, Bakir PG, Sauvage C (2007) Damage identification on the Tilff bridge by vibration monitoring using optical fiber strain sensors. J Eng Mech ASCE 133(2):185–193
4. Moaveni B, He X, Conte JP, De Callafon RA (2008) Damage identification of composite beam using finite element model updating. Comput Aided Civ Infrastruct Eng 23(5):339–359
5. Moaveni B, He X, Conte JP, Restrepo JI (2010) Damage identification study of a seven-story full-scale building slice tested on the UCSD-NEES shake table. Struct Saf 32(5):347–356
6. Ntotsios E, Papadimitriou C, Panetsos P, Karaiskos G, Perros K, Perdikaris PC (2009) Bridge health monitoring system based on vibration measurements. Bull Earthq Eng 7:469–483
7. Yuen KV, Beck JL, Au SK (2004) Structural damage detection and assessment by adaptive Markov chain Monte Carlo simulation. Struct Control Health Monit 11:327–347
8. Simoen E, Moaveni B, Conte JP, Lombaert G (2013) Uncertainty quantification in the assessment of progressive damage in a seven-story full-scale building slice. J Eng Mech ASCE, in press, doi: 10.1061/(ASCE)EM.1943-7889.0000610
9. Moser P, Moaveni B (2011) Environmental effects on the identified natural frequencies of the Dowling Hall Footbridge. Mech Syst Signal Process 25(7):2336–2357
10. Moser P, Moaveni B (2013) Design and deployment of a continuous monitoring system for the Dowling Hall Footbridge. Exp Tech 37(1):15–26
11. Moaveni B, Behmanesh I (2012) Effects of changing ambient temperature on finite element model updating of the Dowling Hall Footbridge. Eng Struct 43:58–68
12. Metropolis N, Rosenbluth AW, Rosenbluth MN, Teller AH, Teller E (1953) Equations of state calculations by fast computing machines. J Chem Phys 21:1087–1092
13. Hasting WK (1970) Monte Carlo sampling method using Markov Chains and their applications. Biometrika 57:97–109
14. Chib S, Greenberg E (1995) Understanding the Metropolis-Hasting algorithm. Am Stat 49(4):327–335
15. Haario H, Saksman E, Tamminen J (2001) An adaptive Metropolis algorithm. Bernouli 7:223–242
16. Bowman J (2003) Vibration testing and modal identification of the Dowling Hall Footbridge at Tufts University [Master's thesis]. Department of Civil and Environmental Engineering, Tufts University, Medford
17. Van Overscheen P, De Moore B (1996) Subspace identification for linear systems. Kluwer Academic, Boston
18. Filippou FC, Constantinides M (2004) FEDEASLab getting started guide and simulation examples. Technical report NEESgrid-2004-22. Available from: http://fedeaslab.berkeley.edu
19. Beck JL, Au SK (2002) Bayesian updating of structural models and reliability using Markov Chain Monte Carlo simulation. J Eng Mech 128(4):380–391
20. Beck JL, Katafygiotis LS (1998) Updating models and their uncertainties. I: Bayesian statistical framework. J Eng Mech 124(4):455–461
21. Sohn H, Law HK (1997) A Bayesian probabilistic approach for structure damage detection. Earthq Eng Struct Dyn 26:1259–1281
22. Katafygiotis LS, Beck JL (1998) Updating models and their uncertainties. II: model identifiability. J Eng Mech 124(4):463–467
23. Beck JL, Au SK (2001) Monitoring structural health using a probabilistic measure. Comput Aided Civ Infrastruct Eng 16:1–11
24. Ching J, Chen YC (2007) Transitional Markov Chain Monte Carlo method for Bayesian model updating, model class selection, and model averaging. J Eng Mech 133(7):816–832

Chapter 6
Considering Wave Passage Effects in Blind Identification of Long-Span Bridges

S. Farid Ghahari, M. Ali Ghannad, James Norman, Adam Crewe, Fariba Abazarsa, and Ertugrul Taciroglu

Abstract Long-span bridges usually experience different input excitations at their ground supports that emanate from differences in wave arrival times, and soil conditions, as well as loss of coherency in arriving waves. These spatial variations can drastically influence the dynamic response; hence, this phenomenon must be considered in any vibration-based identification method. There are numerous Multi-Input Multi-Output (MIMO) identification techniques that may be applied to data recorded at long-span bridges that experience spatial variations in their input motions. However, inertial soil-structure interaction effects severely reduce the accuracy of these techniques because the actual Foundation Input Motion (FIM) cannot be recorded during earthquakes. In this study, we present an extension to a novel blind identification method that we had developed earlier, which enables the method to handle multiple input motions. For the sake of simplicity, we only consider wave passage effects—that is, all unknown input motions are assumed to be identical except for a known/unknown phase-delay. This method comprises two steps. In the first step, the spatial time-frequency distributions of recorded responses are used for extracting the mode shapes and the modal coordinates. This is achieved through a Blind Source Separation (BSS) technique. In the second step, cross relations among the extracted modal coordinates are used for identifying the natural frequencies, damping ratios, modal contribution factors, along with the unknown input motions through a least-squares technique. Both simulated and experimental examples are provided, which suggest that the method is capable of accurately identifying the dynamic characteristics of long-span bridges from recorded response signals without the knowledge of input motions, even in the presence of wave passage effects due to phase-delays.

Keywords Blind system identification • Long-span bridges • Multiple input motions • Wave passage effects • Spatial time-frequency distribution

S.F. Ghahari
Ph.D. Candidate, Civil Engineering Department, Sharif University of Technology, Azadi Ave., Tehran, Iran
e-mail: ghahari@mehr.sharif.edu

M.A. Ghannad
Associate Professor, Civil Engineering Department, Sharif University of Technology, Azadi Ave., Tehran, Iran
e-mail: ghannad@sharif.edu

J. Norman
Lecturer, Department of Civil Engineering, University of Bristol, Queen's Building, University Walk, Clifton, Bristol, UK
e-mail: james.norman@ramboll.co.uk

A. Crewe
Senior Lecturer, Department of Civil Engineering, University of Bristol, Queen's Building, University Walk, Clifton, Bristol, UK
e-mail: a.j.crewe@bristol.ac.uk

F. Abazarsa
Ph.D. Candidate, Structural Engineering Department, International Institute of Earthquake Engineering and Seismology, P.O. Box 3913/19395, Tehran
e-mail: f.azarsa@iiees.ac.ir

E. Taciroglu (✉)
Associate Professor, Civil and Environmental Engineering Department, University of California, Los Angeles, 5731E Boelter Hall, Los Angeles, CA 90095, USA
e-mail: etacir@ucla.edu

T. Simmermacher et al. (eds.), *Topics in Model Validation and Uncertainty Quantification, Volume 5*,
Conference Proceedings of the Society for Experimental Mechanics Series 41,
DOI 10.1007/978-1-4614-6564-5_6,

6.1 Introduction

Modal identification of civil structures subjected to strong ground shaking has been a subject of intense research for more than four decades (see, e.g., [1] and references contained therein). Consequently, many different identification approaches have been developed (e.g., [2]). System identification research is a multi-disciplinary field; and therefore, many of the existing methods are common among mechanical, electrical, and structural engineering disciplines. Although there are a few studies in which identification is carried out without the knowledge of input excitations [3,4]; all of the traditional techniques used for identification of civil structures require the input motions to be known/measured. Yet, frequently, true input excitations are not known, for example, due to Soil-Structure Interaction (SSI), or they are unmeasured because of inadequate instrumentation. For structures affected by SSI, the Foundation Input Motions (FIMs) are significantly different from both Free-Field Motions (FFMs)—which are recorded far from the structure—and from foundation responses [5]. Kinematic Interaction (KI)—which is due to the difference between the stiffness of a nearly rigid foundation and the surrounding soil—causes the motion experienced by the foundation (i.e., FIM) to differ significantly from the FFM. Inertial Interaction (II) due to the structure's mass and flexibility as well as the attenuating effects of soil also render the foundation responses to be different from FIMs. Even in the absence of soil-structure interaction, in many real-life cases, the foundation responses are not recorded at all, or recorded with a very low Signal-to-Noise Ratio (SNR).

Recently, two new identification methods have been developed with which the modal properties of a system can be identified by using only its response (output) signals recorded during ambient vibrations [6, 7] or earthquakes [8]. The method, which can be used for earthquake response signals [8], works in two steps: first the Spatial Time-Frequency Distributions (STFDs) of the response signals are used to blindly identify the mode shapes and the modal coordinate signals. Then, cross relations among the modal coordinates are employed to determine the system's natural frequencies and damping ratios. An extension of this technique for systems subject to multiple input motions (e.g., bridges) is presented here.

As indicated in many prior studies, the spatial variation of input motions—due to wave arrival times, soil conditions, and coherency-losses—can have a significant effect on the seismic response of both long and short span bridges [9–12]. For such cases, output-only MIMO identification techniques must be used. In the present study, we only consider wave passage effects—that is, all unknown input motions are assumed to be identical except for a known/unknown phase-delay. This new method shares the first step of our previously proposed technique [8] to extract mode shapes and modal coordinates from recorded acceleration response signals. In the second step, the extracted modal coordinates are estimated through a nonlinear least-squares technique from which natural frequencies, damping ratios, modal contribution factors, as well as input motions can be identified. If the phase-delay is unknown, then its value can also be estimated by repeating the said process with several phase-delays and by determining the one that minimizes the final value of the residual norm in the least-squares analysis.

6.2 Proposed Identification Technique

Consider the idealized lumped mass system shown in Fig. 6.1, which provides a schematic representation of typical multiple-span bridges with multiple piers and two abutments. Assuming stiffness-proportional damping, or by neglecting the effective damping forces relative to inertial forces, the equation of dynamic equilibrium for this system can be written as [13]

$$\mathbf{M}\ddot{\mathbf{x}}(t) + \mathbf{C}\dot{\mathbf{x}}(t) + \mathbf{K}\mathbf{x}(t) = -\sum_{l=1}^{n_g} \mathbf{M}\mathbf{l}_l \ddot{x}_{gl}(t) \tag{6.1}$$

where $\mathbf{M}$, $\mathbf{C}$, and $\mathbf{K}$ are mass, damping and stiffness matrices of the system, respectively. $\mathbf{x}(t)$ is a vector that contains the internal displacement response of all Degrees Of Freedom (DOFs), while $\dot{\mathbf{x}}(t)$, and $\ddot{\mathbf{x}}(t)$ are their corresponding velocity and acceleration vectors. l_l is a vector that assigns the influence of each input acceleration, $\ddot{x}_{gl}(t)$, on the system's response; and is a function of internal and boundary stiffnesses. n_g is the number of all existing input motions. Using the modal expansion theorem, the recorded internal accelerations can be stated as,

$$\ddot{\mathbf{x}}(t) = -\sum_{i=1}^{n_g} \phi_i \ddot{q}_i(t) \tag{6.2}$$

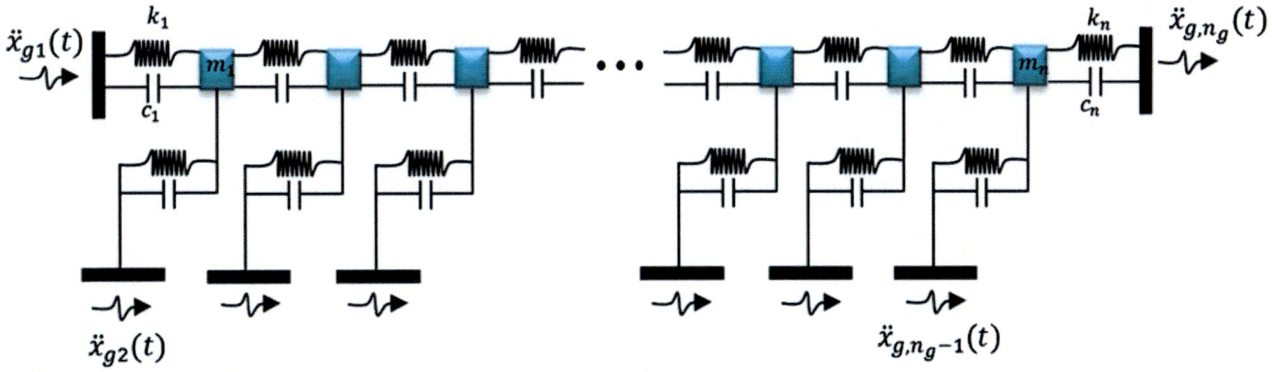

Fig. 6.1 Lumped model of a multi-span bridge

where ϕ_i and $\ddot{q}_i(t)$ denote the ith mode shape vector and modal acceleration, respectively; and n is the number of modes. $\ddot{q}_i(t)$ is the relative acceleration response of a Single DOF (SDOF) system subjected to the excitation $\ddot{\mathbf{x}}(t) = -\sum_{i=1}^{n_g} \phi_i \ddot{q}_i(t)$; and it can be expanded as,

$$\ddot{q}_i(t) = \sum_{l=1}^{n_g} \ddot{q}_i^l(t) \tag{6.3}$$

where $\ddot{q}_i^l(t)$ is the relative acceleration of the SDOF system corresponding to the ith mode subjected to $\beta_i^l \ddot{x}_{gl}(t)$ where $\beta_i^l = \phi_i^T \mathbf{M} \mathbf{l}_l / \phi_i^T \mathbf{M} \phi_i$ is the ith modal participation factor for the lth input acceleration. It follows that the absolute acceleration of the ith modal coordinate for the lth input can be stated as,

$$\ddot{q}_i^{t,l}(t) = \ddot{q}_i^l(t) + \beta_i^l \ddot{x}_{gl}(t) \tag{6.4}$$

Thus, the absolute acceleration response of the system can be written as,

$$\ddot{\mathbf{x}}^t(t) = \sum_{i=1}^{n} \varphi_i \sum_{l=1}^{n_g} \ddot{q}_i^l(t) + \sum_{l=1}^{n_g} \mathbf{l}_l \ddot{x}_{gl}(t), \tag{6.5}$$

or

$$\ddot{\mathbf{x}}^t(t) = \sum_{i=1}^{n} \varphi_i \left[\sum_{l=1}^{n_g} \ddot{q}_i^{t,l}(t) - \sum_{l=1}^{n_g} \beta_i^l \ddot{x}_{gl}(t) \right] + \sum_{l=1}^{n_g} \mathbf{l}_l \ddot{x}_{gl}(t) \tag{6.6}$$

Rearranging Eq. (6.6) yields

$$\ddot{\mathbf{x}}^t(t) = \sum_{i=1}^{n} \varphi_i \sum_{l=1}^{n_g} \ddot{q}_i^{t,l}(t) + \sum_{l=1}^{n_g} \left(\mathbf{l}_l - \sum_{l=1}^{n} \varphi_i \beta_i^l \right) \ddot{x}_{gl}(t) \tag{6.7}$$

The second term in Eq. (6.7) is a zero vector, because for each input motion $\mathbf{l}_l = \sum_{l=1}^{n} \varphi_i \beta_i^l$. Therefore,

$$\ddot{\mathbf{x}}^t(t) = \sum_{i=1}^{n} \varphi_i \sum_{l=1}^{n_g} \ddot{q}_i^{t,l}(t) = \sum_{i=1}^{n} \phi_i \ddot{q}_i^t(t) \tag{6.8}$$

where $\ddot{q}_i^t(t) = \sum_{l=1}^{n_g} \ddot{q}_i^{t,l}(t)$. It is possible, then, to state the modal coordinate $\ddot{q}_i^{t,l}(t)$ as a discrete Auto-Regressive model with eXternal input (ARX) [8], as in

$$\ddot{q}_i^{t,l}[k] = A_i \ddot{q}_i^{t,l}[k-1] + B_i \ddot{q}_i^{t,l}[k-2] + \beta_i^l C_i \ddot{x}_{gl}[k] + \beta_i^l D_{igl} \ddot{x}[k-1], \tag{6.9}$$

where k is the time index. The coefficients in Eq. (6.9) are given by

$$A_i = 2e^{-\xi_i \omega_{ni} T} \cos(\omega_{di} T), \tag{6.10}$$

$$B_i = -e^{-2\xi_i \omega_{ni} T}, \tag{6.11}$$

$$C_i = 2\xi_i \omega_{ni} T, \tag{6.12}$$

$$D_i = \omega_{ni} T e^{-\xi_i \omega_{ni} T} \left[\frac{\omega_{ni}}{\omega_{di}} (1 - 2\xi_i^2) \sin(\omega_{di} T) - 2\xi_i \cos(\omega_{di} T) \right], \tag{6.13}$$

where ξ_i, ω_{ni}, and $\omega_{di} = \omega_{ni}\sqrt{1-\xi_i^2}$ are the damping ratio, and the undamped and damped natural frequencies of ith mode, respectively; and T is sampling time. Thus, the absolute acceleration of the ith modal coordinate can be written as summation of $\ddot{q}_i^{t,l}[k]$ for all input motions, as in

$$\ddot{q}_i^{t,l}[k] = A_i \ddot{q}_i^{t,l}[k-1] + B_i \ddot{q}_i^{t,l}[k-2] + C_i \sum_{l=1}^{n_g} \beta_i^l \ddot{x}_{gl}[k] + D_i \sum_{l=1}^{n_g} \beta_i^l \ddot{x}_{gl}[k-1], \tag{6.14}$$

Having the absolute acceleration modal coordinates, the unknown parameters (ξ_i, ω_{ni}, and β_i^l), the input motions ($\ddot{x}_{gl}$) can be identified through a nonlinear least-square technique. To wit, consider the minimization problem

$$\min_{\mathbf{z}} \|\mathbf{G}(\mathbf{z})\|_2^2 \tag{6.15}$$

where $\|\cdot\|_2$ indicates the Euclidean norm of the matrix in the bracket, and

$$\mathbf{z} \equiv \left[\xi_1, \ldots, \xi_n, \omega_{n1}, \ldots, \omega_{nn}, \beta_1^1, \ldots, \beta_n^1, \ldots, \beta_n^{n_g}, \ddot{x}_{g1}, \ldots, \ddot{x}_{gn_g} \right] \tag{6.16}$$

$$[\mathbf{G}(\mathbf{z})]_{k,i} = \ddot{q}_i^t[k] - A_i \ddot{q}_i^t[k-1] - B_i \ddot{q}_i^t[k-2] - C_i \sum_{l=1}^{n_g} \beta_i^l \ddot{x}_{gl}[k] - D_i \sum_{l=1}^{n_g} \beta_i^l \ddot{x}_{gl}[k-1], \tag{6.17}$$

Here, $\ddot{x}_{gl}$is a vector that contains the time history of the lth unknown input motion; and $\mathbf{G}(\mathbf{z})$ is an $N \times n$ matrix where N denotes number of time samples. As long as the number of equations ($N \times n$) is greater than the number of unknowns ($n + n + n_g \times n + n \times N$), a nonlinear least-squares method can be employed to solve minimization problem posed in Eq. (6.15). This condition is mostly satisfied for practical cases because

$$n + N > n_g \times N + 2 \times n + n \times n_g, \tag{6.18}$$

or

$$N > n(2 + n_g)/(n - n_g) \tag{6.19}$$

However, as Eq. (6.14) is recursive, it does not generate N completely independent equations for each mode. Moreover, in order to obtain reliable results in least-squares techniques, it is important to have the number of equations to be far greater than the number of unknowns. As such, only the wave passage effect will be considered in this study, which means that all unknown input motions are assumed to be identical except for a known or unknown phase-delay. It is useful to note here that the phase-delay parameter can be easily estimated based on the soil shear-wave velocity and the distance between piers. On account of this simplifying assumption, Eq. (6.14) is rewritten as follows,

$$\ddot{q}_i^{t,l}[k] = A_i \ddot{q}_i^t[k-1] + B_i \ddot{q}_i^t[k-2] + C_i \sum_{l=1}^{n_g} \beta_i^l \ddot{x}_{gl}[k-\delta_l] + D_i \sum_{l=1}^{n_g} \beta_i^l \ddot{x}_g[k-\delta_l-1], \tag{6.20}$$

where $\ddot{x}_g[k]$ is the input acceleration and δ_l (with $l = 1, \ldots, n_g$) denotes the time delay for each pier. Assuming a known phase-delay, Eq. (6.17) can be rewritten as,

$$[\mathbf{G}(\mathbf{z})]_{k,i} = \ddot{q}_i^t[k] - A_i \ddot{q}_i^t[k-1] - B_i \ddot{q}_i^t[k-2] - C_i \sum_{l=1}^{n_g} \beta_i^l \ddot{x}_{gl}[k-\delta_l] + D_i \sum_{l=1}^{n_g} \beta_i^l \ddot{x}_g[k-\delta_l-1], \tag{6.21}$$

where

$$\mathbf{z} \equiv \left[\xi_1, \rightleftharpoons, \xi_n, \omega_{n1}, \rightleftharpoons, \omega_{nn}, \beta_1^1, \rightleftharpoons, \beta_n^1, \rightleftharpoons, \beta_n^{n_g}, \ddot{x}_g\right]. \tag{6.22}$$

For cases with unknown phase-delay, this parameter can be also identified using *iterative* nonlinear least-squares analyses. Indeed, the phase-delay cannot be identified through direct nonlinear least-squares analysis, as it is an index; however, it may be estimated iteratively by minimizing the final residual norm of several nonlinear least-squares analyses, as will be shown later through both synthetic and experimental examples. The aforementioned identification technique is based on having the modal coordinate signals, which are obtained through the first step of the identification method recently proposed by Ghahari et al. [8]. The said method is based on a time-frequency Blind Source Separation (BSS) method [14, 15], and is briefly described below.

6.2.1 Time-Frequency Blind Source Separation

Modal summation presented in Eq. (6.8) can be represented in matrix form as follows,

$$\ddot{\mathbf{x}}^t(t) = \mathbf{\Phi}\ddot{\mathbf{q}}^t(t) \tag{6.23}$$

where $\mathbf{\Phi}$ is the mode shape matrix, and $\ddot{\mathbf{q}}^t(t)$ is a vector of time that contains the absolute acceleration modal coordinates. Spatial Time Frequency Distribution (STFD) of Eq. (6.23) can be written using any quadratic time-frequency method. The particular technique adopted here is the *Smoothed Pseudo Wigner-Ville Distribution* (SPWVD) [16] that yields the following representation of the recorded signals

$$\mathbf{D}_{\ddot{\mathbf{x}}^t\ddot{\mathbf{x}}^t}(t,f) = \mathbf{\Phi}\mathbf{D}_{\ddot{\mathbf{q}}^t\ddot{\mathbf{q}}^t}(t,f)\mathbf{\Phi}^H \tag{6.24}$$

where the superscript H denotes a Hermitian transpose. At each time-frequency point (t,f), $\mathbf{D}_{\ddot{\mathbf{x}}^t\ddot{\mathbf{x}}^t}$ and $\mathbf{D}_{\ddot{\mathbf{q}}^t\ddot{\mathbf{q}}^t}$ are STFD matrices of response and modal coordinate signals, respectively. The absolute acceleration modal coordinates are disjoint (or quasi-disjoint) signals in the time-frequency domain; that is, they have distinct time-frequency representation. Time-frequency points at which only one modal coordinate is present are called auto-source points, and at such points $\mathbf{D}_{\ddot{\mathbf{q}}^t\ddot{\mathbf{q}}^t}(t,f)$ is a diagonal matrix that theoretically possesses only one nonzero diagonal element. The following criterion can be used to identify such points,

$$\left|\frac{\lambda_{\max}\left[\mathbf{D}_{\ddot{\mathbf{x}}^t\ddot{\mathbf{x}}^t}(t,f)\right]}{\|\mathbf{D}_{\ddot{\mathbf{x}}^t\ddot{\mathbf{x}}^t}(t,f)\|_F} - 1\right| > \varepsilon, \tag{6.25}$$

where $\|\cdot\|_F$ denotes the Frobenius norm, $\lambda_{\max}[.]$ represents the largest eigenvalue of its argument matrix, and ε is a small positive scalar (typically, $\varepsilon = 0.001$). Applying a Joint Approximate Diagonalization (JAD) [17, 18] on all $\mathbf{D}_{\ddot{\mathbf{x}}^t\ddot{\mathbf{x}}^t}$ matrices that are calculated at all auto-source points, the mode shape matrix can be estimated. Having identified the mode shape matrix, the absolute acceleration modal coordinates are easily calculated by simple matrix inversion at each time instant, as in

$$\ddot{\mathbf{q}}^t(t) = \mathbf{\Phi}^{-1}\ddot{\mathbf{x}}^t(t) \tag{6.26}$$

6.3 Application Examples

6.3.1 Synthetic Simulation

A 10 DOF linear finite element model, shown in Fig. 6.2a, is used for assessing the proposed identification technique's performance and accuracy. This model, incidentally, is the same model that was previously used by Papadimitriou et al. [19]. Soil stiffness at both abutments is modeled by rotational and translational springs; however, rotational springs are neglected here, as their stiffnesses are small in comparison to the bending stiffness of the deck. Values of the system parameters are $k_{yA} = 1.1k_y, k_{yE} = 0.9k_y$ where $k_y = 10^7\,\mathrm{N/m}$. Bending rigidities of the deck elements are $0.95EI$, $1.05EI$, $0.9EI$, and $0.95EI$ for elements AB, BC, CD, and DE, respectively. Mass of the bridge is modeled by five lumped masses positioned at points A–E with values $m_A = m_E = 8\times10^3\,\mathrm{kg}$, and $m_B = m_C = m_D = 16\times10^3\,\mathrm{kg}$. This model is reduced using static condensation as shown in Fig. 6.2b. Moreover, Rayleigh damping is considered with the damping ratios of the first and the last modes set at 3 %. In addition to vertical deformation of the bridge under vertical input motions, this model can be also applied to

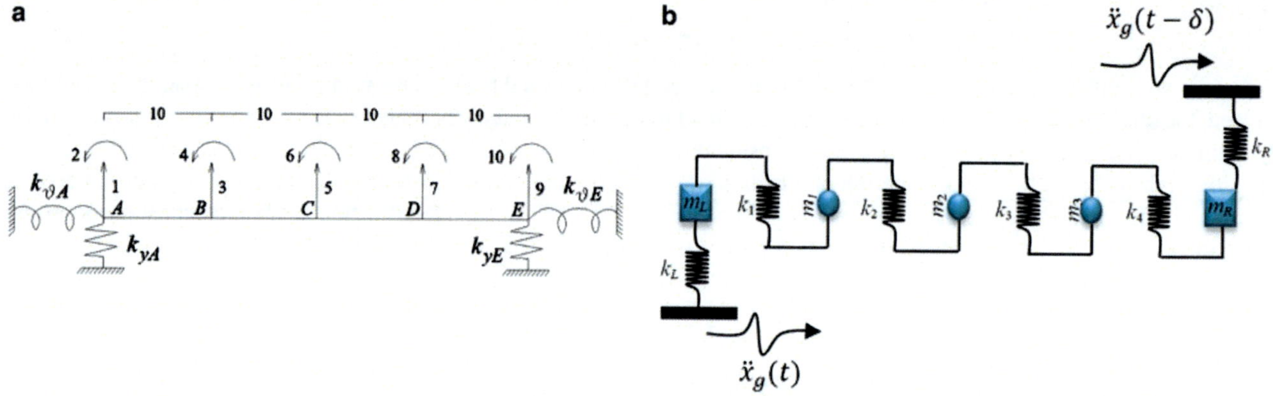

Fig. 6.2 Single-span bridge used in this study **(a)** finite element model [19], and **(b)** condensed model

transverse motions under horizontal input. Natural frequencies of the system for modes one through five are 4.66, 17.39, 32.24, 38.65, and 45.14 rad/s, respectively. Damping ratios for modes one through five are 3, 1.78, 2.34, 2.66, and 3 %. Finally, modal contribution factors related to the left and right abutments for modes one through five are (0.985, 1), (0.511, -0.516), (0.581, 0.814), (0.750, -0.524), and (0.356, 0.176), respectively.

Horizontal accelerogram recorded in El Centro Array #9 during Imperial Valley earthquake, 1940 [20] is used as input motion for summating dynamic response of the system. This dynamic analysis is carried out using the lsim command in MATLAB [21] with a 50 Hz sampling frequency (the original accelerogram is digitized at a 100 Hz rate, but it was down-sampled to reduce the computing time). To consider the wave passage effect, the right abutment is assumed to experience the input acceleration 1 s after the left abutment. Considering the span length shown in Fig. 6.2a, this phase-delay corresponds to a wave propagation velocity of 40 m/s. While this velocity is not quite realistic, it is used to demonstrate the phase-delay effects on identification results. SPWVD of the input accelerations along with absolute acceleration responses of all DOFs are shown in Fig. 6.3.

Application of the first step of the identification procedure yields the mode shapes and the modal coordinates. Identified mode shapes are highly accurate, as evidenced by the Modal Assurance Criterion (MAC) values computed between the identified and the analytical mode shapes, whose values are 0.990, 0.990, 0.997, 0.995, 0.955 for the first to the last mode, respectively. This index is defined as $\text{MAC}_i = (\phi_i^a \cdot \phi_i^i)^2 / \left[\|\phi_i^a\|^2 \|\phi_i^i\|^2 \right]$ for the ith mode, where $\phi_i^i$and ϕ_i^a denote the ith identified and analytical mode shapes, respectively. Provided that the phase-delay is known, application of the nonlinear least-squares procedure described earlier yields the natural frequencies, damping ratios, modal contribution factors as well as the input excitation time history. However, proper bounds and initial values can improve the accuracy of results. Table 6.1 displays the values set for initial values, lower, and upper bounds of natural frequencies, damping ratios, and modal contribution factors of all modes. Also, a zeromean random signal with limited amplitude in $[-1, 1]$ is considered as the initial estimation of the unknown input motion. In the second column of this table, the numbers in parentheses indicate the mode number.

The dynamic parameters identified through the nonlinear least-squares procedure using a trust-region reflective algorithm [22] are presented in Table 6.2. These values are obtained using 50 iterations, after which significant changes are not observed in $\|\mathbf{G}(\mathbf{z})\|_2^2$ as shown in Fig. 6.4. Note that the phase-delay is assumed as a known parameter in this example. As seen, while identified values are acceptably accurate; the identified damping ratios are not exact. This is mainly due to errors accrued during the first step (BSS), because the second step (least-squares) can be shown to produce exact results for exact modal coordinates. Moreover, due to the scaling ambiguity, it is not possible to exactly recover the input motion with its original Peak Ground Acceleration (PGA), because input motion depends on the modal coordinates used, which can be varied by mode shape scaling. Figure 6.5 displays the initial guess, identified, and original input motions. After fixing the scaling ambiguity, it is clearly observed that the proposed identification method is also able to extract the exact input motion with very good accuracy.

To see difference between exact response of the system and response calculated by identified parameters, i.e. mode shapes, natural frequencies, damping ratios, modal contribution factors, and input motions, Fig. 6.6 is presented. As it can be seen, response signals reconstructed by identified parameters match well with their exact corresponding values; that is, parameters cannot be identified more accurately as current values can almost exactly reconstruct response signals.

As mentioned earlier, phase-delay can also be identified by repeating the least-squares analyses with different phase-delay values and by determining the particular value that corresponds to the minimum converged residual norm. Figure 6.7 displays

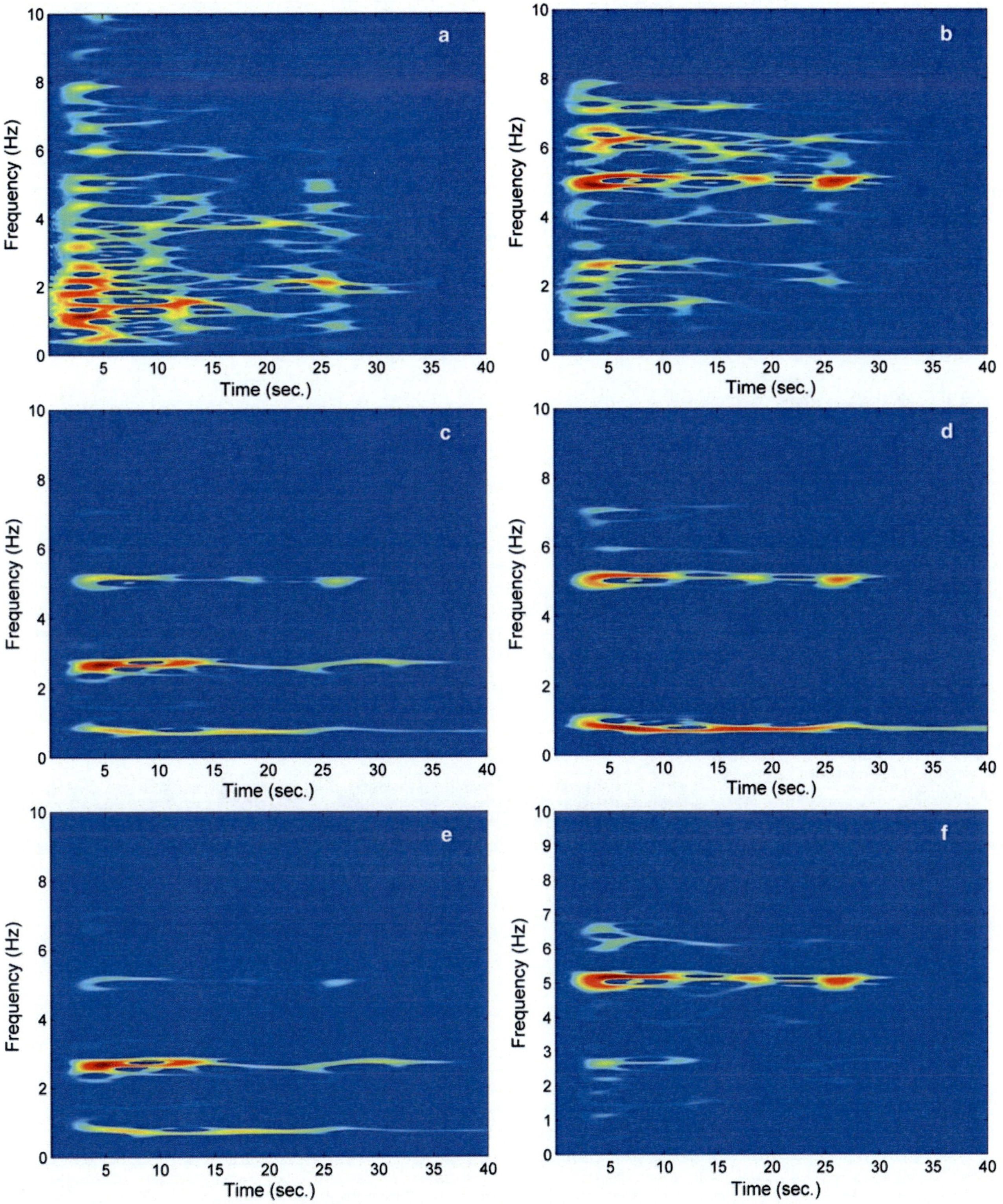

Fig. 6.3 Time-frequency representations of (**a**) input acceleration, (**b**) DOF 1's, (**c**) DOF 2's, (**d**) DOF 3's, (**e**) DOF 4's, and (**f**) DOF 5's absolute acceleration

Table 6.1 Parameters used in the least-squares analysis

	Natural frequency (rad/s)	Damping ratio (%)	Modal contribution factor
Lower bound	1	0.1	−1, −1
Initial guess	10(1), 20(2), 30(3), 40(4), 50(5)	5	0.5, 0.5
Upper bound	50	10	1, 1

Table 6.2 Parameters identified through the least-squares analysis

Mode no	Natural frequency (rad/s)	Damping ratio (%)	Modal contribution factor
1	4.77	2.09	0.965, 1.000
2	17.09	1.15	0.513, −0.513
3	32.27	1.53	0.571, 0.774
4	38.56	3.22	0.739, −0.522
5	45.66	2.46	0.425, 0.221

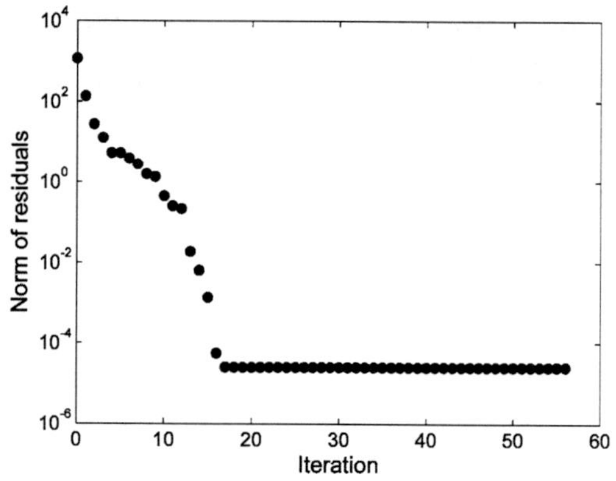

Fig. 6.4 Residual norm $\|\mathbf{G}(\mathbf{z})\|_2^2$ vs. iterations

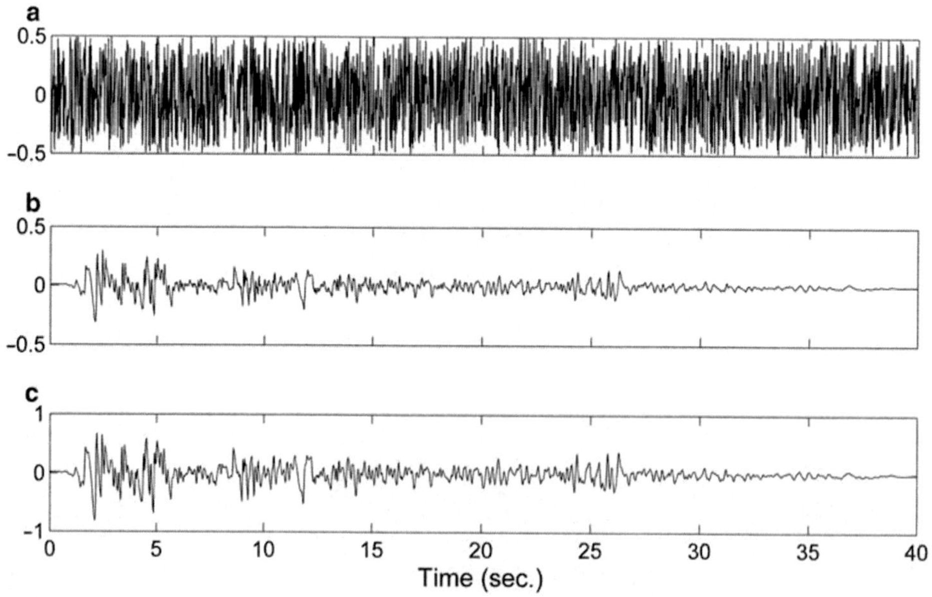

Fig. 6.5 (**a**) Initial guess, (**b**) original, and (**c**) identified input motions

the values of the final/converged residual norm for ten different values of phase-delays. To obtain comparable results, all least-squares parameters—i.e. initial values, lower bounds, and upper bounds—were the same for all these ten cases; and the residual norm values after 20 iterations were used. As seen, the minimum occurs for the 1-s phase-delay which is identical to the exact value considered in this example.

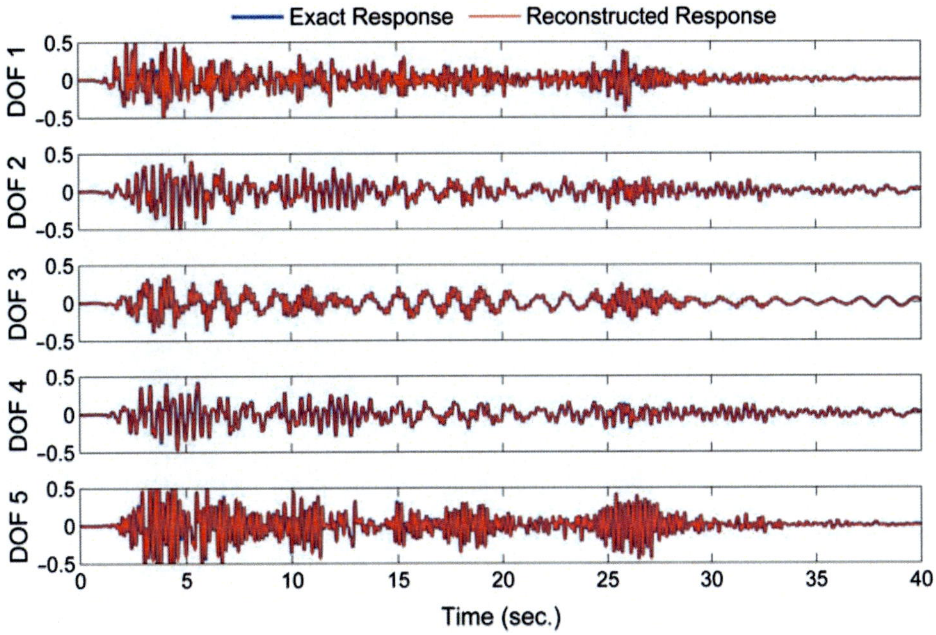

Fig. 6.6 Comparison between the exact response of the bridge and the reconstructed response

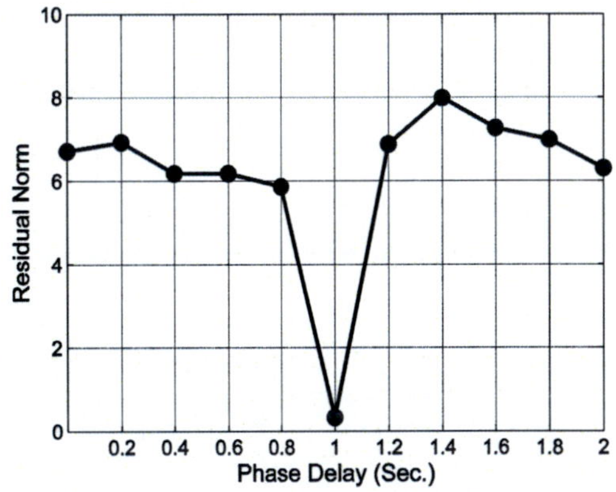

Fig. 6.7 Variation of final residual norm versus phase-delay

6.3.2 Experimental Data: The Prototype Viaduct Bridge

In this section, application of the proposed identification technique is shown for a small-scale viaduct bridge. The prototype bridge is 200 m long with three piers at equal spacing. This structure has been previously tested in other experimental works with synchronous [23, 24] and asynchronous [25] inputs. The physical tests were performed on a specially designed multiple support excitation test bed, which comprises of a set of five independent single axis shaking tables, specifically designed to allow simulation of any type of multiple support excitation (left side of Fig. 6.8). The model bridge can be seen mounted on the MSE experimental test bed in right side of Fig. 6.8.

This experimental bridge is a 1:50 scale model of the bridge prototype, in which there are four 50 m spans as shown in Fig. 6.9. More details of this model can be found in [26]. The first three modes of vibration of this model has been numerically calculated and shown in Fig. 6.10.

To excite the test model, an artificially generated time-history displacement was used, which was response spectrum compatible in accordance with Eurocode 8 Part 1 (2004). For considering wave passage effects, this input motion is applied with a 7 ms phase-delay at each pier from 1 to 5. Figure 6.11 displays the time histories of two input motions recorded at left and right abutments. As seen, there is approximately a 29 ms phase-delay between these two points, which is very close to $4 \times 7 = 28$ ms. This phase-delay is chosen based on a surface velocity of 1,000 m/s for firm ground. As there are only three

Fig. 6.8 Experimental test bed (*left*), and the model bridge on the test bed (*right*)

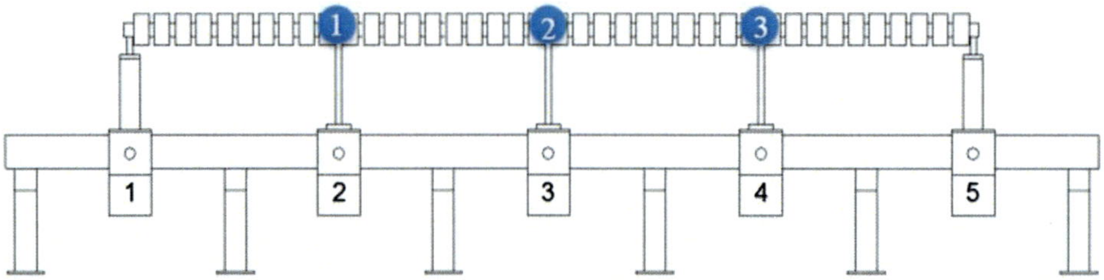

Fig. 6.9 Schematic representation of experimental test bed

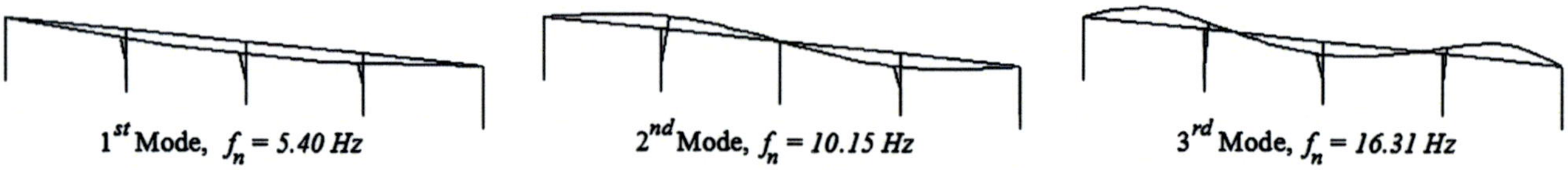

Fig. 6.10 The first three analytical mode shapes of the test model

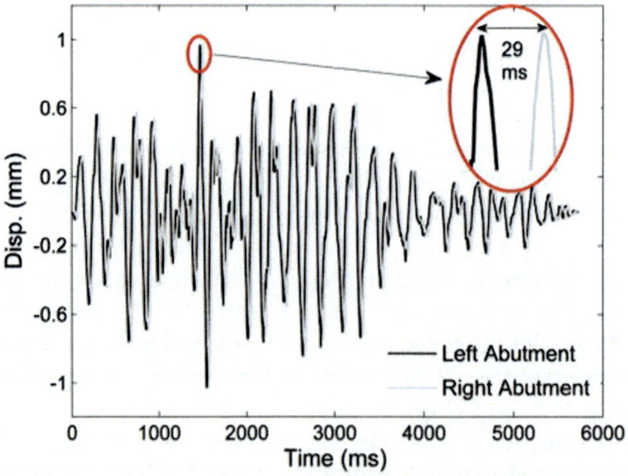

Fig. 6.11 Phase-delay measured by difference between input motions recorded at two abutments

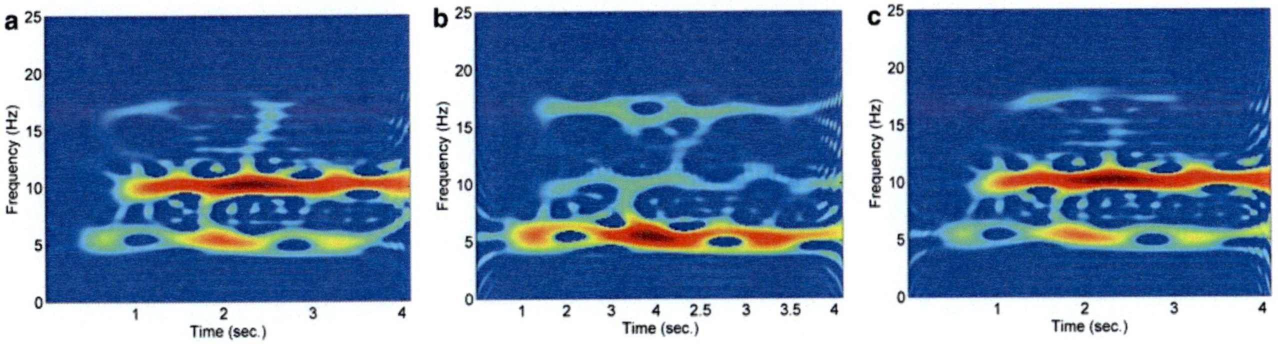

Fig. 6.12 Time-frequency representation of recorded acceleration response signals at points (**a**) 1, (**b**) 2, and (**c**) 3 on the deck

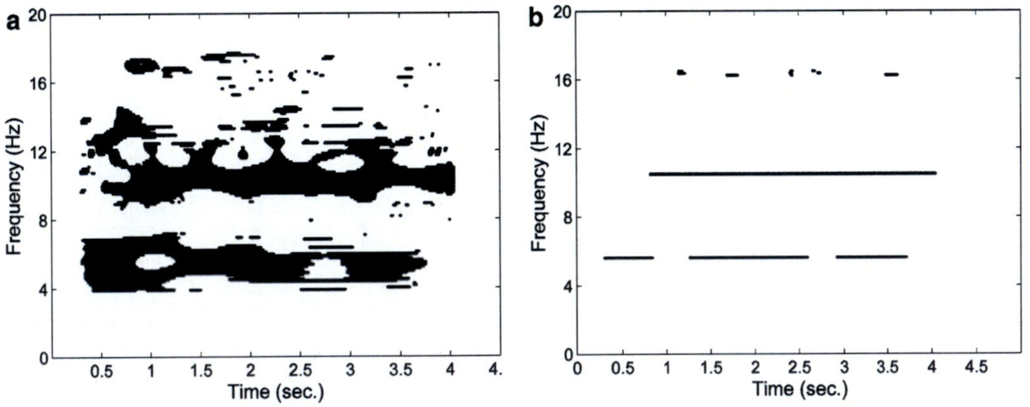

Fig. 6.13 (**a**) First and (**b**) second estimates of the auto-source points

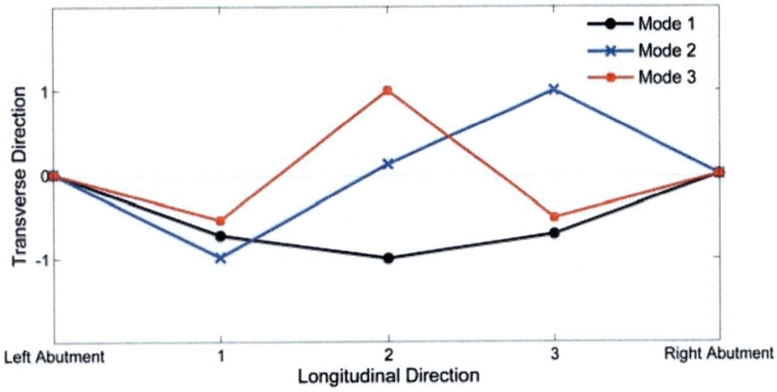

Fig. 6.14 Identified first three mode shapes

sensors to record the response signals, recorded displacement signals are filtered with a low-pass Butterworth filter with a cut-off frequency of 25 Hz, so that only the first three modes have a contribution. Figure 6.12 displays the time-frequency representations of the acceleration response signals.

To extract mode shapes and modal coordinates, the separation approach proposed by Ghahari et al. [8] is employed on the acceleration response signals—i.e., signals recorded at points 1, 2, and 3 on the deck. Figure 6.13a displays the auto-source points selected using the criterion introduced in Eq. (6.25) with $\varepsilon = 0.001$. In order to increase the accuracy of the identified mode shapes, it is expedient to limit the auto-source points around the dominant frequencies identified using extracted modal coordinates. Thus, by repeating the auto-source selection with the same $\in$ around $\pm\%1$ of each dominant frequency, new auto-source points are chosen, which are shown in Fig. 6.13b. By employing the JAD technique on STFD matrices corresponding to these newly chosen auto-source points, mode shapes are identified. Figure 6.14 shows the plan

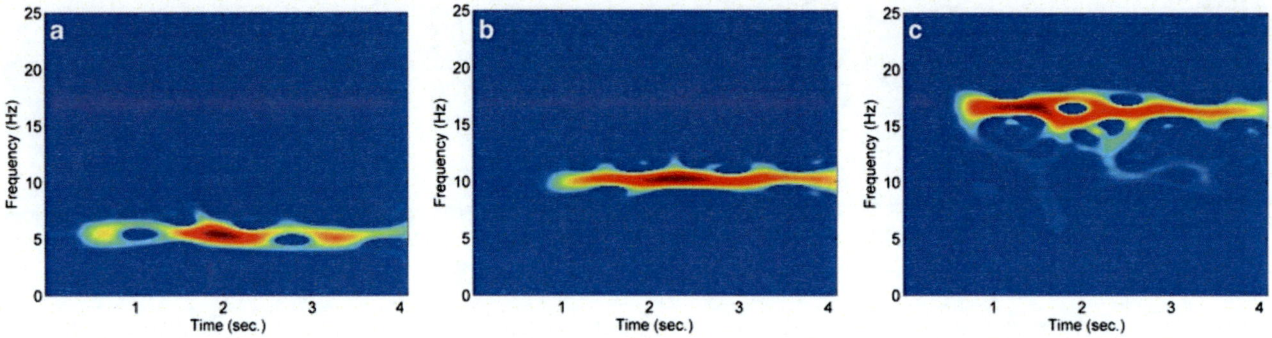

Fig. 6.15 Time-frequency representation of extracted modal coordinates of (**a**) first mode, (**b**) second mode, and (**c**) third mode

Fig. 6.16 Variation of final residual norm versus phase-delay

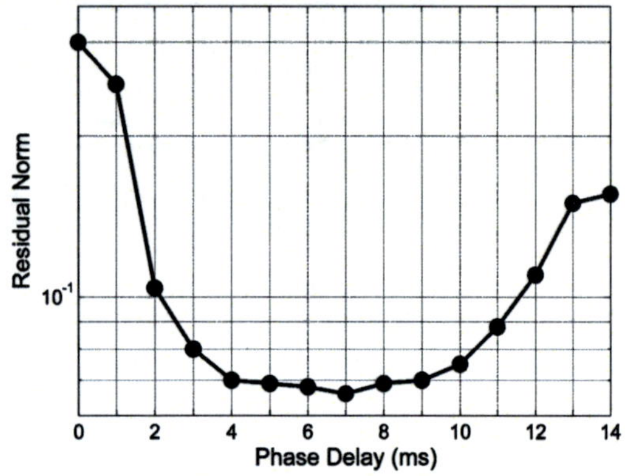

Table 6.3 Identified parameters from least-squares analysis

Mode no	Natural frequency (Hz)	Damping ratio (%)
1	5.62	0.41
2	10.50	0.70
3	16.36	1.35

view of identified mode shapes. In this graph, two ends of the mode shapes are virtually connected to zero, as abutment response signals were not used for the identification process. As seen, the identified mode shapes are almost symmetric, which is expected due to the geometry of the structure.

After identifying the mode shapes, modal coordinates are easily extracted through Eq. (6.26). Time-frequency representations of the extracted modal coordinates are shown in Fig. 6.15. As seen, a dominant frequency is observed that depicts the natural frequency of each mode. These are subsequently used as initial estimates of the natural frequencies for the minimization process, which yields the natural frequencies, damping ratios, modal contribution factors, and the input motions. Similar to the previous example, the optimization process is repeated several times with a different assumed value of phase-delay in order to find the delay that yields the minimum value of the residual norm. Figure 6.16 displays the converged residual norm for 15 different values of phasedelays that range from 0 to 14 ms. As seen, the minimum residual occurs at 7 ms phase-delay, which is equal to the original value. The natural frequencies and damping ratios corresponding to this phase-delay are displayed in Table 6.3. The identified natural frequencies are identical to those identified from the ridges of time-frequency graphs shown Fig. 6.15. For verification of the damping ratios, an input-output MIMO identification technique must be used, which is omitted here. However, a comparison between the estimated and the exact input motions is presented, this is—albeit indirect—an evidence of the accuracy of the identified parameters. As it can be seen in Fig. 6.17, while the identified input motion is not exact, it is very similar to its exact counterpart. Errors observed in this figure are mainly due the following reasons:

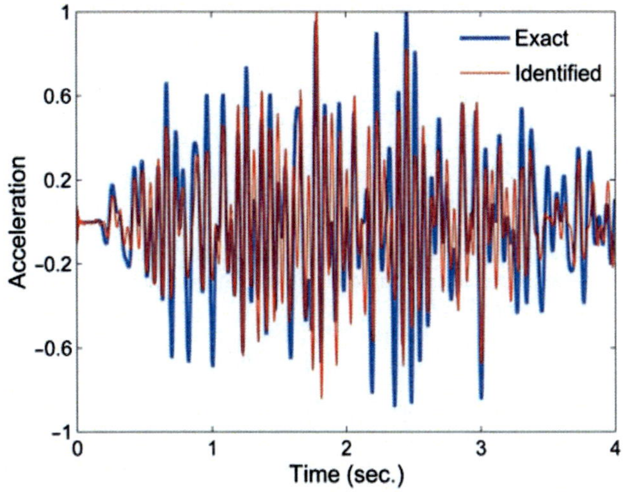

Fig. 6.17 Comparison between exact and identified input motions

- The identified mode shapes are not exact as higher modes may have contributions to the response signals; and their better resolution is limited because only three sensors are used.
- The number of unknown parameters in comparison to the known parameters is high, because only three modes have been used in the optimization process.

6.4 Conclusions

Due to differences in wave arrival times, soil conditions, and loss of coherency in arriving waves, long-span bridges usually experience different input excitations at their ground supports. Thus, identification of dynamic characteristics of these types of structures cannot be accurate if they are not considered as multiple input systems. Although there are numerous Multi-Input Multi-Output (MIMO) identification techniques, they are not applicable to structures that are interacting with their surrounding soil media, because signals recorded on foundations are not the actual input motions due to feedback from the superstructure's response (dubbed inertial interaction). Recently, a new system identification technique was developed by Ghahari et al. [8] to overcome this problem, which is based on Blind Source Separation (BSS) techniques. The said technique works in two steps—*viz.*, (i) time-frequency separation, in which mode shapes and modal coordinates are decomposed from response signals; and (ii) estimation of natural frequencies and damping ratios, in which cross-relations among modal coordinates are employed to identify the dynamic properties.

In this study, the second step of the said new BSS technique was extended to treat systems subject to multiple input motions. To that end, a nonlinear least square technique was used to estimate the system's dynamic characteristics and the unknown input motions form extracted modal coordinates. For the sake of simplicity, only wave passage effects with known phase-delays were studied here, and potential extensions in this regard are left for future studies. Both simulated and experimental examples were provided, which demonstrated that the new method is capable of accurately identifying the dynamic characteristics of long-span bridges from recorded response signals without the knowledge of input motions, even in the presence of wave passage effects stemming from known/unknown phase-delays.

References

1. Doebling SW, Farrar CR (1999) The state of the art in structural identification of constructed facilities. Report by the ASCE committee on structural identification of constructed facilities. Los Alamos National Laboratory, Los Alamos, NM, USA
2. Maia NMM, Silva JMM (2001) Modal analysis identification techniques. Philos Trans R Soc Lond A 359:29–40
3. Pridham BA, Wilson JC (2004) Identification of based-excited structures using output-only parameter estimation. Earthq Eng Struct Dyn 33:133–155
4. Lin CC, Hong LL, Ueng JM, Wu KC, Wang CE (2005) Parametric identification of asymmetric buildings from earthquake response records. Smart Mater Struct 14:850–861
5. Wolf JP (1985) Dynamic soil-structure interaction. Prentice-Hall, Englewood Cliffs

6. Abazarsa F, Ghahari SF, Nateghi F, Taciroglu E (2012) Response-only modal identification of structures using limited sensors. Struct Control Health Monit. doi:10.1002/stc.1513
7. Abazarsa F, Nateghi F, Ghahari SF, Taciroglu E (2013) Blind identification of non-classically damped systems from free or ambient vibration records. Earthq Spectra 29(3). In press
8. Ghahari SF, Abazarsa F, Ghannad MA, Taciroglu E (2012) Response-only modal identification of structures using strong motion data. Earthq Eng Struct Dyn. doi:10.1002/eqe.2268
9. Werner SD, Lee LC, Wong HL, Trifunac MD (1977) An evaluation of the effects of traveling seismic waves on the three-dimensional response of structures. Agbabian Associates, El Segundo, Report No. R-7720-4514
10. Abdel-Ghaffar AM, Rubin LI (1982) Suspension bridge response to multiple support excitations. J Eng Mech 108:419–435
11. Zerva A (1990) Response of multi-span beams to spatially incoherent seismic ground motions. Earthq Eng Struct Dyn 19:819–832
12. Lupoi A, Franchin P, Pinto PE, Monti G (2005) Seismic design of bridges accounting for spatial variability of ground motion. Earthq Eng Struct Dyn 34:327–348
13. Chopra A (1995) Dynamics of structures: theory and applications to earthquake engineering. Prentice Hall Inc., Upper Saddle River, NJ
14. Belouchrani A, Amin MG (1998) Blind source separation based on time-frequency signal representations. IEEE Trans Signal Process 46(11):2888–2897
15. Belouchrani A, Abed-Meriam K, Amin MG, Zoubir AM (2004) Blind separation of nonstationary sources. IEEE Signal Process Lett 11(7):605–608
16. Auger F, Flandrin P (1995) Improving the readability of time-frequency and time-scale representations by the reassignment method. IEEE Trans Signal Process 43:1068–1089
17. Cardoso JF (1994) Perturbation of joint diagonalizers. Telecom Paris, Signal Department, Technical Report 94D023
18. Cardoso JF, Souloumiac A (1994) Jacobi angles for simultaneous diagonalization. J Matrix Anal Appl 17(1):161–164
19. Papadimitriou C, Beck JL, Katafygiotis LS (2001) Updating robust reliability using structural test data. Probab Eng Mech 16:103–113
20. Pacific Earthquake Engineering Research Centre (PEER), http://peer.berkeley.edu/smcat/
21. MATLAB, version 7.0 (2004) The Math Works Inc, Natick
22. Coleman TF, Li Y (1996) An interior, trust region approach for nonlinear minimization subject to bounds. SIAM J Optim 6:418–445
23. Pinto AV (1996) Pseudodynamic and shaking tables on RC bridges. European Consortium of Earthquake Shaking Tables (ECOEST). Prenormative research in support of eurocode 8, Report No. 5. Bristol University, Bristol, UK
24. Zapico JL, Gonzalez MP, Friswell MI, Taylor CA, Crewe AJ (2003) Finite element model updating of a small scale bridge. J Sound Vib 268(5):993–1012
25. Crewe AJ, Norman AJP (2006) Experimental modelling of multiple support excitations of long span bridges. In: 4th international conference on earthquake engineering, Taipei, Paper No. 127
26. Norman, AJP, Crewe AJ (2008) Development and control of a novel test rig for performing multiple support testing of structures. In: 14th world conference on earthquake engineering, Beijing

Chapter 7
Quantification of Parametric Model Uncertainties in Finite Element Model Updating Problem via Fuzzy Numbers

Yildirim Serhat Erdogan, Mustafa Gul, F. Necati Catbas, and Pelin Gundes Bakir

Abstract Analytical and numerical models that simulate the physical processes inevitably contain errors due to the mathematical simplifications and the lack of knowledge about the physical parameters that control the actual behavior. In this sense, parametric identification of civil engineering structures using uncertain numerical models should be subject to a particular interest in terms of accuracy and reliability of identified models. In this study, model uncertainties are modeled by fuzzy numbers and quantified using fuzzy model updating approach. In order to find the possible variation range of the response parameters (e.g. natural frequencies, mode shapes and strains) using uncertain finite element model, successive updating is employed. A simplified approach is proposed in order to facilitate the time consuming successive model updating phase. The identified variation range of the response parameters is employed to construct the fuzzy membership functions for each response parameter. Finally, fuzzy finite element model updating method (FFEMU) is used to obtain the membership functions of the model parameters. Different sets of model parameters are chosen to represent different models in terms of accuracy and these parameters are identified in the same way to investigate the model complexity. A two span laboratory grid structure developed for simulating bridge structures is used to validate and demonstrate the proposed approaches. The results show that the proposed approaches can efficiently be utilized to quantify the modeling uncertainties for more realizable and quantitative condition assessment and decision making purposes.

Keywords Finite element model updating • Fuzzy numbers • Model uncertainties • Optimization • Inverse fuzzy problems

7.1 Introduction

Identification of civil structures using structural health monitoring data has gained considerable attention with the development of measurement technologies [1–3]. Model based methods such as Finite Element Model Updating (FEMU) provide calibrated models, which better represent the actual behavior of structures [4, 5]. These models might be used for different analysis like damage detection, reliability analysis, load rating calculation etc. in order to make reliable decisions for the remaining life of the structure or retrofitting purposes [6–8]. However, the model and the measurement uncertainties that directly affect the updating process have still to be considered in order to obtain reliable updated models, which provide a variation range for model prediction in probabilistic sense. Quantification of these uncertainties is not a straightforward task, especially when the degree of freedom of the structure is too high since it requires too many model calculations [9, 10].

Y.S. Erdogan • P.G. Bakir
Department of Civil Engineering, Istanbul Technical University, Maslak, 34469, Istanbul, Turkey
e-mail: yserdogan@itu.edu.tr; gundesbakir@yahoo.com

M. Gul
Department of Civil and Environmental Engineering, University of Alberta, Edmonton, AB, Canada
e-mail: mustafa.gul@ualberta.ca

F.N. Catbas (✉)
Department of Civil, Environmental and Construction Engineering, University of Central Florida, Orlando, FL, 32816-2450, USA
e-mail: catbas@ucf.edu

T. Simmermacher et al. (eds.), *Topics in Model Validation and Uncertainty Quantification, Volume 5*,
Conference Proceedings of the Society for Experimental Mechanics Series 41,
DOI 10.1007/978-1-4614-6564-5_7, © The Society for Experimental Mechanics, Inc. 2013

The main objective of this study is to investigate and then demonstrate a model uncertainty quantification approach first in the laboratory assuming that the measurement noise is time invariant and also less than certain real life civil engineering structures since the experiments are conducted in a controlled laboratory environment. The impact of modeling uncertainties on the response predictions are quantified by Fuzzy Finite Element Model Updating (FFEMU) method. Several measurement data sets (e.g. strain readings, modal parameters) and updating model parameter sets are generated and used to update the baseline FEM model of the structure. By this way, uncertainty amount in structural responses are compared for different data sets in order to quantify the uncertainty effect on model predictions. A benchmark grid structure designed for investigating structural health monitoring technologies and St-Id strategies at the University of Central Florida is used as the test structure [11, 12]. A damage case is adopted in which the boundary conditions are turned to be flexible to simulate the damage at the supports. In order to extract modal parameters for initial and damage cases, Complex Mode Indicator Function (CMIF) is employed [11, 13]. However, many other methods can be used effectively for experimental modal analysis like subspace based system identification techniques [14].

FFEMU method together with some proposed constraints which is required for the uniqueness of the inverse solution is employed to quantify uncertainties in the model. A Gaussian Process model, which is the surrogate for the numerical model is used to tackle with the computational issues. The results show that appropriate measurement sets should be generated instead of involving the complete measurement in order to keep the uncertainty at certain levels. This is due to the fact that the response parameters, which are more affected by the model uncertainties, cause the total uncertainty in the updated model to increase. Hence, the uncertainty quantification methods should be employed for different measurement sets and compared with each other to determine the appropriate data sets required for reliable models, particularly when both static and dynamic data are utilized together.

7.2 Fuzzy Finite Element Model Updating

Fuzzy set theory [15] is one of the efficient ways to quantify parametric uncertainty involved in the input and output parameters. Fuzzy numbers have been used as a tool in order to investigate the effect of uncertainties in different engineering input/output systems [16, 17]. A general illustration of fuzzy forward and inverse analysis for an I/O system is given in Fig. 7.1. In Fig. 7.1, the inputs and the outputs are the fuzzy parameters with different membership functions. Those membership functions can be chosen depending on the assumed and/or the quantified uncertainty in the parameters with the help of expert knowledge and the past experiences.

The equations for objective functions and related constraints that have to be strictly applied in order to make inverse problem to have a unique solution and capture all uncertainty in model responses are given in Eqs. (7.1), (7.2), (7.3), (7.4) and (7.5).

$$\min f\left(\theta^{\text{int}}\right) = \underline{r}\left(\theta^{\text{int}}\right)^{T} \mathrm{W}\underline{r}\left(\theta^{\text{int}}\right) + \bar{r}\left(\theta^{\text{int}}\right)^{T} W\bar{r}\left(\theta^{\text{int}}\right) \tag{7.1}$$

$$\theta^{\text{int}} = \left[\underline{\theta}, \overline{\theta}\right] \tag{7.2}$$

$$\underline{r}\left(\theta^{int}\right) = \frac{\underline{\gamma}\left(\theta^{int}\right) - \underline{\gamma}^{e}}{\underline{\gamma}^{e}} \tag{7.3}$$

$$\bar{r}\left(\theta^{int}\right) = \frac{\bar{\gamma}\left(\theta^{int}\right) - \bar{\gamma}^{e}}{\bar{\gamma}^{e}} \tag{7.4}$$

$$\left[\lambda^{int}, \phi^{int}, \varepsilon^{int}\right] = f^{model}\left(\theta^{int}\right) \tag{7.5}$$

$$\underline{\theta}_{i}^{(j+1)} \leq \underline{\theta}_{i}^{(j)} \tag{7.6}$$

$$\bar{\theta}_{i}^{(j+1)} \geq \bar{\theta}_{i}^{(j)} \tag{7.7}$$

$$\underline{\gamma}_{k}^{num} \leq \underline{\gamma}_{k}^{\exp} \forall j \in [0, 1]\, k = 1, \ldots .nm \tag{7.8}$$

$$\bar{\gamma}_{k}^{num} \geq \bar{\gamma}_{k}^{\exp} \forall j \in [0, 1]\, k = 1, \ldots .nm \tag{7.9}$$

$$f^{\text{infeasible}}(\theta) = f(\theta) + H \tag{7.10}$$

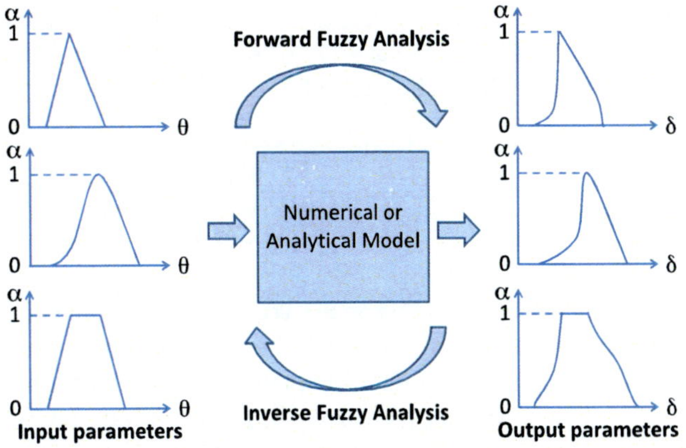

Fig. 7.1 General structure of the forward and the inverse fuzzy problem

where, θ^{int} is the interval valued updating parameter vector, W is the weighting matrix, which might be determined intuitively considering the relative accuracy of the measurements, $\gamma = [\lambda, \phi, \varepsilon]$ is the response vector, $f(\theta)$ is the objective function and $f^{\text{model}}(\theta)$ is the model function that governs the physical process. In Eq. (7.5), frequency, mode shape and strain vectors are denoted by λ, φ and ε, respectively. The bar above and below the response quantities denotes the upper and lower quantities. In addition to the objective function provided by Eqs. (7.1), (7.2), (7.3), (7.4) and (7.5), some additional constraints have to be introduced to preserve the monotonic behavior of the fuzzy set. It should also be noted that the optimization problem given Eqs. (7.1), (7.2), (7.3), (7.4) and (7.5) has to be solved for some specified numbers of α-level in order to capture the nonlinear relationship between inputs and outputs. The equations (7.8) and (7.9) can effectively be satisfied by assigning some penalties to the infeasible regions in the output space domain. In Eq. (7.10), the term H is a number in which its value is very high compared to the objective function value. By this way, infeasible regions can be disregarded.

7.3 Gaussian Process Model for FFEMU

Gaussian Process (GP) models are very efficient to develop approximate I/O computer models which require less computational effort to calculate. Since FFEMU is a computationally expensive method, it is not allow the computer simulation to be used directly due to time and cost constraints. Hence, surrogate models are efficient solutions to this problem with the limitation in the number of input parameters. However, [18] states that the GP can effectively be implemented to problems with 50 input parameters. This number is sufficient in most cases to identify a real life civil engineering structure.

Let x be the vector valued input parameter and the $[Y(X_1), Y(X_2), \ldots Y(X_n)]$ be the outputs based on n observations. In order to choose appropriate observations for the input parameter x, Central Composite Design (CCD) can be utilized as second-order design method [19]. If Y(x) denotes a Gaussian Process the mean and the covariance can be expressed by Eqs. (7.11) and (7.12).

$$E\left[Y(x)\right] = q(x)^T \beta \tag{7.11}$$

$$Cov\left[Y(x), Y(x*)\right] = \sigma^2 C\left(x, x^* \,|\xi\,\right) \tag{7.12}$$

Where (x) are trend functions given by $[1x^T]^T$ for linear trend and 1 for constant trend; β is the vector of regression coefficients, x^* is the untested input, σ^2 and $C(x, x*|\xi)$ are the variance of overall process and the correlation function, respectively. The governing parameters of the correlation function, which have to be determined together with β and σ^2 to create the GP model, are represented by ξ. The correlation function and the joint distribution function for Y expending on n observation are given by Eqs. (7.13) and (7.14).

$$C\left(x, x*\right) = exp\left(-\sum_{i=1}^{m} \xi_i \left(x_i - x_i^*\right)^2\right) \tag{7.13}$$

$$\boldsymbol{Y} \sim \boldsymbol{N}_n\left(q^T(x)\beta, \sigma^2 \boldsymbol{R}\right) \tag{7.14}$$

If we know all the parameters governing the overall Gaussian Process, the mean and the covariance of the output can be calculated by Eqs. (7.15) and (7.16) instead of solving linear system of equations or eigenvalue problems in which large system matrices has to be handled.

$$E\left[\boldsymbol{Y}\left(x^{*}\right)|\boldsymbol{Y}\right]=\boldsymbol{q}^{T}\left(x^{*}\right)\boldsymbol{\beta}+r^{T}\left(x^{*}\right)\boldsymbol{R}^{-1}\left(\boldsymbol{Y}-\boldsymbol{F}\boldsymbol{\beta}\right) \tag{7.15}$$

$$Cov\left[Y(x)Y\left(x^{*}\right)|\boldsymbol{Y}\right]=\sigma^{2}\left(C\left(x,x^{*}\right)-r^{T}\boldsymbol{R}^{-1}\boldsymbol{r}\right) \tag{7.16}$$

In Eqs. (7.15) and (7.16), r is the vector of correlations between x and each of the observation points, r^* is the vector of correlations between x* and each of the observation points. F is a matrix with rows $q^T(x_i)$. In this study, there is no need to calculate covariance of the predictions since this information will not be used in the scope of our methodology. In order to obtain the governing parameters ξ, β and the variance of the process σ^2 of GP, maximum likelihood estimation (MLE) might be employed which is required to solve optimization problem for each response variable only once. More details can be found in [20].

7.4 Numerical Verification

The benchmark grid structure developed for bridge health monitoring studies is used to investigate the model uncertainties for different experimental data sets. The grid structure has been designed to enable researchers to explore the use of different sensor technologies, St-Id and damage detection algorithms under different conditions, which offer promise cases for real life bridge structures. The girders and the columns of the grid have been constructed using steel sections S3 × 5.7 and W12 × 26, respectively. The 3D cad model and plan view of the structure are given in Fig. 7.2. More details about the grid structure can be found in [12].

Several static and dynamic tests have been conducted on the grid structure. The natural frequencies, mode shapes and displacements obtained from different loading cases are involved in FFEMU procedure. The natural frequencies and mode shapes are obtained by using Complex Mode Indicator Function (CMIF) method [11, 13]. In static case, five different loading conditions are considered. Single loads (671.6 N) are applied to different nodes for each static loading case. The nodes where the single static loads are applied are as follows: (1) N3-N6-N9-N12 (2) N3-N9 (3) N6-N12 (4) N3-N6 (5) N9-N12. In order to demonstrate the methodology, a damage case in which two supports located at nodes N7 and N14 are replaced with Duro50 elastomeric pads to simulate flexible boundaries. The experimental and the deterministically updated natural frequencies (the natural frequencies which correspond to the α-cut level 1) are presented in Table 7.1.

The fuzzy response parameters obtained from updated fuzzy model are investigated for two main cases. In the first case, different sets of experimental data which contain several numbers of measured dynamic and static responses are considered. In the second case, different sets of updating parameters are used in fuzzy updating procedure in order to investigate the effect of uncertainties in structural response parameters (e.g. natural frequencies, static displacements). The flexible boundary conditions case with elastomeric pads is considered to be the damage case and use to demonstrate the fuzzy updating and analysis concepts. In the first case, column stiffnesses are chosen as the updating parameters. The data sets used in FFEMU is given in Table 7.2. Only results for the natural frequencies obtained from updated fuzzy model are presented in this paper. However, same results can be deducted for both strain and mode shapes.

After employing Fuzzy FEM to the updated model, fuzzy valued response parameters are obtained for each data sets. Membership functions of four predicted natural frequencies using updated fuzzy model are given in Fig. 7.3. The fourth and the fifth frequencies are not illustrated in order to make the figure more tractable. As apparent in Fig. 7.3, given frequencies are capable of illustrating the trend in uncertainty distribution for the increasing frequencies. The uncertainty in the frequencies is increasing when the amount of data involved in the updating is increased. The deterministic values for the first natural frequency which correspond to α-level 1 are very close to experimental one when the first six natural frequencies and first mode shapes (f + 1 m) are included. The amount of imprecision is also very low for this frequency. This is same for the f + 3 m together with slight increase in the uncertainty. This means that the effect of modeling uncertainties on the first natural frequency is low and this frequency can be predicted more accurate using the updated model as expected. However, the accuracy of predictions decreases and amount of imprecision increases when the strain data is included. This is due to fact that the uncertainty sources coming from dynamic and static tests are different. In the static tests, there might be some additional uncertainties due to loading. As opposed to static case, in dynamic tests there are some additional uncertainties arising from slight nonlinearity in boundaries due to the loss of contact and lack of tension stiffness in supports. In static tests, these problems are surpassed by adding some additional weights on support. However, first two natural frequencies and

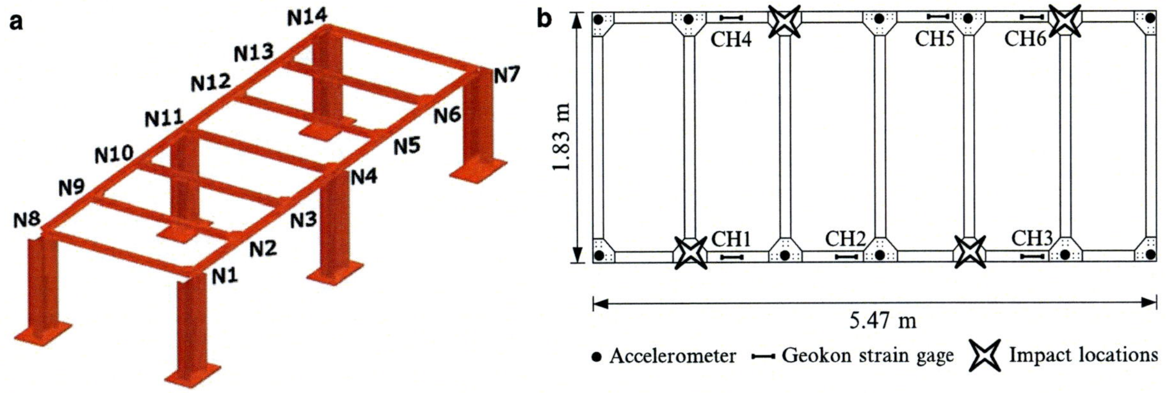

Fig. 7.2 (**a**) 3D view (**b**) plan view and sensor configuration

Table 7.1 Comparison of natural frequencies obtained from updated FEM and experiments

Mode number	Initial FEM (Hz)	Deterministically updated FEM (Hz)	Experimental frequencies (Hz)	
			Initial case	Damage case
1	21.52	22.65	22.72	14.31
2	26.22	27.60	27.67	18.50
3	32.69	34.10	34.13	26.12
4	40.72	42.57	42.15	32.25
5	62.73	64.54	65.18	37.68
6	66.52	67.99	68.27	47.68
7	93.98	94.02	95.36	61.18
8	96.47	96.70	97.90	65.50

Table 7.2 Data sets involved in the updating procedure

Data set	Abbreviation	Explanation
1	$f+1\text{m}$	First six natural frequencies and first mode shape values
2	$f+3\text{m}$	First six natural frequencies and first three mode shape values
3	$f+\varepsilon$ (1 loading)	First six natural frequencies and strain values from first loading
4	$f+\varepsilon$ (3 loading)	First six natural frequencies and strain values from the first three loading
5	ε (3 loading)	Only strain values from first three loadings
6	Full data	First six natural frequencies, first three mode shape values and strain values from first three loadings

mode shapes are affected less from these uncertainties and they can be predicted more precisely by means of FEMU through the first few modal parameters.

The imprecision amounts given in Table 7.4 demonstrate the statements made in the previous paragraph. The imprecision amount is low and the deterministic values are close to measurements for the data set f + 1 m. In Table 7.3, it is also apparent that the imprecision increases dramatically for the first two frequencies with the increase in data due to the uncertainty coming from other response variables. It should also be mentioned that while total imprecision may be less with fewer data and more imprecision with more data, models calibrated with more data better represent the overall structural characteristics.

In the second case three sets of updating parameter are used to update the FEM of the benchmark grid structure. The updating parameter sets are as follows: (1) Axial stiffness of the columns and the moment of inertia of the connections. The connections are grouped as N2,N3 – N5,N6 – N9,N10 – N12,N13 – N11,N4 (Total 11 parameters) (2) Axial stiffness of the columns and moment of inertia of the connections at nodes N2,N3,N4,N5,N6,N9,N10,N11,N12,N13. For this set, the connections are grouped as N2,N3,N4,N5,N6 and N9,N10,N11,N12,N13 (Total 8 parameters) (3) Axial stiffness of the columns (Total 6 parameters). Full data set is used as the measurement data for all updating parameter set. The fuzzy natural frequencies obtained from fuzzy models updated using each sets of updating parameters and amount of imprecision are given in Fig. 7.4 and Table 7.1, respectively.

As apparent in Fig. 7.4 and Table 7.1, the amount of uncertainty is decreased as the number of updating parameters is increased. In addition, the deterministic values of response parameters which correspond to the α-cut level 1 are close to the

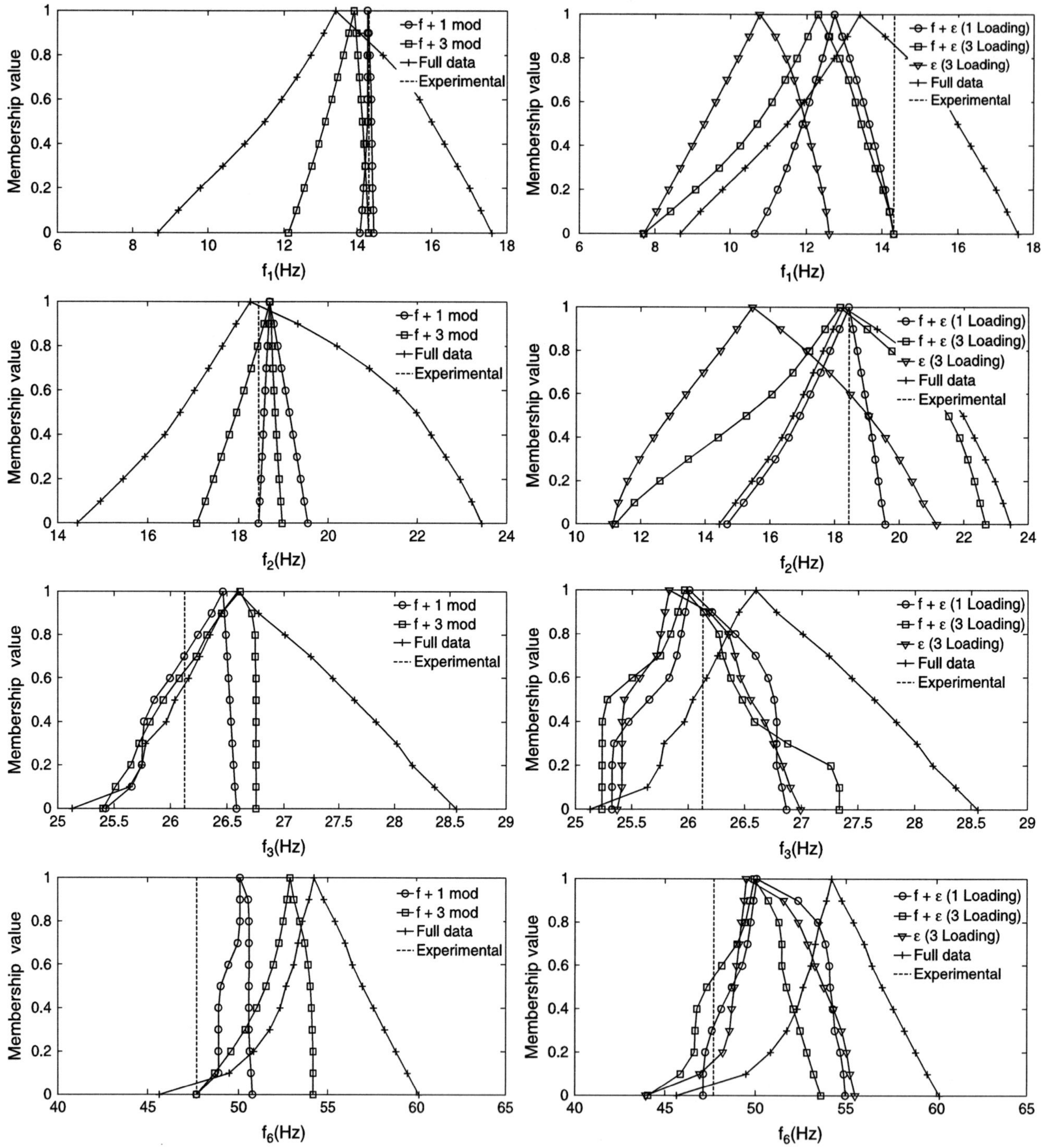

Fig. 7.3 Updated natural frequencies using different SHM data set for the damage case

Table 7.3 Imprecision in the response for different data sets

Response/data set	$f+1$ mode	$f+3$ mode	$f+$ strain (1 loading)	$f+$ strain (3 loading)	Strain (3 loading)	Full data
f_1	0.0101	0.0754	0.1404	0.2422	0.2449	0.3396
f_2	0.0296	0.0471	0.1214	0.3366	0.3724	0.2751
f_3	0.0218	0.0289	0.0380	0.0441	0.0388	0.0595
f_6	0.0254	0.0513	0.1053	0.0852	0.1032	0.0928

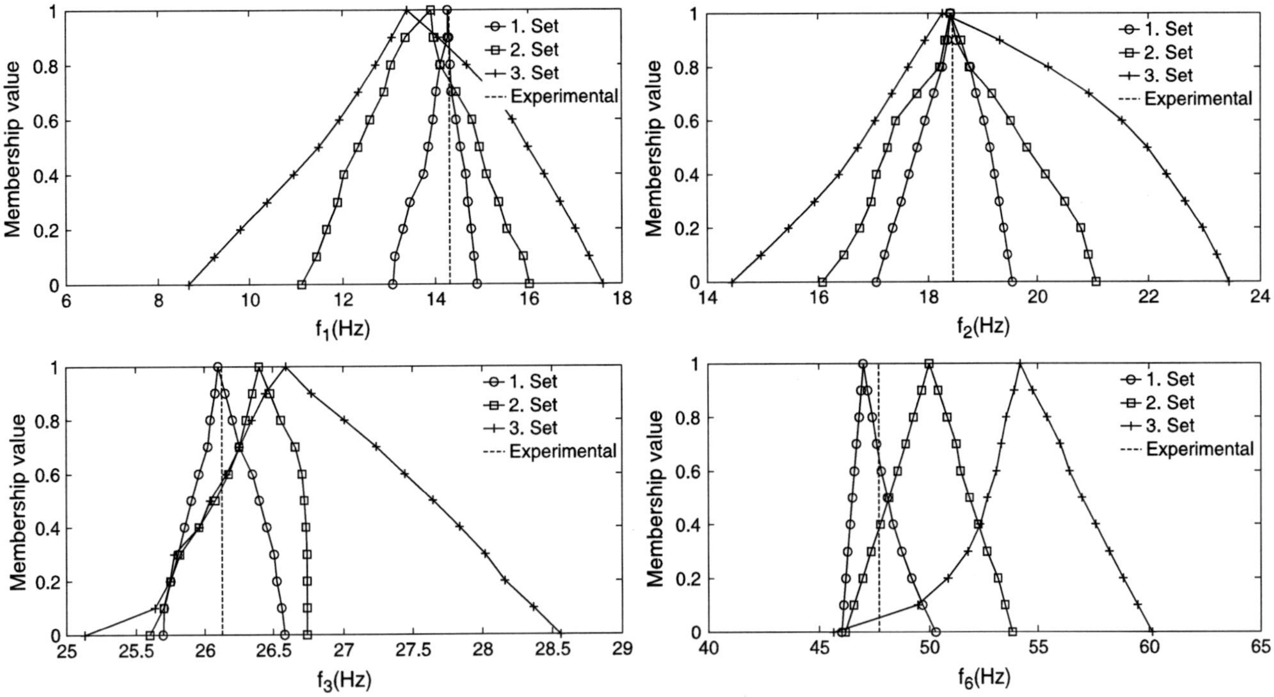

Fig. 7.4 Updated natural frequencies using different updating parameter set

Table 7.4 Imprecision in the response for different updating parameter sets

Response/data set	1. set	2. set	3. set
f_1	0.0511	0.1034	0.3396
f_2	0.0445	0.0814	0.2751
f_3	0.0113	0.0310	0.0595
f_6	0.0333	0.0620	0.0928

experimental values. However, more updating parameters bring some disadvantageous beside low uncertainty and accuracy. The computational complexity increases with higher number of model parameters. In addition, the updated parameters might not provide physically meaningful predictions for different loading conditions. Hence, number of updating parameters should be restricted considering the geometric and material properties of the actual structure.

7.5 Conclusion

Results show that the uncertainty in the response variables increases as the number of experimental data used in FFEMU increases. This can be explained with the fact that the uncertainty in all response variables contributes to total uncertainty in the updated model. This is analogous to applying a high precision curve-fit to very few data points, say two data points. As the number of data points increase, it can be obvious that the imprecision will increase, yet the overall characterization of the entire data set will be better. The imprecision and uncertainty amount in model parameters are observed to increase especially when strain data is used. This increment arises from the difference between the dynamic and the static test setup and the global (modal) and local (strain) responses that these measurements represent. The difference in test setups and responses lead to different uncertainty sources, reducing the reliability of individual parameter predictions. In static tests, additional weights have been used to prevent the loss of contacts in supports while these weights were removed during dynamic tests. Additionally, there is no tension stiffness in boundaries, which introduces nonlinearity in the supports of the test structure. However, similar uncertainties can be expected to exist in real life structures. Hence, special care should be paid for generating appropriate data sets to develop the updated models with the least uncertainty. This can be achieved by ignoring some measurements such as dynamic response at boundaries or strain measurements of "virtually unstressed" members. In addition, low weighting factors can be assigned to the measurements that increase the uncertainty in response predictions. Finally, more complicated models, which are not preferable in most cases, might be another solution in some sense. As demonstrated in the second case, inclusion of more updating parameter lead lower uncertainty amount in the response. However, inclusion of additional model parameters in updating process might not be feasible in all cases since it is always possible to obtain physically inconvenient model parameter values. By this way, the degree of freedom of the updated model is increased which may provide unfeasible response predictions for different loading conditions.

References

1. Catbas FN, Kijewski-Correa TL, Aktan AE (eds) (2012a) Structural identification of constructed systems: approaches, methods and technologies for effective practice of St-Id. ASCE (accepted). ISBN:978-0784411971
2. Xu B, He J, Rovekamp R, Dyke SJ (2012) Structural parameters and dynamic loading identification fromin complete measurements: approach and validation Mech Syst Signal Process 28:244–257
3. Esfandiari A, Bakhtiari-Nejad F, Sanayei M, Rahai A (2010) Structural finite element model updating using transfer function data. Comput Struct 88:54–64
4. Mottershead JE, Michael L, Friswell MI (2011) The sensitivity method in finite element model updating: a tutorial. Mech Syst Signal Process 25(7):2275–2296
5. Bakir PG, Reynders E, De Roeck G (2007) Sensitivity-based finite element model updating using constrained optimization with a trust region algorithm. J Sound Vib 305:211–225
6. Buezas FS, Rosales MB, Filipich CP (2011) Damage detection with genetic algorithms taking into account a crack contact model. Eng Fract Mech 78:695–712
7. Meruane V, Heylen W (2011) An hybrid real genetic algorithm to detect structural damage using modal properties. Mech Syst Signal Process 25:1559–1573
8. Gokce HB, Catbas FN, Gul M, Frangopol DM (2012) Structural identification for performance prediction considering uncertainties: a case study of a movable bridge. J Struct Eng ASCE (accepted). doi:http://dx.doi.org/10.1061/(ASCE)ST.1943-541X.0000601
9. Xiong Y, Chen W, Tsui KL, Apley DW (2009) A better understanding of model updating strategies in validating engineering models. Comput Meth Appl Mech Eng 198:1327–1337
10. Catbas FN, Gokce HB, Frangopol DM (2012b) Predictive analysis by incorporating uncertainty through a family of models calibrated with structural health monitoring data. ASCE J Eng Mech (accepted). doi:http://dx.doi.org/10.1061/(ASCE)EM.1943-7889.0000342
11. Catbas FN, Aktan AE, Brown DL (2004) Parameter estimation for multiple input multiple output analysis of large structures. J Eng Mech ASCE 130(8):921–930
12. Catbas FN, Gul M, Burkett J (2008) Damage assessment using flexibility and flexibility-based curvature for structural health monitoring. Smart Mater Struct 17(1):015–024
13. Catbas FN, Brown DL, Aktan AE (2006) Use of modal flexibility for damage detection and condition assessment: case studies and demonstrations on large structures. J Struct Eng ASCE 132(11):1699–1712
14. Bakir PG (2011) Automation of the stabilization diagrams for subspace based system identification. Expert Syst Appl 38(12):14390–14397
15. Zadeh LA (1965) Fuzzy sets. Inf Control 8(3):338–353
16. Massa F, Ruffin K, Tison T, Lallemand B (2005) A complete method for efficient fuzzy modal analysis. J Sound Vib 309:63–85
17. Nicolai BN, Egea JA, Scheerlinck N, Banga JR, Datta AK (2011) Fuzzy finite element analysis of heat conduction problems with uncertain parameters. J Food Eng 103(1):38–46
18. O'Hagan A (2006) Bayesian analysis of computer code outputs: a tutorial. Reliab Eng Syst Saf 91:1290–1300
19. Box GEP, Wilson KB (1951) On the experimental attainment of optimum conditions (with discussion). J R Stat Soc B 13(1):1–45
20. Mc Farland JM (2008) Uncertainty analysis for computer simulations through validation and calibration. Department of Mechanical Engineering, Vanderbilt University

Chapter 8
Quantifying Maximum Achievable Accuracy of Identified Modal Parameters from Noise Contaminated Free Vibration Data

Eric M. Hernandez

Abstract This paper derives exact results for the maximum achievable accuracy when estimating modal parameters from free vibration signals contaminated by Gaussian white noise. These limits are found through the Cramer-Rao lower bound. The paper compares the exact findings with previous approximate results found in the literature. Comparisons are drawn with results from stochastic simulations of a Fourier domain approach for estimation of natural frequency and damping in single degree of freedom system.

Keywords Estimation • Cramer-Rao • Fisher information • Uncertainty quantification • Damping

8.1 Introduction

Identification of modal properties of structures based on their measured response is essential for many applications in structural dynamics, vibration based structural health monitoring and control. In some cases noise contaminated free vibration response measurements are available and one wishes to identify modal parameters from these. Free vibration response of a classically damped multi-degree of freedom linear structural system is given as

$$z(x,t) = \sum_i \phi_i(x) y_i(t) \tag{8.1}$$

where

$$y_i(t) = e^{-\xi_i \omega_{n,i} t} \left[\left(\frac{\dot{y}_{o,i} + \xi_i \omega_{n,i} y_{o,i}}{\omega_{d,i}} \right) \sin \omega_{d,i} t + y_{o,i} \cos \omega_{d,i} t \right] \tag{8.2}$$

and $\phi_i(x)$ is the modal amplitude of the ith mode at location x. ξ_i and $\omega_{n,i}$ are the modal damping ratio and modal circular frequency of the ith mode and $\omega_{d,i} = \omega_i \sqrt{1-\xi_i^2}$. The main objective of this paper is to obtain an expression for the maximum achievable accuracy when estimating modal parameters based on analysis of measurement signals given by

$$s(x,t) = z(x,t) + \nu(t) \tag{8.3}$$

where $\nu(t)$ is the measurement noise. This problem has been studied previously, although not with the specific interest in structural dynamics by Wigren and Nehorai [3]. In their study some assumptions were made, specifically (A1) noise is a zero mean white Gaussian random process, (A2) the signal is measured until it fades away, (A3) sampling is sufficient (Nyquist frequency is higher than signal frequency), (A4) low damping in the signal and (A5) low spectral interference. Gersch [1] also studied a similar problem, however, with emphasis in the case of a system excited by a white noise, and no free vibration present. In our case, we are studying the complementary problem, namely no excitation, only free vibration. In this paper

E.M. Hernandez (✉)
Assistant Professor, University of Vermont, 301 Votey Hall, 33 Colchester Ave., Burlington, VT 05405, USA
e-mail: eric.hernandez@uvm.edu

T. Simmermacher et al. (eds.), *Topics in Model Validation and Uncertainty Quantification, Volume 5*,
Conference Proceedings of the Society for Experimental Mechanics Series 41,
DOI 10.1007/978-1-4614-6564-5_8,

we relax the assumptions imposed in [3] and derive exact expressions for the Cramer-Rao lower bound for modal frequency, damping ratio and initial conditions.

8.2 Fisher Information

The Fisher information is a metric on the amount of information that an observable random variable X carries about an unknown parameter θ upon which the probability of X depends. The probability function for X, which is also the likelihood function for θ, is a function $f(X;\theta)$; it is the probability mass (or probability density) of the random variable X conditional on the value of θ. The Fisher information matrix is defined as

$$I(\theta) = -E\left[\frac{\partial^2 \ln f(X,\theta)}{\partial \theta^2}\right] \tag{8.4}$$

It can be shown [2] that the Fisher information matrix of a noise corrupted scalar signal $s(t)$ dependent on multiple parameters θ_i is given by

$$\mathbf{I}(\theta) = \frac{1}{\sigma^2}\sum_{n=0}^{N-1} \frac{\partial s(n;\theta)}{\partial \theta_i}\frac{\partial s(n;\theta)}{\partial \theta_j} \tag{8.5}$$

where σ^2 is the variance of the additive zero mean noise.

8.3 Cramer-Rao Lower Bound

If $f(X;\theta)$ satisfies the regularity condition

$$E\left[\frac{\partial \ln(f(X;\theta))}{\partial \theta}\right] = 0 \quad \forall\theta \tag{8.6}$$

where the expectation is taken with respect to $f(X;\theta)$, then the variance of any unbiased estimator $\hat{\theta}$ must satisfy

$$Var(\hat{\theta}) \geq CRLB = \frac{1}{I(\theta)} \tag{8.7}$$

This lower limit is known as the Cramer-Rao lower bound (CRLB). For the case of multiple parameters, the Fisher information becomes a matrix and the CRLB is given by

$$CRLB = \mathbf{I}(\theta)^{-1} \tag{8.8}$$

8.4 Cramer-Rao Lower Bound for a SDoF

For the special case of Eq. 8.3

$$s(t) = e^{-\xi\omega_n t}\left[\left(\frac{\dot{x}_o + \xi\omega_n x_o}{\omega_d}\right)\sin\omega_d t + x_o\cos\omega_d t\right] + \nu(t) \tag{8.9}$$

the exact expression for the derivatives necessary to compute Eq. 8.5 are given by

$$\frac{\partial s}{\partial \omega} = e^{-\xi\omega_n t}[-\xi t\left(\frac{\dot{x}_o + \xi\omega_n x_o}{\omega_d}\right)\sin\omega_d t + t\left(\frac{\dot{x}_o + \xi\omega_n x_o}{\omega_d}\right)\sqrt{1-\xi^2}\cos\omega_d t - \frac{\dot{x}_o}{\omega_n\omega_d}\sin\omega_d t$$
$$-x_o\xi t\cos\omega_d t - x_o t\sqrt{1-\xi^2}\sin\omega_d t] \tag{8.10}$$

$$\frac{\partial s}{\partial \xi} = e^{-\xi\omega_n t}[-\omega_n t\left(\frac{\dot{x}_o + \xi\omega_n x_o}{\omega_d}\right)\sin\omega_d t + \left(\xi\left(\frac{\dot{x}_o}{\omega_n} + \xi x_o\right)(1-\xi^2)^{-\frac{3}{2}} + \frac{x_o}{\sqrt{1-\xi^2}}\right)\sin\omega_d t$$

$$-\left(\frac{\dot{x}_o + \xi\omega_n x_o}{\omega_d}\right)\left(\frac{\omega_n \xi t}{\sqrt{1-\xi^2}}\right)\cos\omega_d t - x_o\omega_n t\cos\omega_d t + x_o\left(\frac{\omega_n \xi t}{\sqrt{1-\xi^2}}\right)\sin\omega_d t] \tag{8.11}$$

$$\frac{\partial s}{\partial \dot{x}_o} = \frac{e^{-\xi\omega_n t}}{\omega_d}\sin\omega_d t \tag{8.12}$$

$$\frac{\partial s}{\partial x_o} = \frac{\xi}{\sqrt{1-\xi^2}}e^{-\xi\omega_n t}\sin\omega_d t + e^{-\xi\omega_n t}\cos\omega_d t \tag{8.13}$$

to the best knowledge of the author this is the first time these expressions are derived explicitly in the literature. In [3], it was shown that given some limiting simplifying assumption, namely, (A1) noise is a zero mean white gaussian random process, (A2) the signal is measured until it fades away, (A3) sampling is sufficient (Nyquist frequency is higher than signal frequency), (A4) damping in the signal tends to zero and (A5) low spectral interference, the CRLB of the parameters $[\alpha, \beta, \omega, \phi]$ of a signal described by

$$y(t) = \alpha e^{-\beta t}\sin(\omega t + \phi) \tag{8.14}$$

can be compactly expressed

$$CRLB = \sigma^2 \begin{bmatrix} 8\beta & \frac{8\beta^2}{\alpha} & 0 & 0 \\ \frac{8\beta^2}{\alpha} & \frac{16\beta^3}{\alpha^2} & 0 & 0 \\ 0 & 0 & \frac{16\beta^3}{\alpha^2} & -\frac{8\beta^2}{\alpha^2} \\ 0 & 0 & -\frac{8\beta^2}{\alpha^2} & \frac{8\beta}{\alpha^2} \end{bmatrix} \tag{8.15}$$

with

$$\alpha = \sqrt{y_o^2 + \left(\frac{\xi\omega_n y_o + \dot{y}_o}{\omega_d}\right)^2} \tag{8.16}$$

$$\beta = \xi\omega_n \tag{8.17}$$

$$\phi = \arctan\left(\frac{y_o\omega_d}{\xi\omega_n y_o + \dot{y}_o}\right) \tag{8.18}$$

the extent to which the exact expressions derived by the author and the limiting simplified expressions derived in [3] differ will be illustrated in the numerical section of the paper.

8.5 Fourier Domain Identification

In this section we derive Fourier domain expressions to estimate the frequency and damping of SDOF. The Fourier transform of the free vibration response of a single mode in Eq. 8.2 is given by

$$Y(\omega) = \mathcal{F}(y(t)) = \int_{-\infty}^{\infty} y(t)e^{-i\omega t}\,dt = a\frac{\omega_d}{s^2+\omega_d^2} + b\frac{s}{s^2+\omega_d^2} \tag{8.19}$$

where

$$a = \frac{\dot{y}_o + \xi\omega_n y_o}{\omega_d} \tag{8.20}$$

$$b = y_o \tag{8.21}$$

$$s = \xi\omega_n + i\omega \tag{8.22}$$

Then

$$Y(\omega) = \frac{a\omega_d + b(\xi\omega_n + i\omega)}{(\xi\omega_n + i\omega)^2 + \omega_d^2} \tag{8.23}$$

and

$$|Y(\omega)|^2 = \frac{(a\omega_d + b\xi\omega_n)^2 + (b\omega)^2}{(\omega_n^2 - \omega^2)^2 + (2\xi\omega\omega_n)^2} \tag{8.24}$$

To determine abscissa (which we denote ω_*) corresponding to the maximum value of Eq. 8.24 we solve the following equation

$$\frac{\partial |Y(\omega)|^2}{\partial \omega} = 0 \tag{8.25}$$

which after some algebra results in the real and positive solution to the following fourth order polynomial equation:

$$-2b^2\omega^4 - 4\beta\omega^2 + (2b^2\omega_n^4 - \beta\alpha) = 0 \tag{8.26}$$

where

$$\beta = (a\omega_d + b\xi\omega_n)^2 \tag{8.27}$$

$$\alpha = 8(\xi\omega_n)^2 - 4\omega_n^2 \tag{8.28}$$

For the special (and very useful) case of $y(0) = 0$ and $\dot{y}(0) = v_o$ the solution of Eq. (8.26) is given by

$$\omega_* = \omega_n\sqrt{1 - 2\xi^2} \tag{8.29}$$

By examining the ratio

$$r = \frac{|Y(\omega_*)|^2}{|Y(0)|^2} = \frac{1}{4\xi^2} \tag{8.30}$$

one can express the damping ratio ξ without prior knowledge of the undamped frequency as

$$\hat{\xi} = \frac{1}{2\sqrt{r}} \tag{8.31}$$

Substituting this result into Eq. 8.29 one can estimate the undamped frequency as

$$\hat{\omega}_n = \frac{\omega_*}{\sqrt{1 - \hat{\xi}^2}} = \frac{2\omega_*}{\sqrt{\frac{4r-1}{r}}} \tag{8.32}$$

Note that are independent of the magnitude of the initial velocity.

8.6 Numerical Illustration

In this section we proceed to illustrate the theoretically exact CRLB for frequency and damping and compare it with results from stochastic simulations on a SDoF subject to a unit initial velocity given noise contaminated measurements. The noise is Gaussian with zero mean and standard deviation $\sigma = 0.01$. Figure 8.1 illustrates the CRLB induced coefficient of variation versus the coefficient of variation obtained from implementation of Eqs. 8.31 and 8.32 in the context of stochastic simulations. Every point in the figure represents 1,000 stochastic simulations (this number was found to be satisfactory and convergent in the mean and variance). The time step used for the simulations was $\Delta t = 0.01$ s and the total simulation time $t = 10$ s.

Figure 8.2 illustrates, for a range of natural frequencies and damping ratios, the ratio between the CRLB induced coefficient of variation for natural frequency and damping ratio. As can be seen it is clearly verified that the damping ratios are significantly more difficult to estimate than natural frequencies for all values examined. Also it is interesting to note that the ratio is practically independent of frequency.

To conclude, consider Fig. 8.3 which illustrates the difference between the approximation provided in [3] and the exact results derived in this paper for the natural circular frequency. As can be seen the difference is not negligible.

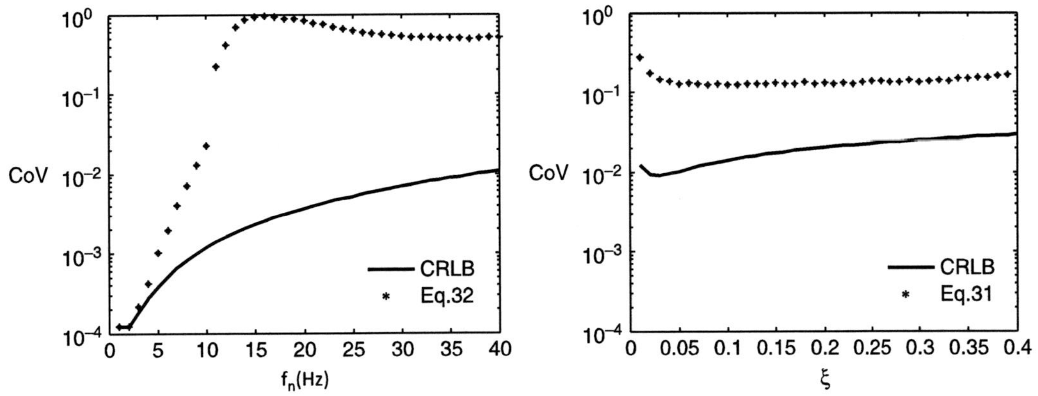

Fig. 8.1 Comparison of CRLB induced coefficient of variation versus coefficient of variation obtained from implementing frequency domain estimation of damping and undamped frequency. The figure on the *left* corresponds to the marginal at a damping ratio of 0.01, the figure on the *right* corresponds to the marginal at a natural frequency of 1 Hz

Fig. 8.2 Ratio of CRLB induced coefficient of variation as a function of fundamental frequency and damping ratio

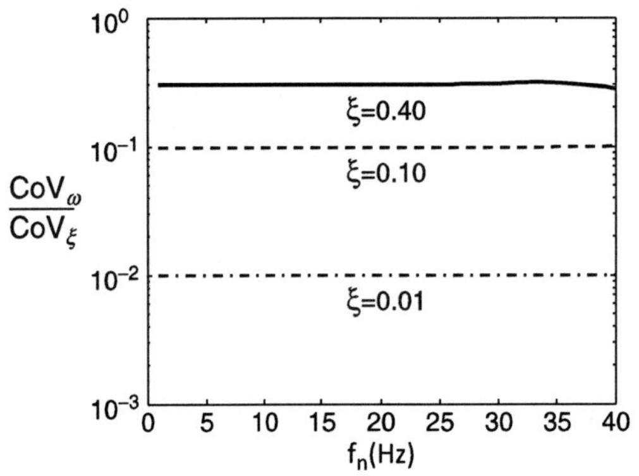

Fig. 8.3 Comparison of exact and approximate CRLB for the natural circular frequency. Results are shown for a damping ratio of 0.01

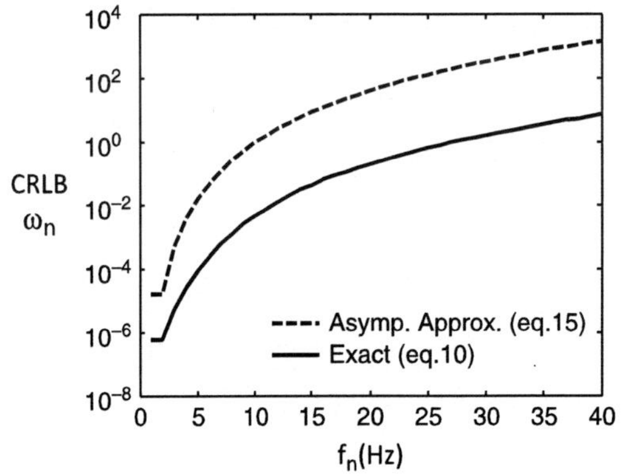

8.7 Conclusions

The paper derives exact expressions for the Cramer-Rao lower bound for natural frequency, damping ratio, and initial conditions of a single degree of freedom system subject to initial conditions. To the best knowledge of the author, this is the first time these results are explicitly presented in the literature. It was shown that, similarly to the case of random vibrations [1], noise contaminated signals contain significantly less information about damping in comparison to the information contained about natural frequency. This was quantified by comparing the respective coefficients of variation induced by the CRLB.

References

1. Gersch W (1974) On the achievable accuracy of structural system parameter estimates. J Sound Vib 34(1):63–79
2. Kay SM (1993) Fundamentals of statistical signal processing: estimation theory. Prentice Hall, Upper Saddle River
3. Wigren T, Nehorai A (1991) Asymptotic Cramer-Rao bounds for estimation of the parameters of damped sine waves in noise. IEEE Trans Signal Process 39(4):1017–1020

Chapter 9
Using P-Box and PiFE to Express Uncertainty in Model Updating

Ramin Madarshahian, Juan M. Caicedo, and Boris A. Zárate

Abstract This paper proposes the use of probability bounds with the Pseudo-inverse Finite Element (PiFE) method for structural model updating. The technique estimates the probability bound of structural parameters based on dynamic or static features such as modal parameters or static displacements. Two methods are explored for the calculation of the probability bounds: (i) Naïve method and (ii) all possible combinations. The capabilities of the technique are explored using a two degree of freedom structural system where the stiffness is considered uncertain. Results indicate that both the Naïve and all possible combination techniques are applicable with PiFE and produce bounds that include the cumulative distribution function of the structural parameters. The probability bounds found with the all possible combinations method was narrower for this particular example.

Keywords Uncertainty • Model updating • Interval method • P-box • PiFE

9.1 Introduction

Mathematical models of structural systems such as finite elements are frequency used to predict the structure's behavior. Updating of these models based on experimental data is important when the structural system being modeled is already existing and some of the characteristics of the system are unknown [1–3]. Dynamic tests are an attractive alternative as a source of data for model updating because these tests can be performed without the need to close the structure [4,5]. The data obtained from these tests provides modal information of a structure [6–8] that can be then used to improve the characteristics of a numerical model. There are different model updating approaches available on the literature depending on the type of data available and the type of structural model [9–13]. Some of these techniques recognize that the need to include uncertainty on their approach by using probability methods [14, 15]. For example, Beck and Katafygiotis [16] and Katafygiotis and Beck [17] proposed a Bayesian framework for model updating that has been used and extended by many other researchers. These methods seek to describe a particular structural parameter (or model) in terms of a probability distribution [18]. However, one can argue that even defining the type of probability density function is difficult in many cases. Many applications assume a distribution because that is the best that can be done based on the information available. An alternative to express uncertainty is to define boundaries (or interval) of input data regardless of knowing the type of distribution [19, 20]. This open the possibility for experienced analyst to consider an interval that expresses uncertainty without the need to define a distribution. For example, when measuring the length of an object with a meter stick marked off in millimeters, a person can define an interval between 9.99 and 10.1 cm even if that person does not have enough information or experience to make a judgment

R. Madarshahian (✉)
Doctoral Candidate, Department of Civil and Environmental Engineering, University of South Carolina, 300 Main, Columbia, SC 29208, USA
e-mail: mdrshhn@email.sc.edu

J.M. Caicedo
Associate Professor, Department of Civil and Environmental Engineering, University of South Carolina, 300 Main, Columbia, SC 29208, USA

B.A. Zárate
Research Scientist, Mistras Group, 195 Clarksville Road, Princeton Junction, NJ 08550, USA

T. Simmermacher et al. (eds.), *Topics in Model Validation and Uncertainty Quantification, Volume 5*, Conference Proceedings of the Society for Experimental Mechanics Series 41,
DOI 10.1007/978-1-4614-6564-5_9,

about the probability distribution of the numbers in this interval. All in all considering a variable as an interval is good alternative, when we have lack of information about the data distribution of particular parameter.

This paper presents the expansion of the Pseudo-inverse Finite Element (PiFE) to identify stiffness characteristics using intervals. The method does not assume any distribution for the structural parameters nor the features characterizing the structure (i.e. natural frequencies and mode shapes). Rather an interval for them is assumed. The method generates a probability box (P-box) bounding the cumulative distribution function of the structural parameters [21]. The methodology is tested using a 2 DOF numerical model. A probability density function is defined for the stiffness of the structure to generate natural frequencies and mode shapes used as input data for the proposed technique. The distribution of the parameters is considered unknown in the updating process and it is only used for comparison with the resulting P-box.

9.2 Pseudo-inverse Finite Element (PiFE)

PiFE is an inverse method to identify parameters of structural systems (e.g. mass density, moment of inertia, etc.) based on static or modal data. The method is called PiFE because it is applied in a similar way to the finite element method. The following paragraphs describe the general idea of the method and the derivation for a 2 DOF system using a set of identified natural frequencies and mode shapes. Modal data can be found using one of the many available modal identification techniques such as the Stochastic Subspace Identification [22] or the Eigensystem Realization Algorithm [6, 23, 24]. More information about how to expand PiFE to complex structures can be found in [25–27]. The eigenvalue problem for a system with no damping is [28].

$$\left(K - M\omega_i^2\right)\phi_i = 0 \tag{9.1}$$

where K, and M are stiffness and mass matrix respectively, and ω_i, and ϕ_i are i-th natural frequency and mode shape of the structure respectively found experimentally. Equation (9.1) can be rewritten such that all unknown parameters are located in new defined vector Θ and then other known parameters are gathered in a matrix Δ_i, called identification matrix.

$$\Gamma_i = \Delta_i\Theta \tag{9.2}$$

where the matrix Γ_i is a vector with the same numbers of rows as degrees of freedom (DOF) calculated as $\Gamma_i = M\omega_i\phi_i$. This formulation is attractive because the identification matrix Δ_i in Eq. (9.2) because can be written for each element of the structure, rotated from local to global coordinates, assembled to a global identification matrix for the whole structure in a similar fashion as the application of the finite element method. Furthermore, it is possible to extend Eq. (9.2) to contain all n available modes obtained experimentally:

$$\Gamma = \Delta\Theta \tag{9.3}$$

where $\Gamma = [\Gamma_1^T, \ldots, \Gamma_n^T]^T$, $\Delta = [\Delta_1^T, \ldots, \Delta_n^T]$, and the superscript "$T$" represents transpose. The n available modes do not need to be in order. For example, if PiFE was applied to a 4 DOF structure, it is possible to apply the technique with only the second and fourth modes of a particular structure (i.e. there is no requirement to experimentally identify the first and third mode). The vector Θ can be obtained after Γ is multiplied by pseudo-inverse of identification matrix Δ.

$$\Theta = \Delta^{\dagger}\Gamma \tag{9.4}$$

where the superscript $\dagger$ denotes pseudo-inverse.

9.3 Interval Arithmetic and Associated Challenges

PiFE has been previously applied in a deterministic fashion where a set of identified natural frequencies and mode shapes are used to estimate a set of structural parameters. However, performing a deterministic identification does not provide any information about the related uncertainty on the identified parameters. One way common method to express uncertainty is to probability distributions. Another possibility is to express the structural parameters by an interval which tells us about its bounds. For instance $\overleftrightarrow{X}$ is a quantity between its lower bound $\overleftarrow{x}$ and upper bound $\overrightarrow{x}$ which are real numbers and can be represented by $\overleftrightarrow{x}$. In a similar fashion that we deal with numbers we can define operators for intervals. For instance $\overleftrightarrow{x} \times \overleftrightarrow{y}$ can be obtained by considering the maximum and minimum of $\overleftarrow{x} \times \overrightarrow{y}$, $\overleftarrow{x} \times \overrightarrow{y}$, $\overrightarrow{x} \times \overleftarrow{y}$ and $\overrightarrow{x} \times \overrightarrow{y}$. Also for more complicated operators like inverse of a matrix which its arrays are interval several algorithm are proposed [29–31].

Ignoring the dependency of variables of a particular function could result in overestimating the interval of the resulting function. For example, consider the variables $\overleftrightarrow{x} = [1,5]$ and $\overleftrightarrow{y} = [1,5]$ and the function $f(\overleftrightarrow{x}, \overleftrightarrow{y}) = \overleftrightarrow{x} - \overleftrightarrow{y}$. If x and y are independent and $f(\overleftrightarrow{x}, \overleftrightarrow{y}) = \overleftrightarrow{x} - \overleftrightarrow{y} = [-4,4]$. However, if they are completely correlated such that $\overleftrightarrow{x} = \overleftrightarrow{y}$, the resulting interval should be $f(\overleftrightarrow{x}, \overleftrightarrow{y}) = f(\overleftrightarrow{x}, \overleftrightarrow{x}) = \overleftrightarrow{x} - \overleftrightarrow{x} = [0,0]$. Although the resulting interval considering the two variables independent still bounds the values of the correlated variables the interval could be so wide that it is not useful anymore. These interval obtained by ignoring the dependency between the variables is called the *naïve* solution. One approach to improve the results is to simplify function before calculating the interval. Following the example described above this corresponds to simplifying the function $f(\overleftrightarrow{x}, \overleftrightarrow{x}) = 0$. An alternative to the naïve solution is to consider all possible combinations of our interval bounds in our solution using real numbers arithmetic instead of interval arithmetic. It should be noted there are associated challenges with this technique. For example, intervals in dominators should not change sign. In addition, this is only applicable to monotonic functions. The application of intervals presented here all intervals are selected such that they do not change sign, i.e. there is no zero in them, and all relationships between variables are constructed using four common arithmetic operators and are monotonic. PiFE provide us suitable framework to apply interval arithmetic.

9.4 Validation Example

The 2 DOF structure shown in Fig. 9.1 is used to explore the capabilities of the proposed technique. The stiffness of the two floors are considered the unknown parameters to be estimated using PiFE. Modal parameters (natural frequencies and mode shapes) are required to implement PiFE. In an experimental setting these parameters are obtained by applying modal identification techniques to acceleration records. The identification of the parameters to a series of acceleration records would result in samples for the natural frequencies and mode shapes. In this particular example the samples are obtained by assuming a probability distribution function (PDF) for the k_1 and k_2, sample for values of k_1 and k_2 and solving the eigenvalue problem. The validation of the technique is performed by comparing the P-box from PiFE with the initial probability distribution.

9.4.1 Modal Parameter Intervals

One could make an assumption for the probability of the bounds to implement the proposed technique. However, we decided not to make this assumption and sample an assumed PDF for the spring constants to obtain samples for the bounds. The PDF of the stiffness of the springs is considered as a normal distribution based on reported variability of materials due to process of fabrication [32]. The parameters of the PDF are $\mu_{k_1} = 3\frac{N}{m}$, $\mu_{k_2} = 2\frac{N}{m}$ and standard of deviation of 1% of mean value as shown in Fig. 9.2.

The samples for the bounds are calculated using a two step approach. First, 100 samples of the stiffness constants are considered to calculate 100 set of modal parameters and estimate one set of intervals. The modal parameters of the structure might not follow a normal distribution because the function that relates the spring constants to modal data is not linear (i.e. eigenvalue problem). However, assuming a unimodal distribution (i.e. a distribution with only one peak) is reasonable.

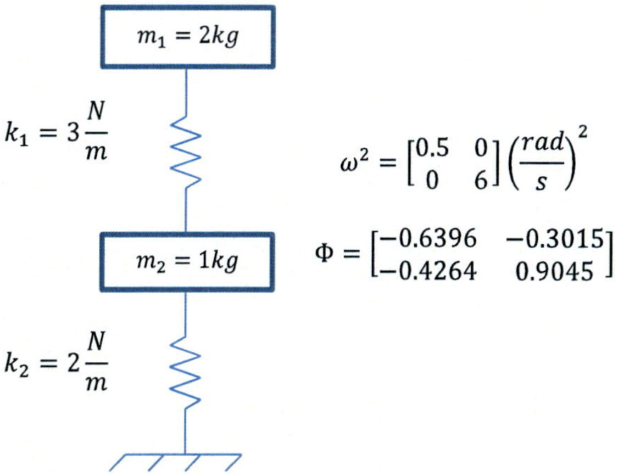

Fig. 9.1 Two DOF model and its modal information

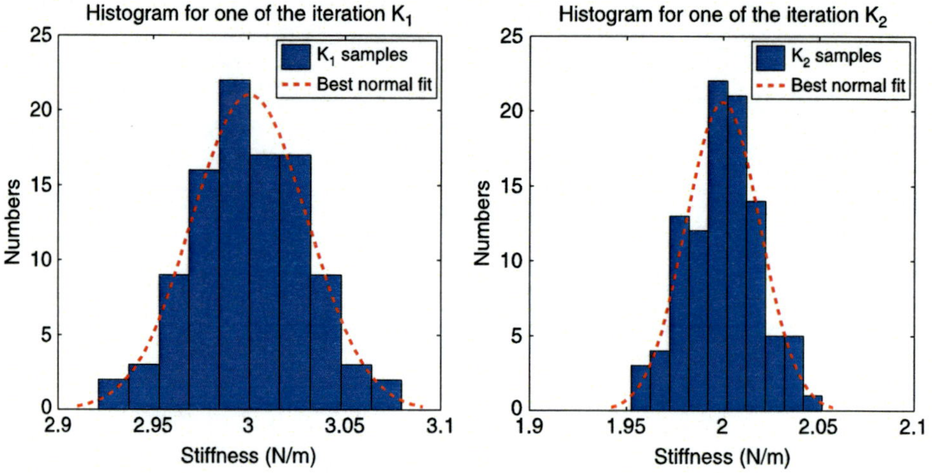

Fig. 9.2 Histogram of normal distributed samples for stiffness with standard of deviation of 0.01 of mean

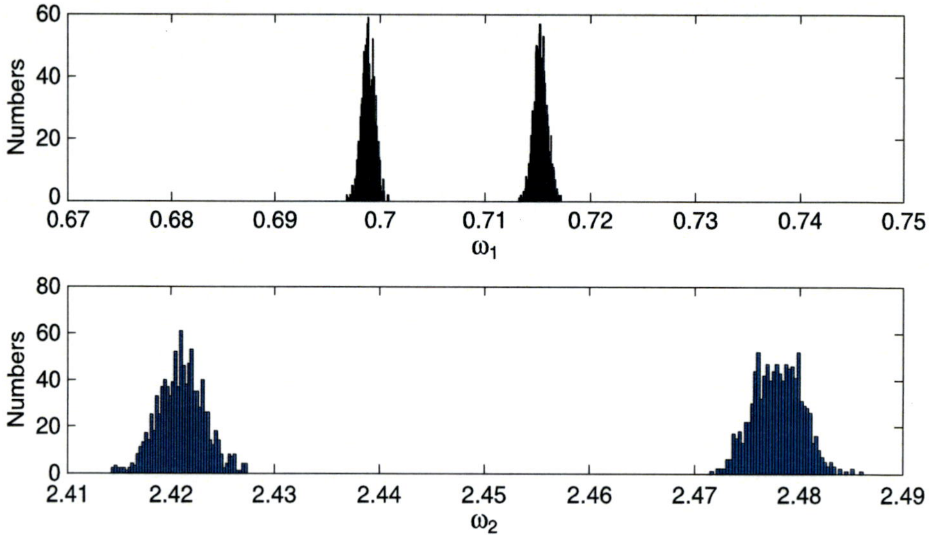

Fig. 9.3 Histogram of intervals for natural frequencies

Therefore, 90 % of the probable values will be located between -3σ to $+3\sigma$ from the mean (where σ is the standard deviation) based on the Vysochanskiï–Petunin inequality [33]. The process is repeated for a total of 1,000 set of intervals.

Figures 9.3, 9.4 and 9.5 show the histograms of the intervals for the obtained natural frequencies and modal coordinates for each mass normalized mode. The X-axis of each figure represents the same range of values for easy comparison. The obtained intervals are different for each natural frequency. Furthermore, they are different for different modal coordinates (even for the same mode). One possible explanation for this behavior is the nonlinear relationship between the modal parameters, stiffness and mass. The data described on these three figures is used as the input parameters for PiFE as discussed in the following section. Similar plots could be found experimentally by analyzing different acceleration records and performing modal analysis.

9.4.2 Application of PiFE and Naïve Method

Solving Eq. (9.4) analytically for the 2 DOF model results in

$$\begin{cases} k_1 = C_{11} \times m_1 + C_{12} \times m_2 \\ k_2 = C_{21} \times m_1 + C_{22} \times m_2 \end{cases} \tag{9.5}$$

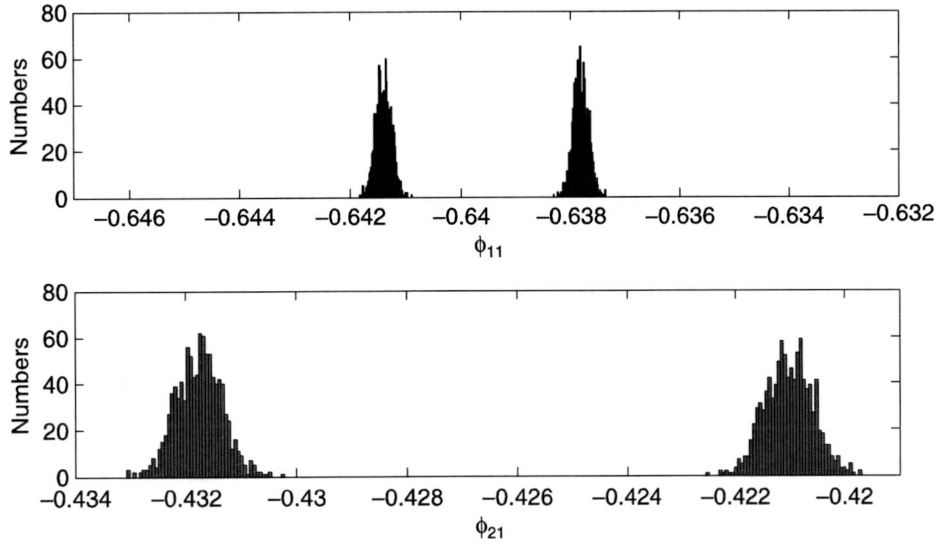

Fig. 9.4 Histogram of intervals for modal coordinates of the first mode

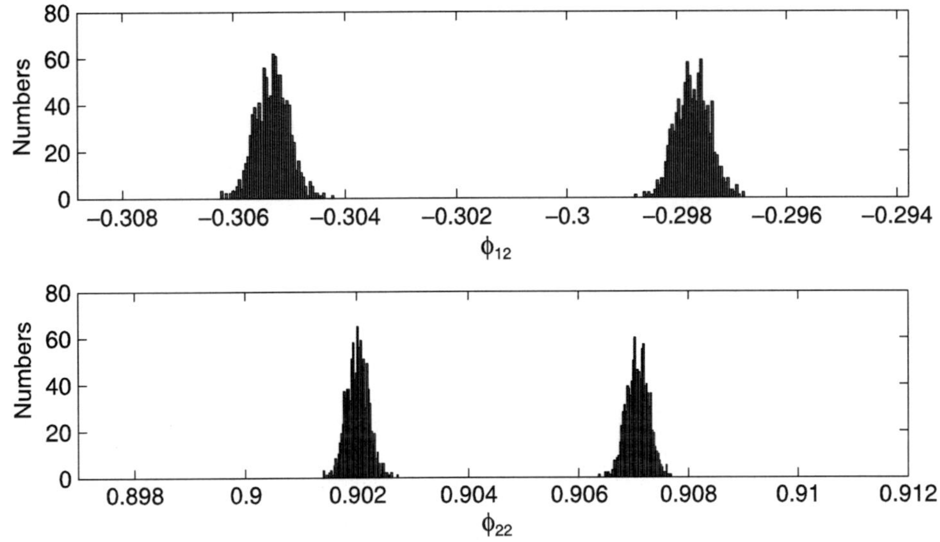

Fig. 9.5 Histogram of intervals for modal coordinates of the second mode

where C_{11}, C_{12}, C_{21}, and C_{22} are functions of the modal coordinates and natural frequencies. These functions should be analytically simplified to apply the naïve method resulting in

$$C_{11} = \frac{\left(\phi_{11}\omega_1^2\left(\phi_{11}-\phi_{21}\right)+\phi_{12}\omega_2^2\left(\phi_{12}-\phi_{22}\right)\right)\left(\phi_{21}^2+\phi_{22}^2\right)}{\phi_{11}^2\left(\phi_{21}^2+2\phi_{22}^2\right)-2\phi_{12}\phi_{22}^3+\phi_{21}^4+\phi_{22}^4-\phi_{11}\left(2\phi_{21}^3+4\phi_{21}\phi_{22}^2\right)+\phi_{12}^2\phi_{22}^2+2\phi_{21}^2\phi_{22}^2} \tag{9.6a}$$

$$C_{12} = \frac{-\phi_{21}\phi_{22}\left(\omega_1^2-\omega_2^2\right)\left(\phi_{11}\phi_{22}+\phi_{12}\phi_{21}-2\phi_{21}\phi_{22}\right)}{\phi_{11}^2\phi_{21}^2+2\phi_{11}^2\phi_{22}^2-2\phi_{11}\phi_{21}^3-4\phi_{11}\phi_{21}\phi_{22}^2+\phi_{12}^2\phi_{22}^2-2\phi_{12}\phi_{22}^3+\phi_{21}^4+2\phi_{21}^2\phi_{22}^2+\phi_{22}^4} \tag{9.6b}$$

$$C_{21} = \frac{\left(\phi_{11}\omega_1^2\left(\phi_{11}-\phi_{21}\right)+\phi_{12}\omega_2^2\left(\phi_{12}-\phi_{22}\right)\right)\left(-\phi_{21}^2+\phi_{11}\phi_{21}-\phi_{22}^2+\phi_{12}\phi_{22}\right)}{\phi_{11}^2\left(\phi_{21}^2+2\phi_{22}^2\right)-2\phi_{12}\phi_{22}^3+\phi_{21}^4+\phi_{22}^4-\phi_{11}\left(2\phi_{21}^3+4\phi_{21}\phi_{22}^2\right)+\phi_{12}^2\phi_{22}^2+2\phi_{21}^2\phi_{22}^2} \tag{9.6c}$$

$$\begin{aligned} C_{22} = {} & \frac{\omega_1^2\phi_{21}^2+2\omega_2^2\phi_{22}^2}{\phi_{21}^2+2\phi_{22}^2} \\ & - \frac{\phi_{21}\phi_{22}\left(\phi_{12}\phi_{22}\right)\left(\omega_1^2\omega_2^2\right)\left(-\phi_{21}^3\phi_{11}\phi_{21}^2\phi_{21}\phi_{22}^2\phi_{12}\phi_{21}\phi_{22}\phi_{11}\phi_{22}^2\right)}{\left(\phi_{21}^2\phi_{22}^2\right)\left(\phi_{11}^2\left(\phi_{21}^2+2\phi_{22}^2\right)-2\phi_{12}\phi_{22}^3+\phi_{21}^4+\phi_{22}^4-\phi_{11}\left(2\phi_{21}^3+4\phi_{21}\phi_{22}^2\right)+\phi_{12}^2\phi_{22}^2+2\phi_{21}^2\phi_{22}^2\right)}. \end{aligned} \tag{9.6d}$$

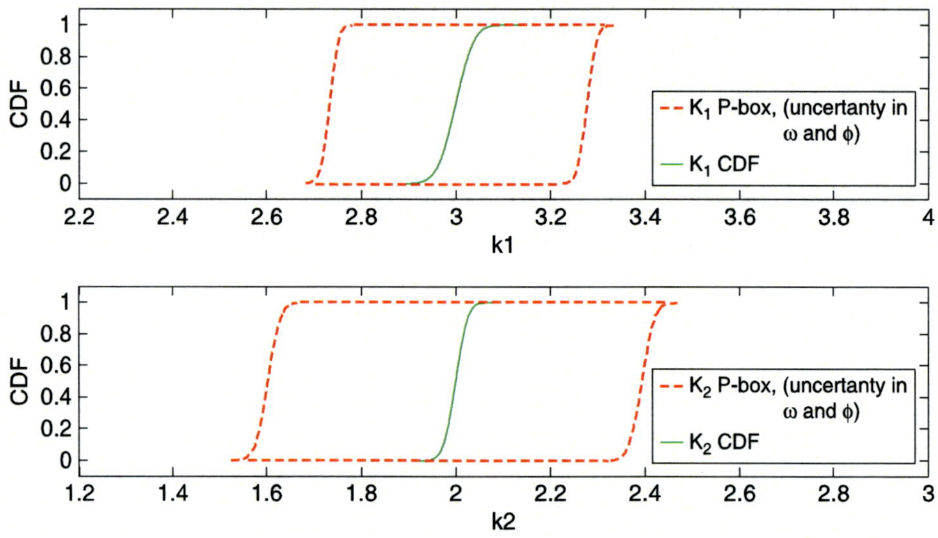

Fig. 9.6 P-box for spring constants using the naïve method

As it is clear from Eqs. (9.5) and (9.6a) even for a 2 DOF system the relationship between modal data and spring constants is not simple. A P-box is calculated using Eq. (9.5) with the modal data presented in Figs. 9.3, 9.4 and 9.5. Each upper and lower bound for the modal coordinates will result in a distribution for the upper and lower bounds of the spring constants. Figure 9.6 shows the resulting P-box where the dashed lines represent the lower and upper bounds of probabilities for the spring constants. The solid line indicates the original CDF. As expected, this CDF is bounded by the P-box. Notice that there is no need to assume a distribution for the parameters being identified in the pcess of estimating the P-box. Furthermore, the P-box does not indicate a specific probability distribution but rather an area that contains the CDF.

9.4.3 Application of PiFE and All Possible Combination Method

This method explores all the possible combinations of the bounds of the input parameters (i.e. modal parameters) to determine the bounds of the unknowns (i.e. spring constants). For example, in the case of the 2 DOF problem there are two natural frequencies and four modal coordinates. Therefore, there are six lower bounds and six upper bounds for a total of $n = \sum_{i=0}^{6} \binom{6}{i} = \sum_{i=0}^{6} \frac{6!}{(6-i)!i!} = 64$ possible combinations for each sample set. Each of these n combinations are introduced to Eq. (9.5) and using algebraic arithmetic (not interval arithmetic) n values of the parameters are obtained. The maximum and minimum values are selected as upper bound and lower bound of the unknown parameters. This process is repeated for each set of intervals considered. This method is only applicable to monotonic functions within the interval because non-monotonic functions could result in a higher or lower value with a value of the parameter inside the bounds.

Figure 9.7 shows the resulting P-box (dashed line) and the original CDF (solid line) for each spring constant. The P-box contains the original CDF indicating that all possible combinations method was effective in estimating the bounds of the CDF.

9.5 Conclusions

This paper presents the implementation of two interval methods to the Pseudo-inverse Finite Element method (PiFE). The proposed technique estimates the bounds of the cumulative distribution function of the unknown structural parameters without the need of assuming any distribution. The method provides a region that contains the CDF of the structural parameters. The Naïve method uses the interval arithmetic to find the unknown parameters while all possible combinations applies algebraic arithmetic. The methods were tested with a 2 DOF structure where the stiffness was considered uncertain. Figures 9.6 and 9.7 shows the results for each method. As illustrated, the techniques estimated a probability box (P-box) that

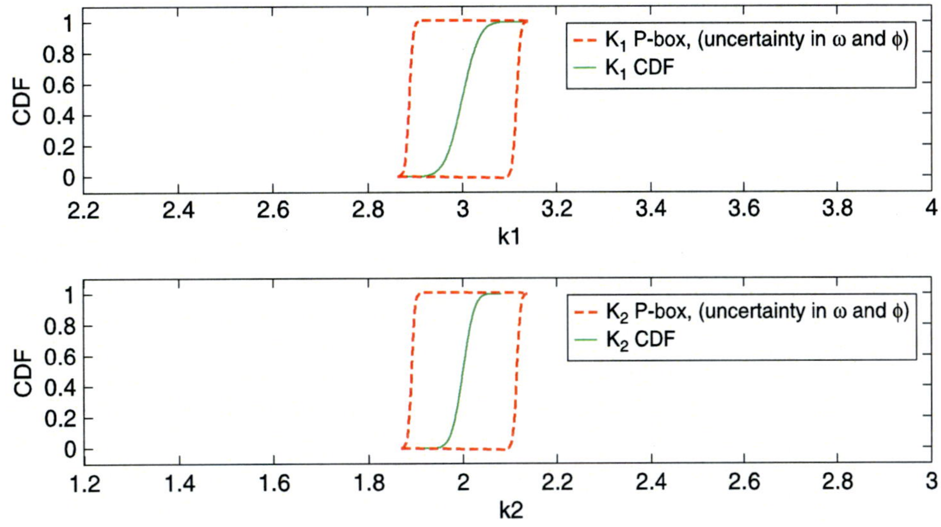

Fig. 9.7 P-box for spring constants using all possible combinations

contained the underlain CDF of these parameters. The all possible combinations method provided a narrower P-box than the Naïve approach. A possible explanation for this is that the Naïve approach does not consider the dependency between values.

Acknowledgements This material is based upon work supported by the National Science Foundation under Grant No.CMMI-0846258.

References

1. Ewins DJ (2007) Modal analysis and modal testing. In: Crocker MJ (ed) Handbook of noise and vibration control. John Wiley & son, Hoboken, NJ, USA, pp. 565–574, doi:10.1002/9780470209707.ch47
2. Sirca GF Jr, Adeli H (2012) System identification in structural engineering. Scientia Iranica. 19(6), pp. 1355–1364, http://www.sciencedirect.com/science/article/pii/S1026309812001940
3. Franco G, Betti R, Lu H (2004) Identification of structural systems using an evolutionary strategy. J Eng Mech 130(10):1125–1139
4. Caicedo J (2011) Practical guidelines for the Natural Excitation Technique (NExT) and the Eigensystem Realization Algorithm (ERA) for modal identification using ambient vibration. Exp Tech 35(4):52–58
5. James GH et al (1992) Modal testing using natural excitation. In: 10th international modal analysis conference, San Diego
6. Caicedo JM, Dyke SJ, Johnson EA (2003) Natural excitation technique and eigensystem realization algorithm for phase I of the IASC-ASCE benchmark problem: simulated data. J Eng Mech 130(1):49–60
7. Alvin KF et al (2003) Structural system identification: from reality to models. Comput Struct 81(12):1149–1176
8. Yuen K-V (2012) Updating large models for mechanical systems using incomplete modal measurement. Mech Syst Signal Process 28(0):297–308
9. Kenigsbuch R, Halevi Y (1998) Model updating in structural dynamics: a generalised reference basis approach. Mech Syst Signal Process 12(1):75–90
10. Jaishi B, Ren W-X (2007) Finite element model updating based on eigenvalue and strain energy residuals using multiobjective optimisation technique. Mech Syst Signal Process 21(5):2295–2317
11. Arora V, Singh SP, Kundra TK (2010) Further experience with model updating incorporating damping matrices. Mech Syst Signal Process 24(5):1383–1390
12. Lin RM, Lim MK, Ong JH (1993) Improving finite element models in the higher frequency range using modified frequency response function sensitivity method. Finite Elem Anal Des 15(2):157–175
13. Moaveni B et al (2012) Finite element model updating for assessment of progressive damage in a three-story infilled RC frame. J Struct Eng 442
14. Nasrellah HA, Manohar CS (2011) Finite element method based Monte Carlo filters for structural system identification. Probab Eng Mech 26(2):294–307
15. Moaveni B, Conte JP, Hemez FM (2009) Uncertainty and sensitivity analysis of damage identification results obtained using finite element model updating. Comput Aided Civ Infrastruct Eng 24(5):320–334
16. Beck JL, Katafygiotis LS (1998) Updating models and their uncertainties. I: Bayesian statistical framework. J Eng Mech 124(4):455–461
17. Katafygiotis LS, Beck JL (1998) Updating models and their uncertainties. II: model identifiability. J Eng Mech 124(4):463–467
18. Zhang H (2012) Interval importance sampling method for finite element-based structural reliability assessment under parameter uncertainties. Struct Saf 38(0):1–10
19. Konijn HS (1987) Distribution-free and other prediction intervals. Am Stat 41(1):11–15
20. Sellke T (1996) Generalized gauss-chebyshev inequalities for unimodal distributions. Metrika 43(1):107–121
21. Zhang H, Mullen RL, Muhanna RL (2010) Interval Monte Carlo methods for structural reliability. Struct Saf 32(3):183–190

22. Zárate BA, Caicedo JM (2008) Finite element model updating: multiple alternatives. Eng Struct 30(12):3724–3730
23. Li P, Hu SLJ, Li HJ (2011) Noise issues of modal identification using eigensystem realization algorithm. Procedia Eng 14(0):1681–1689
24. Siringoringo DM, Fujino Y (2008), System identification of suspension bridge from ambient vibration response. Eng Struct 30(2):462–477
25. Madarshahian R et al (2012) Direct inverse finite element model updating. In: 2012 joint conference of the engineering mechanics institute and the 11th ASCE joint specialty conference on probabilistic mechanics and structural reliability, Notre Dame
26. Caicedo JM (2003) Structural health monitoring of flexible civil structures. In: Civil engineering. Washington University in Saint Louis, Saint Louis, p 160
27. Caicedo JM, Dyke SJ (2002) Determination of member stiffnesses for structural health monitoring. In: 3rd world conference in structural control, Como
28. Chopra AK (2012) Dynamics of structures: theory and applications to earthquake engineering, 4th edn. Prentice Hall, Upper Saddle River, p 944. xxxiii
29. Chen S-H, Yang X-W (2000) Interval finite element method for beam structures. Finite Elem Anal Des 34(1):75–88
30. Muhanna R, Zhang H, Mullen R (2007) Interval finite elements as a basis for generalized models of uncertainty in engineering mechanics. Reliab Comput 13(2):173–194
31. Zhang H, Muhanna RL (2009) Interval approach for nondeterministic linear static finite element method in continuum mechanics problems. Int J Reliab Saf 3(1):201–217
32. Hess PE et al (2002) Uncertainties in material and geometric strength and load variables. Nav Eng J 114(2):139–166
33. Pukelsheim F (1994) The three sigma rule. Am Stat 48(2):88–91

Chapter 10
Robust Model Calibration with Load Uncertainties

D. Pereiro, S. Cogan, E. Sadoulet-Reboul, and F. Martinez

Abstract The goal of this work is to propose a model calibration strategy for an industrial problem consisting in a MW class geared wind turbine power train subjected to uncertain loads. Lack of knowledge is commonplace in this kind of engineering system and a realistic model calibration cannot be performed without taking into account this type of uncertainty. The question at stake in this study is how to perform a robust predictive model of a dynamic system given that the excitations are poorly known. The uncertainty in the latter will be represented with an info-gap model. The tradeoff between fidelity to data and robustness to uncertainty is then investigated in order to maximize the robustness of the prediction error at a given horizon of uncertainty. This methodology is illustrated on a simple academic model and on a more complex engineering system representing a wind turbine geared power train.

Keywords Robust calibration • Model updating • Load uncertainty • Model fidelity • Transient analysis • Wind turbine

10.1 Introduction

Validating structural dynamic numerical models is a common engineering task. Confidence in simulation results is critical for product development and risk management, and the preferred framework to quantify this confidence is model verification and validation (V&V) [1]. The confrontation between numerical simulations and experimental observations often indicates that the model is unsatisfactory and four different paradigms have been developed over the years to improve fidelity to data based on model calibration strategies, namely:

1. Reference basis methods
   A class of nonparametric identification methods that solve a constrained optimization problem to obtain perturbations of the stiffness and mass matrices for the nominal finite element model based on a set of measured eigensolutions. Fidelity to data is optimized [2–4].
2. Local deterministic methods
   A class of parametric identification methods that solve a nonlinear constrained optimization problem to obtain deterministic estimates of local stiffness and mass properties based on a set of measured eigensolutions, frequency response functions, or time domain responses. Fidelity to data is optimized over the feasible design space [5–7].
3. Local stochastic methods
   A class of parametric identification methods that solve a nonlinear constrained optimization problem to obtain statistical estimates of local stiffness and mass properties based on a set of measured eigensolutions, frequency response functions, or time domain responses. Here again, fidelity to data is optimized over the feasible design space [8, 9].

D. Pereiro (✉) • F. Martinez
IK4-Ikerlan, Pº J.M. de Arizmendiarrieta 2, 20500 Arrasate-Mondragón, Guipuzkoa, Spain
e-mail: dpereiro@ikerlan.es

S. Cogan • E. Sadoulet-Reboul
Department of Applied Mechanics, University of Franche-Comté, 24 rue de l'Epitaphe, 25000 Besançon, France
e-mail: scott.cogan@univ-fcomte.fr; emeline.sadoulet-reboul@univ-fcomte.fr

T. Simmermacher et al. (eds.), *Topics in Model Validation and Uncertainty Quantification, Volume 5*,
Conference Proceedings of the Society for Experimental Mechanics Series 41,
DOI 10.1007/978-1-4614-6564-5_10,

4. Local robust methods
   A class of parametric identification methods that solve a nonlinear constrained optimization problem to obtain either deterministic or statistical estimates of local stiffness and mass properties based on a set of measured eigensolutions, frequency response functions, or time domain responses. A robust satisficing approach is adopted to find an adequate compromise between fidelity to data and robustness of to lack of knowledge in the system and its environment [10].

The first three paradigms have been studied extensively on both academic and industrial examples. As different as these approaches are, they share the same fundamental hypothesis, namely that the best solution to the calibration problem is the one that minimizes the defined test-analysis metric. While this assumption is commonplace, it fails to recognize that most real world problems are plagued by various sources of lack of knowledge due to poorly understood physics. The fourth paradigm has received relatively little attention by the structural dynamics community. It attempts to bring the impact of lack of knowledge into the picture by analyzing the three fundamental components inherent in model-based prediction, namely:

1. Fidelity to data
   Quantifies the test-analysis distances based on either deterministic or stochastic metrics.
2. Robustness to lack of knowledge
   Quantifies the impact of gaps in knowledge on the model prediction error.
3. Predictive looseness
   Quantifies the confidence that can be accorded to the calibrated model given the range of model responses which are consistent with the lack of knowledge.

The inherent antagonistic nature of these components leads to fundamental tradeoffs that forms the basis of the new model calibration paradigm.

The present work is based on the concepts presented in [10] and serves to illustrate the relevance of robust parameter calibration under load uncertainties. The focus in this paper will be on the tradeoff between fidelity to data and robustness and the strategy is applied first on a single degree of freedom (SDOF) damped spring mass system under uncertain excitation. Secondly, a simple but realistic wind turbine power train model is calibrated with respect to the measured transient response. Wind turbine power trains are a suitable example to illustrate both the problem faced and the approach. The problem because in the wind energy booming market the pressure on releasing newer, more powerful and more reliable machines is very high, and the design requirements of a 20 year and over 90 % availability [11] must be assessed. A wind turbine is also a type of machine which has to operate in a wide range of load conditions that are difficult, expensive or even impossible to measure in detail. This type of device is suitable to illustrate the approach because in the authors opinion the robust parameter calibration strategy presented here can be a step forward on the process of assessing models prediction credibility.

10.2 Robust Parameter Calibration

A numerical model can be denoted by:

$$y = \mathcal{M}(q) \tag{10.1}$$

The model $\mathcal{M}$ defines a relationship between y, the response feature of interest and q, model input parameters or decision variables. The impact of lack of knowledge on model responses will be studied using info-gap models of uncertainty [12]. The uncertain variable is denoted by u, and the info gap model of uncertainty is defined as $U(u_0;\alpha)$, hence:

$$y = \mathcal{M}(q;u),\ u \in U(u_0;\alpha) \tag{10.2}$$

where the parameter α is known as the horizon of uncertainty, while u_0 is the nominal value of the uncertain variable. The objective here is to study the robustness of the model fidelity to experimental data given uncertainty loads. The discrepancy between reference and simulation data can be assessed with a metric $D(q;u)$ defined as the norm of the difference between the test data y^{Test} with the results obtained from the model y:

$$D(q;u) = \left\| y^{Test} - y \right\|_2 \tag{10.3}$$

where y^{Test} ; y will be assumed here to be real vectors. Furthermore, let D_c denote the greatest level of discrepancy that can be tolerated:

$$D(q;u) \leq D_c \tag{10.4}$$

The robustness of the discrepancy with respect to uncertainty can now be formally defined as:

$$\hat{\alpha} = \underset{\alpha \geq 0}{Argmax} \underset{U(u;\alpha)}{max} \{D(q;u)/D(q;u) \leq D_c\} \tag{10.5}$$

The robustness function assesses the immunity of the discrepancy to uncertainty in u. Large $\hat{\alpha}$ means that discrepancy is relatively insensitive to variations in the uncertain quantities u, while small $\hat{\alpha}$ means that small variations in the uncertain quantities lead to large discrepancy.

Robust parameter calibration searches for a model design q which maximizes the robustness of the discrepancy for a given horizon of uncertainty:

$$\hat{\alpha}(D_c, q) = max\left(\alpha : \left\{\left(\underset{u \in U(\alpha)}{max} D(q;u)\right) \leq D_c\right\}\right) \tag{10.6}$$

According to Eq. 10.6 the robust calibrated parameter can be defined as:

$$\hat{q}(D_c) = \underset{q}{argmax}(\hat{\alpha}(D_c, q)) \tag{10.7}$$

10.3 Application

10.3.1 SDOF System

The robust parameter calibration strategy is first illustrated on a single degree of freedom dumped spring mass system, where the transient force applied is set to be uncertain.

System model The SDOF spring damper system under load can be described by the equation:

$$M \cdot \ddot{x}(t) + R \cdot \dot{x}(t) + K \cdot x(t) = F(t) \tag{10.8}$$

The system and the reference values used in the model are described in Fig. 10.1. The system is assumed to be at rest at $t = 0$.

Uncertainty model The load uncertainty is represented by an info-gap envelope bound model [12]. The amplitude of the force step is allowed to vary by a specified fractional error from the estimated load. The estimated shape is defined as a rectangular step function over a known time interval:

$$\tilde{F}(t) = \begin{cases} F_0 & t \in (0,T) \\ 0 & t \notin (0,T) \end{cases} ; T \geq 0; t > 0; F_0 > 0 \tag{10.9}$$

The info-gap model of uncertainty can be defined as:

$$U(\alpha) = \left\{F(t) : \left|\frac{F(t) - \tilde{F}(t)}{S(t)}\right| \leq \alpha;\right\}, \alpha \geq 0 \tag{10.10}$$

The time dependent function $S(t)$ normalizes the load:

$$S(t) = \begin{cases} F_0 & t \in (0,T) \\ 0.2 \cdot F_0 & t \notin (0,T) \end{cases} ; T \geq 0; t \geq 0 \tag{10.11}$$

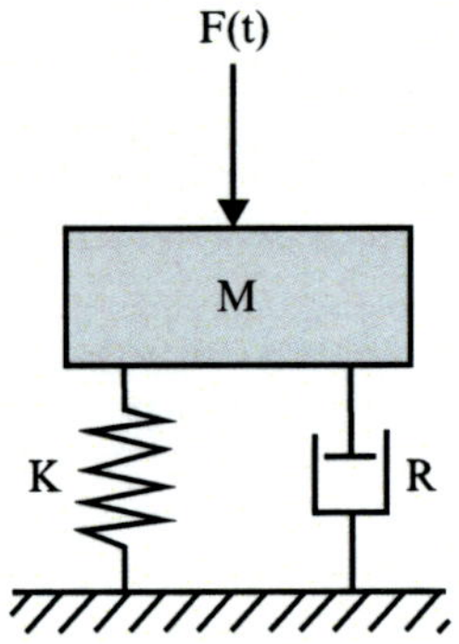

M [Kg]	R [Ns/m]	K [N/m]
1	0.5	10

Fig. 10.1 Single degree of freedom system

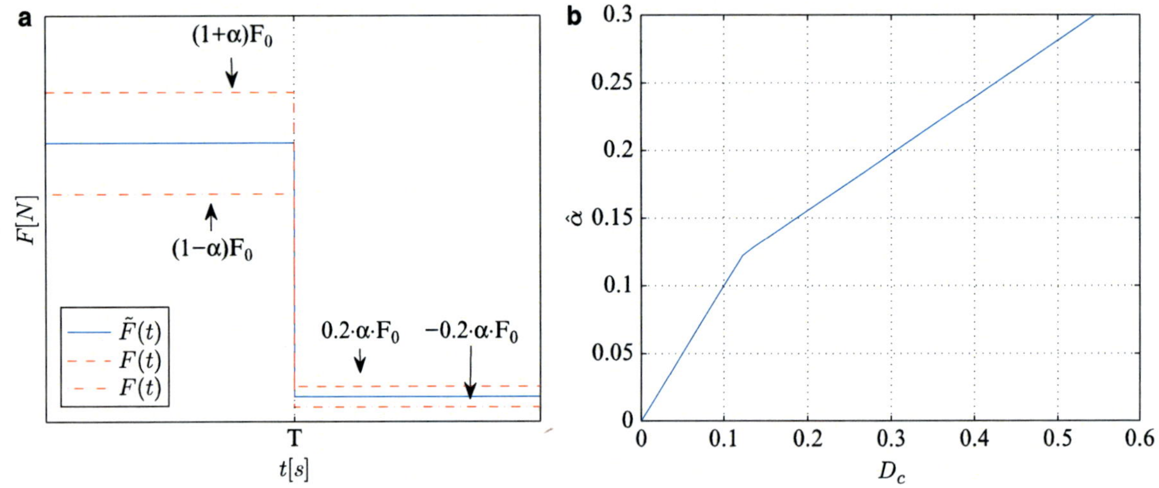

Fig. 10.2 **(a)** Envelope bound uncertainty model. **(b)** Robustness function

Regarding Eqs. 10.9–10.11, $F(t)$ can be expressed as:

$$F(t) = \begin{cases} (1-\alpha)F_0 \leq F(t) \leq (1+\alpha)F_0 & t \in (0,T) \\ -0.2\alpha F_0 \leq F(t) \leq 0.2\alpha F_0 & t \notin (0,T) \end{cases} ; T \geq 0; t \geq 0 \tag{10.12}$$

The uncertainty model is graphically represented in Fig. 10.2a. The value of F_0 is set to 4 N.

Performance requirement The performance requirement is defined as the fractional error of the system maximum response amplitude at an uncertain load $F(t)$ which is not greater than a specified critical discrepancy D_c:

$$|\rho(t)| = \left| \frac{\max_t (x_F(t)) - \max_t (x_{\tilde{F}}(t))}{\max_t (x_{\tilde{F}}(t))} \right| \leq D_c \tag{10.13}$$

Robustness function The robustness function is defined as:

$$\hat{\alpha}(D_c) = \max\left(\alpha : \left\{ \left(\max_{F \in U(\alpha)} |\rho(t)| \right) \leq D_c \right\} \right) \tag{10.14}$$

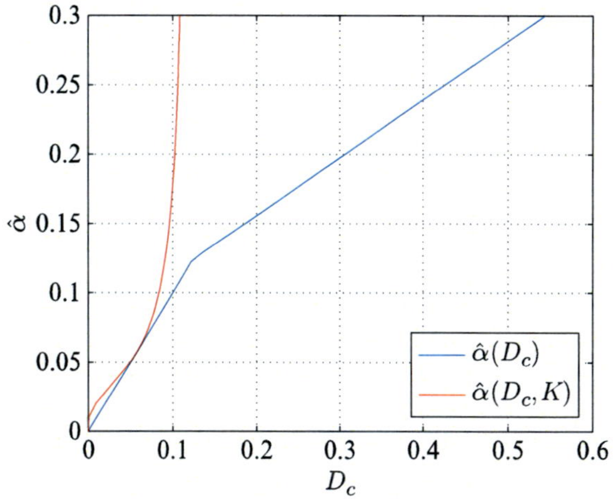

Fig. 10.3 Nominal and calibrated robustness functions

which can be expressed as:

$$\hat{\alpha}(D_c) = max\left(\alpha : \left\{ \left(\max_{F \in U(\alpha)} |\rho(t)| \right) \leq D_c \right\}\right) = max\left(\alpha : \left\{ \left(\left| \frac{\max_t \max_{F \in U(\alpha)} (x_F(t)) - \max_t (x_{\tilde{F}}(t))}{\max_t (x_{\tilde{F}}(t))} \right| \right) \leq D_c \right\}\right) \quad (10.15)$$

Using the convolution integral $\rho(t)$ [13] can be written as follows:

$$\rho(t) = \frac{\max_t \left(\int_{\tau=0}^{\tau=T} F_0 h(t-\tau) d\tau + \alpha \int_{\tau=0}^{\tau=T} F_0 h(t-\tau) d\tau + 0.2\alpha \int_{\tau=T}^{\tau=t} F_0 h(t-\tau) d\tau \right)}{\max_t \left(\int_{\tau=0}^{\tau=T} F_0 h(t-\tau) d\tau \right)} - 1 \quad (10.16)$$

Since the objective is to maximize $\rho(t)$, this implies calculating both the maximum and minimum of $x_F(t))$ as a function of t and α. Formally:

$$\max_{F \in U(\alpha)} F(t) = \begin{cases} (1+\alpha) \cdot F_0 & t \in (0,T) \\ 0.2\alpha F_0 & t \notin (0,T) \end{cases} if\ h(t-\tau) \geq 0 \qquad \begin{cases} (1-\alpha) \cdot F_0 & t \in (0,T) \\ -0.2\alpha F_0 & t \notin (0,T) \end{cases} if\ h(t-\tau) < 0 \quad (10.17)$$

Maximizing Eq. 10.16, leads to:

$$\max_{F \in U(\alpha)} \rho(t) = \frac{\max_t \left(\int_{\tau=0}^{\tau=T} F_0 h(t-\tau) d\tau + \alpha \int_{\tau=0}^{\tau=T} F_0 |h(t-\tau)| d\tau + 0.2\alpha \int_{\tau=T}^{\tau=t} F_0 |h(t-\tau)| d\tau \right)}{\max_t \left(\int_{\tau=0}^{\tau=T} F_0 h(t-\tau) d\tau \right)} - 1 \quad (10.18)$$

The numerator and denominator can be solved analytically. It should be noted that minimizing instead of maximizing (10.16), only affects on the sign of α. The corresponding robustness function is shown in Fig. 10.2b.

Robust parameter calibration A robust parameter calibration is applied to the present example. Selecting the system stiffness as the decision variable, we have:

$$\hat{\alpha}(D_c, K) = max\left(\alpha : \left\{ \left(\max_{F \in U(\alpha)} D(K; F) \right) \leq D_c \right\}\right) \quad (10.19)$$

The result is plotted in Fig. 10.3.

The robust calibrated model guarantees equal or smaller discrepancies than the nominal model at every horizon of uncertainty.

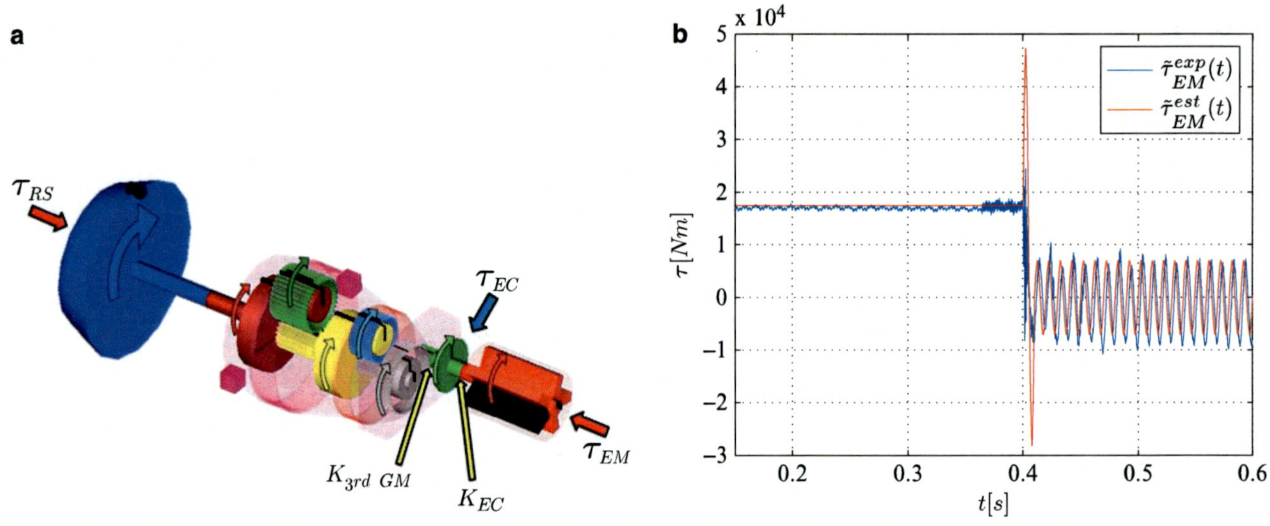

Fig. 10.4 (**a**) Power train model. (**b**) Measured and estimated torque in the generator rotor

10.3.2 Wind Turbine Power Train Model

The robust calibration strategy is now applied to a MW class wind turbine power train model. This case study focuses on a voltage dip which leads to electromagnetic torque variation. The experimentally measured torque in the elastic coupling is used to formulate the discrepancy function.

System model For this application, a discrete model is adopted consisting in a one degree of freedom per body rotational model. The system properties are the inertias of each part of the system, the low speed shaft (LSS) stiffness, the gear teeth stiffness and the high speed shaft (HSS) elastic coupling stiffness. The gearbox support stiffness is modeled as a non linear spring. Such a model is widely used to represent this kind of rotating devices [14, 15]. As for the pitch, the control is assumed to be negligible for the studied load case. The generator torque τ_{EM} is considered as a quadratic generator speed dependent torque after the load case is trigered.

In the present case, a 9 dofs discrete model is considered with one degree of freedom per body rotational model (Fig. 10.4a). The dynamic system to solve can thus be written as:

$$[I] \cdot \{\ddot{\theta}\} + [C] \cdot \{\dot{\theta}\} + [K] \cdot \{\theta\} = \{T\} \tag{10.20}$$

Where $[I]$, $[C]$, and $[K]$ are respectively the inertia matrix, the damping matrix and the stiffness matrix. They depend on inertias, stiffnesses and radii of the various elements of the system. The initial conditions consist in a torque applied to the rotor and the generator. The generator torque depends on the load case rotational speed of the system. The initial positions of the different parts of the system lead to the initial rotation vector. The initial rotational speed is set to a specified value.

The mathematical model is implemented in the Dymola software. Dymola (Dynamic Modeling Laboratory) is suitable for modeling various kinds of physical systems. It supports hierarchical model composition, and acasual connections. Dymola uses a modeling methodology based on object orientation and equations. The usual need for manual conversion of equations to a block diagram is removed by the use of automatic formula manipulation. The equations must be expressed in a neutral form without consideration of the computational order.

Uncertainty model A *voltage dip* is a generator related phenomena [16] and can be produced by a grid oscillation or even manually triggered. It is a sudden and short drop in the generator torque τ_{EM} usually preceded by a sudden torque peak. The torque drops nearly to zero during a period of time and then the control is activated again. A voltage dip is manually triggered in a MW class wind turbine, and the torque in the HSS coupling τ_{EC} is measured. The electromagnetic torque in the generator rotor τ_{EM} is difficult to measure, especially for the short duration torque transient that occur when the voltage dip is triggered. This short transient will be modeled using an info-gap model as a single 100 Hz sinus cycle of uncertain amplitude A_{VD} while the aerodynamic torque input τ_{RS} is assumed to remain constant at its nominal rate on the rotorside. The generator torque input is given by:

$$\tau_{EM}(t) = \begin{cases} \tau_{EM}^{N} & t \in (0, 0.4) \\ \tau_{EM}^{N} \cdot \left(1 - \frac{t-0.4}{0.01}\right) + A_{VD} \cdot e^{-10\cdot(t-0.4)} \cdot sin\left(2 \cdot pi \cdot 100 \cdot (t-0.4)\right) & t \in (0.4, 0.41) \\ \tau_{EM}^{N} \cdot 0.4 \cdot sin\left(2 \cdot pi \cdot 100 \cdot (t-0.4)\right) & t \in (0.41, 0.65) \\ \tau_{EM}^{N} \cdot \left(\frac{\omega_{gen}(t)}{\omega_{gen}^{N}}\right)^2 & t > 0.65 \end{cases} ; t > 0 \tag{10.21}$$

The info-gap model for the uncertain torque transient sinus amplitude:

$$U(\alpha) = \left\{ A_{VD} : \left| \frac{A_{VD} - \tilde{A_{VD}}}{\tilde{A_{VD}}} \right| \leq \alpha ; \right\}, \alpha \geq 0 \tag{10.22}$$

A comparison between the estimated generator torque input $\tilde{\tau}_{EM}^{est}(t)$ (estimated) and the measured power output recorded in the converter and divided by the rotational speed of the system $\tilde{\tau}_{EM}^{exp}(t)$ (experimental) is shown in Fig. 10.4b.

Performance criteria The performance criteria is defined as the fractional error between the ratio of τ_{EC}^{N} and the maximum torque values during the voltage dip for both measured and simulated responses in the HSS elastic coupling τ_{EC}.

$$|\rho(t)| = \left| \frac{\left(\frac{\max_t(\tau_{EC}(t))}{\tau_{EC}^{N}}\right) - \left(\frac{\max_t(\tilde{\tau}_{EC}(t))}{\tilde{\tau}_{EC}^{N}}\right)}{\left(\frac{\max_t(\tilde{\tau}_{EC}(t))}{\tilde{\tau}_{EC}^{N}}\right)} \right| \leq D_c \tag{10.23}$$

The torque increase ratio is of a capital importance for the integrity of the gearbox. A large and sudden increase of the torque in the gearbox generatorside output could lead to gear teeth or bearing failure [17].

Deterministic parameter calibration Before the robustness curve is calculated a deterministic parameter calibration is conducted. Two model parameters (HSS elastic coupling stiffness K_{EC} and third stage gear teeth contact stiffness $K_{3rd\,GM}$ see Fig. 10.4a) are selected as calibration candidates attending to their influence in the torque response. The responses are calculated for a 30 % parameter variations. The error for each point is calculated using the performance criteria defined in Eq. 10.23. The result is plotted in Fig. 10.5a where the color scale represents the performance criteria value for each point of the grid. The parameters which correspond to the better performance (lowest value of $|\rho(t)|$) are selected as reference values. As can be seen in Fig. 10.5a, there is a valley defining a set of local minima, hence the deterministic solution is not uniquely defined. Five sets of parameters – S_1, S_2, S_3, S_4 and S_5 – corresponding to the region of minimums will be selected and analyzed.

Robustness function A robustness analysis is conducted for both the baseline model (before calibration) and the deterministic calibrated models, and a robustness curve is plotted for each selected point in Fig. 10.5a. The results are shown in Fig. 10.5b, where the blue curve corresponds to the original model $\hat{\alpha}_{Nom}$ and the red curves corresponds to each of the sets of parameters selected for calibration $\hat{\alpha}_{S_{1-5}}$. The calibrated models show a much better performance for $\alpha = 0$ than the original model. The robustness curves (nominal and calibrated) cross at a certain α value (0.7 approximately), meaning that a preference reversal occurs as the uncertainty horizon increases.

Robust parameter calibration A robust parameter calibration is now conducted for the power train model (Eq. 10.20) and the uncertain torque transient (Eqs. 10.21 and 10.22). The baseline model defined by the point S_3 is selected and the model is robustly calibrated for different horizons of uncertainty using the same decision variables as before, that is to say, the HSS elastic coupling stiffness K_{EC} and the third stage gear teeth contact stiffness $K_{3rd\,GM}$. For each horizon of uncertainty the parameter set which minimizes the maximum discrepancy is calculated. The robust model parameter calibration is formally defined:

$$\hat{\alpha}(D_c, \mathbf{K}) = max\left(\alpha : \left\{ \left(\max_{A_{VD} \in U(\alpha)} D(\mathbf{K}; A_{VD}) \right) \leq D_c \right\}\right) \textbf{ with } \mathbf{K} = \{K_{EC}; K_{3rd\,GM}\} \tag{10.24}$$

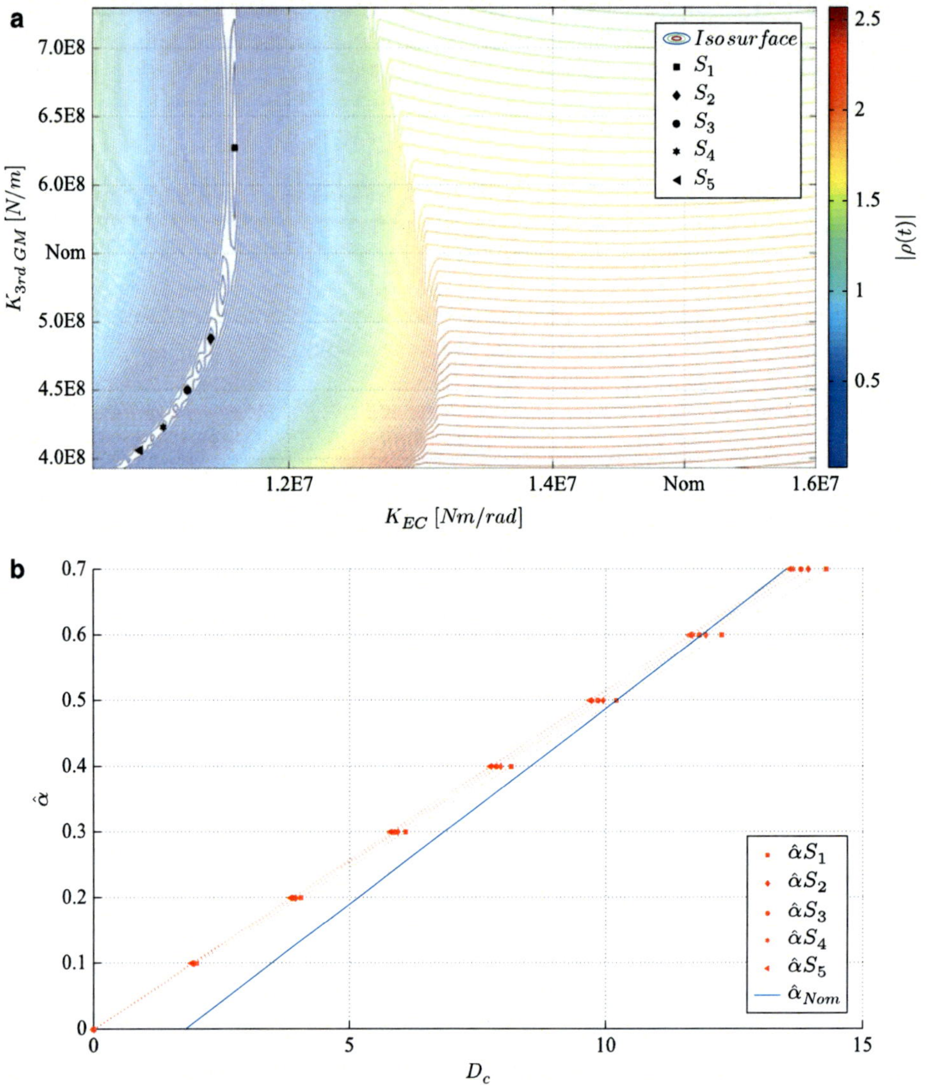

Fig. 10.5 (**a**) Deterministic parameter calibration. (**b**) Robustness curves. Original and calibrated models

The results are plotted in Fig. 10.6. The blue line corresponds to the calibrated model robustness curve $\hat{\alpha}_C$, while the dotted lines correspond to the robustness curves calculated with each set of calibrated parameters for a given α ($\hat{\alpha}_{RC(\alpha=0.1)}$, $\hat{\alpha}_{RC(\alpha=0.2)}$, $\hat{\alpha}_{RC(\alpha=0.3)}$, $\hat{\alpha}_{RC(\alpha=0.4)}$, $\hat{\alpha}_{RC(\alpha=0.5)}$, $\hat{\alpha}_{RC(\alpha=0.6)}$, $\hat{\alpha}_{RC(\alpha=0.7)}$). The red curve links the points of the calibrated model robustness curves at the α value at which each curve is calibrated, it will be denoted as the robust calibrated robustness curve $\hat{\alpha}_{RC}$. This red curve represents the minimum possible maximum discrepancy D_c for different values of α. Once again, the robust calibrated models guarantee lower discrepancies than the deterministically calibrated model.

10.4 Conclusion

Model calibration is an important phase in the overall model validation process and serves to improve fidelity to data. Model calibration methods generally seek to optimize fidelity. Meanwhile, the presence of lack of knowledge in the modeled physics renders fidelity-based approaches suspect given that the purported calibrated performance can no longer be guaranteed. A new robust calibration paradigm was proposed in [10] and forms the basis of ideas explored in this study which focused on the tradeoff between fidelity to data and robustness to uncertainty. The strategy was illustrated on both an academic example and on a more realistic industry related problem with load uncertainty. It was shown that the best deterministic model does not necessarily provide the most robust predictions under lack of knowledge.

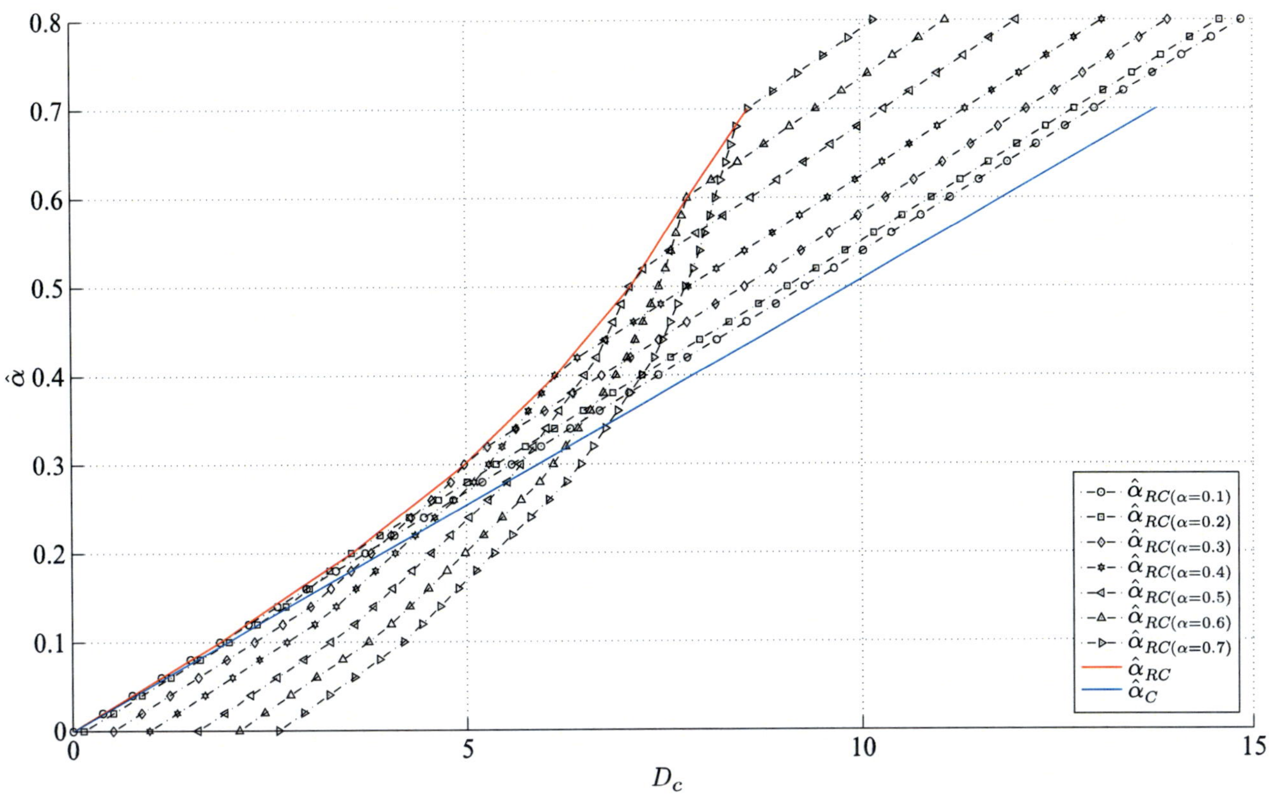

Fig. 10.6 Robustness function for the deterministic and robust calibrated models

Acknowledgements The authors would like to thank Professor Yakov Ben-Haim for his insightful support in the advancement of this study.

References

1. ASME (2006) Guide for verification and validation in computational solid mechanics. American Society of Mechanical Engineers, New York. Tech. Rep. 10
2. Baruch M, Itzhack YB (1978) Optimal weighted orthogonalization of measured modes. AIAA J 16:346–351
3. Berman A, Nagy E (1983) Improvement of a large analytical model using test data. AIAA J 21:1168–1173
4. Halevi Y, Bucher I (2003) Model updating via weighted reference basis with connectivity constraints. J Sound Vib 265:561–581
5. Friswell M, Mottershead JE (1995) Finite element model updating in structural dynamics (Vol. 38). Springer
6. Kerschen G, Worden K, Vakakis A, Golinval J (2006) Past, present and future of nonlinear system identification in structural dynamics. Mech Syst Signal Process 20:505–592
7. Mottershead J, Link M, Friswell M (2011) The sensitivty method in finite element model updating: a tutorial. Mech Syst Signal Process 25:2275–2296
8. Calvi A, Schueller G, Goller B, Broggi M (2011) A stochastic model updating technique for complex aerospace structures. Finite Elem Anal Des 47(7):739–752
9. Link M, Govers Y (2010) Stochastic model updating-covariance matrix adjustment from uncertain experimental modal data. Mech Syst Signal Process 24(3):696–706
10. Ben-Haim Y, Hemez F (2011) Robustness, fidelity and predictions looseness of models. Phys Trans R Soc A 468:227–244
11. IEC 61400-1 (2005) 3rd edn 2005-08 Wind turbines – Part 1: Design requirements, IEC
12. Ben-Haim Y (2006) Info-gap decision theory: decisions under severe uncertainty, 2nd edn. Academic, London
13. Craig RR (1981) Structural dynamics: an introduction to computer methods (Vol. 40). New York: Wiley
14. Lin J, Parker RG (1999) Analytical characterization of the unique properties of planetary gear free vibration. J Vib Acoust 131:316–321
15. Kahraman A (1994) Planetary gear train dynamics. J Mech Des 116:713–720
16. Papathanassiou SA (2001) Mechanical stresses in fixed-speed wind turbines due to network disturbances. IEEE Trans Energy Convers 16(4):361–367
17. Rahimi M, Parniani M (2009) Dynamic behavior and transient stability analysis of fixed speed wind turbines. Renew Energy 34(12):2613–2624

Chapter 11
Simulating the Dynamics of the CX-100 Wind Turbine Blade: Model Selection Using a Robustness Criterion

Kendra L. Van Buren, Sez Atamturktur, and François M. Hemez

Abstract Several plausible modeling strategies are available to develop finite element (FE) models of ever-increasingly complex phenomena. Expert judgment is typically used to choose which strategy to employ, while the "best" modeling approach remains unknown. This paper proposes a decision analysis methodology that offers a systematic and rigorous methodology for comparing plausible modeling strategies. The proposed methodology departs from the conventional approach that considers only test-analysis correlation to select the model that provides the highest degree of fidelity-to-data. The novelty of the methodology lies in an exploration of the trade-offs between robustness to uncertainty and fidelity-to-data. Exploring robustness to model imprecision and inexactness, in addition to fidelity-to-data, lends credibility to the simulation by guaranteeing that its predictions can be trusted even if some of the modeling assumptions and input parameters are incorrect. To demonstrate this approach, an experimental configuration is analyzed in which large masses are used to load the CX-100 wind turbine blade in bending during vibration testing. Two plausible simulations are developed with differing strategies to implement these large masses using (i) a combination of point-mass and spring elements or (ii) solid elements. In this paper, the authors study the ability of the two FE models to predict the experimentally obtained natural frequencies, and the robustness of these competing models to uncertainties in the input parameters. Considering robustness for model selection provides the extent to which prediction accuracy deteriorates as the lack-of-knowledge increases. Therefore, the preferable modeling strategy is the one that offers the best compromise between fidelity-to-data and robustness to uncertainty. To predict the bending vibration of the CX-100 wind turbine blade, it is observed that the modeling strategy with solid elements is far superior to the other one in its ability to provide a compromise between fidelity-to-data and robustness to the modeling assumptions.

Keywords Info-gap decision theory • Model selection • Wind turbine blade • Model complexity • Test-analysis correlation

11.1 Introduction

11.1.1 Motivation

The wind energy industry in the United States has consistently observed the design of larger wind turbines, with the largest ones in 2011 produced with blades 61.5 m in length. In anticipation of this continued trend, blades 100 m in length are already being pursued for future wind turbines [1]. Modeling and simulation techniques can be used to economically and efficiently study the behavior of wind turbines produced at this massive scale, which enables designers to consider both aerodynamic

K.L. Van Buren (✉) · S. Atamturktur
Clemson University, Clemson, SC 29634, USA
e-mail: klvan@clemson.edu; sez@clemson.edu

F.M. Hemez
Los Alamos National Laboratory, Los Alamos, NM 87545, USA
e-mail: hemez@lanl.gov

T. Simmermacher et al. (eds.), *Topics in Model Validation and Uncertainty Quantification, Volume 5,*
Conference Proceedings of the Society for Experimental Mechanics Series 41,
DOI 10.1007/978-1-4614-6564-5_11, © The Society for Experimental Mechanics, Inc. 2013

and structural concerns early in the design process [2], and mitigate the increasing costs of full-scale testing [3]. Finite element (FE) models calibrated against experimental data have gained acceptance for routine use in studying the static and dynamic responses of wind turbine blades, as demonstrated by its inclusion in wind turbine design standards [4]. Further, FE models are advantageous to study complex load cases that arise from in-service wind loading [5, 6], as compared to the idealized loads that are implemented in full-scale experiments [7].

Assumptions and simplifications are routinely implemented in FE models to mitigate our lack-of-knowledge about the underlying physics and to reduce the computation time needed to simulate complex phenomena. For example, it has been proposed to simulate wind turbines using geometrically non-linear, one-dimensional beam elements when coupling FE models of the wind turbine structural response with computational fluid dynamics models of the surrounding airflow [8]. Here, expert judgment is used to identify the need for geometric non-linearity (as opposed to material non-linearity), which can account for large angle displacements as wind turbine blades are produced at larger scales. The use of one-dimensional beam elements is necessary to reduce computation time and make the simulation of wind turbines at the plant scale feasible. Current computing resources and code capabilities prevent the simulation of plant performance based on full-physics, full-coupling, three-dimensional representations of the structural response and air flow, which renders these simplifications necessary. As seen, FE models are only able to provide an approximation of reality due to the need for assumptions and simplifications. The main concern that arises is the extent to which predictions of a numerical simulation can be trusted, given that the models implemented rely on assumptions and simplifications.

Even though there may be a severe lack-of-knowledge about the best modeling strategy, recent wind turbine studies continue to consider a model good quality when its predictions match physical experiments after calibration or validation exercises. It is important to note that the model complexity, as influenced by the model form and number of parameters used to define the model, will also affect the quality of model output [9, 10]. For example, when comparing the output of four different wind turbine drive train models, Martins et al. [11] found that the model that accounted for mechanical damping achieved the best agreement between measurements and simulations, while the less complex models provided poorer agreement with the experimental data. Another study found it necessary to include non-linearity in the structural response of a wind turbine blade to better match the deformation observed during experimental static testing [5]. However, when the quality of numerical models are determined solely by their fidelity to experimental data, the modeling preference strategy will typically lean towards FE models that are overly complex at the risk of over-fitting the experimental data and at the cost of a poor generalization to other, non-tested settings [9]. This paradigm in modeling and simulation has been formally recognized, and it can be shown that fidelity-to-data, robustness to assumptions and predictive capability are antagonistic attributes of any family of models [12].

To quantify the effect on predictions of lack-of-knowledge introduced by modeling assumptions and simplifications, this manuscript proposes a novel, non-probabilistic decision analysis framework rooted in info-gap decision theory (IGDT). IGDT has surfaced as a useful method to study the influence of our ignorance on numerical predictions and the decisions that they support [13]. The methodology proposed herein deviates from other model selection methods (see Sect. 11.1.2), because it is non-probabilistic in nature, and is performed by assessing the trade-offs of fidelity-to-data and robustness of predictions to our lack-of-knowledge. The basic premise is that a good-quality model, while it should be able to reproduce the available measurements, should also provide predictions that are as insensitive as possible to the modeling assumptions and simplifications. In the proposed methodology, model selection is achieved through a rigorous exploration of robustness versus accuracy of predictions. Understanding these trade-offs is important for the development of robust numerical models because it is the very mechanism through which the trustworthiness of predictions can be established.

The methodology proposed in this paper is demonstrated on the bending vibration of the CX-100 wind turbine blade developed at the Sandia National Laboratories (SNL). In two earlier studies, the FE model of the CX-100 blade has undergone rigorous Verification and Validation (V&V) assessments to ensure the credibility of predictions (see Mollineaux et al. [14] for verification studies and Van Buren et al. [15] for the validation studies). These earlier studies are briefly summarized in Sect. 11.2 for completeness. More recently, the CX-100 blade was dynamically tested at the National Renewable Energy Laboratory (NREL) with large masses used to load the blade in bending. The added masses are represented in two alternative configurations, using (i) point masses and stiffening springs or (ii) high-fidelity solid elements. The ability of these two competing modeling strategies to replicate the experimentally obtained natural frequencies is discussed in Sect. 11.3. In Sect. 11.4, the fundamental principles behind IGDT are presented, and the extent to which predictions of these two competing models are robust to uncertainties in the model input parameters is quantified. The strategy that implements high-fidelity solid elements is found to be both more accurate and more robust than the use of point masses. The practical implication of these findings is that predictions, and their accuracy, can be trusted even if some of the modeling assumptions upon which the solid-element FE model relies upon are incorrect.

11.1.2 Related Literature

Model selection has been a widely pursued topic due to the lack-of-knowledge that arises in identifying an appropriate modeling strategy [16]. The importance of model selection arises from the fact that different modeling strategies will affect the quality of predictions. As early as the 1930s, it was observed that using the same data to train an algorithm and evaluate its performance can lead to a false sense of confidence in the results [17]. One method to mitigate this shortcoming in model selection is cross-validation, where experimental data are divided into at least two sets: (i) a calibration set, and (ii) at least one (or more) hold-out sets. The calibration set of experimental data is used to update the model, and the model that is able to best replicate the hold-out sets of data is considered to be the best choice. The data-splitting technique used to divide the available experimental data into different sets is known to affect the model selection process [18].

In the last decade, Bayesian methods have been widely pursued for model selection, with several approaches rooted in its theory, such as the intrinsic Bayes factor [19], fractional Bayes factor [22], deviance information criterion [23], and asymptotic approaches [24]. One advantage of Bayesian approaches is that they will naturally take structural uncertainty into account. Further, when sufficient experimental data are available, the "true" model will always be chosen by Bayesian model selection [25]. One drawback, however, is that the development of defensible prior uncertainties is often difficult and can affect the model selection outcome [26]. Despite the drawbacks to Bayesian approaches, many studies have successfully demonstrated its use in model selection [27, 28].

Several other methods have also been proposed to address the question of model selection, for example the Akaike information criterion [29], minimum description length [30], likelihood ratio tests [31], and information-theoretic measure of complexity [29, 32]. Myung [9] compares the performance of several model selection techniques, with the conclusion that model complexity must be taken into account to ensure that an overly complex model is not selected. Recent studies have also acknowledged the importance of robustness in model selection [27, 33, 34].

The current investigation proposes to address the question of model selection using IGDT, which requires the definition of a family of models to demonstrate the trade-off of fidelity-to-data and robustness to our lack-of-knowledge [13]. Such a family of models can originate from either a probabilistic or non-probabilistic description of the uncertainty and it is referred to as an info-gap model. Previous studies have successfully demonstrated the usefulness of treating uncertainties in an info-gap context, such as exploring the robustness of model predictions to uncertainties in connection stiffnesses [35] and industrial applications [36]. Herein, IGDT will be applied to model parameters to answer the question of model selection. The non-probabilistic treatment of uncertainty limits the assumptions that are applied in the analysis while accounting for model complexity. In addition, explicitly addressing the robustness to uncertainty of model predictions establishes confidence despite the lack-of-knowledge about the modeling assumptions and parameter values used in the simulation.

11.2 Model Development and Experimental Campaign

11.2.1 Development of the CX-100 FE Model

This section provides a review of the development process of the FE model of the CX-100 blade, as discussed in-depth by Mollineaux et al. [14] and Van Buren et al. [15]. The model is developed using an accurate description of the geometry, obtained from design specifications of the wind turbine blade in NuMAD, preprocessor developed at SNL, and imported into ANSYS version 12.1 with Shell-281 finite elements. The mesh discretization is based on an element size of $\Delta x = 8\,\text{cm}$, which produces an overall solution uncertainty of 1.78% for the prediction of low-frequency bending modes of the blade. This mesh size is chosen because the 1.78% numerical uncertainty is comparable to a 3-σ experimental variability of 1.62%, estimated from free-free modal testing performed at the Los Alamos National Laboratory (LANL) [37].

Six independent sections are defined in the development of the FE model: shear web, root, spar cap, trailing edge, leading edge with balsa, and leading edge without balsa. With the exception of the shear web, because it is located inside the cross-section of the blade, these sections are illustrated in Fig. 11.1. Isotropic materials with smeared cross-sectional properties are used to define the material sections.

Modal testing performed at LANL in two configurations is used for calibration and validation studies: (i) free-free, where the blade is suspended with straps, and (ii) clamped-free, where a 250-kg steel bookend fixture is used to fix the base of the blade. The model parameters are calibrated to the free-free and clamped-free experimental natural frequencies in a two-step procedure [15]. To mitigate the uncertainty in the fixity at the base of the blade, fictitious springs are introduced in an attempt

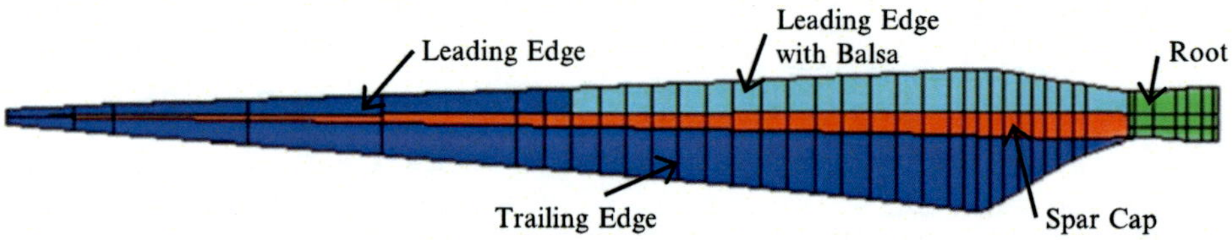

Fig. 11.1 Illustration of the ANSYS model showing different sections of the blade

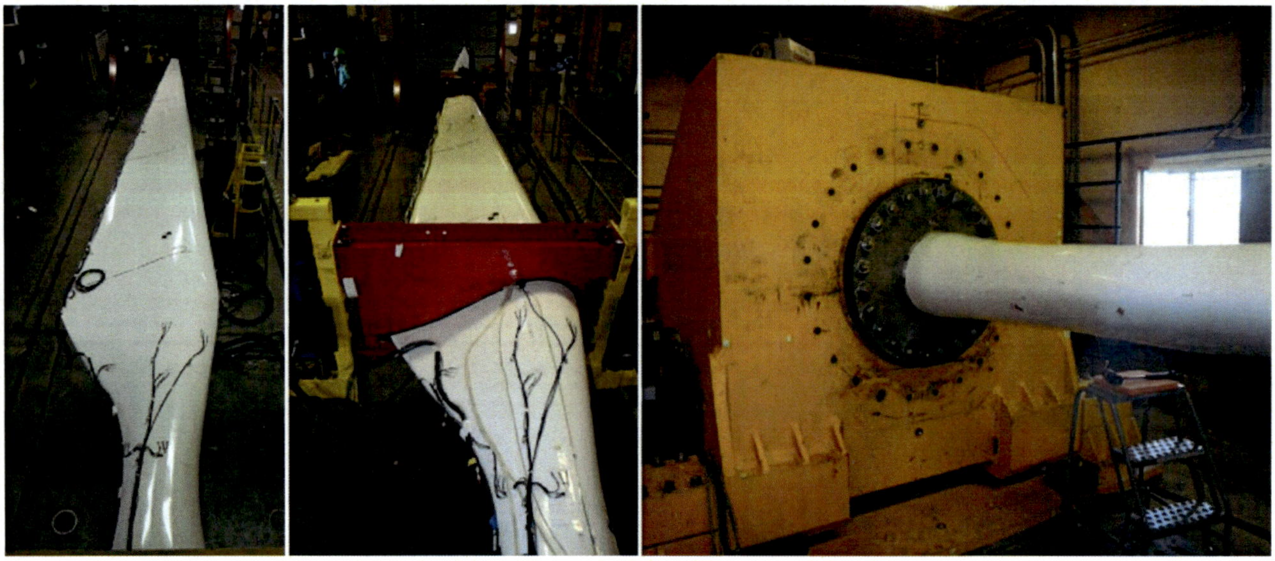

Fig. 11.2 Experimental fixed-free (*left*) configuration, mass-added (*middle*) configuration, and base fixture (*right*)

Table 11.1 Results of the experimental modal analysis

Mode	Fixed-free frequency (Hz)	Mass-added frequency (Hz)
First flap bending	4.35	1.82
Second flap bending	11.51	9.23
Third flap bending	20.54	12.72

to implement a boundary condition that is between the ideal "fixed" and "free" cases. The mode shape vectors are used to validate the FE model, in which the modal assurance criterion is estimated to quantify the agreement of simulation results to the experimental data. An overall correlation of 84% is observed for the free-free modes and 94% for the clamped-free modes.

11.2.2 NREL Modal Testing of the CX-100 Wind Turbine Blade

The CX-100 wind turbine blade is attached to a 6,300 kg (7-ton) steel frame, effectively allowing for a fixed-free boundary condition. Modal testing is performed using a roving impact hammer test procedure under two different setups: first, in a fixed-free condition, and second with large masses clamped to the blade. A 582- and 145-kg mass are added on the blade at the 1.60- and 6.75-m locations, respectively. Four uni-axial accelerometers and one tri-axial accelerometer are used to collect data for hammer impacts at 65 locations: 47 in the flapwise directions, and 18 in the edgewise directions. Three test replicates are performed with a linear average and 150 Hz sampling frequency. The acceleration response is collected with 4,096 sampling points without a window function due to the relatively long sampling period of 11 s [38].

The experimental setups and base fixture are shown in Fig. 11.2, and the first three flapwise frequencies are listed in Table 11.1. It is important to note that the boundary condition provided in the NREL testing is different from the boundary condition used in previous modal testing performed at LANL [15, 37]. The mass-added configuration, in which there is significant mass loading and a change in the compliant boundary condition to a more rigid fixture, defines a different configuration of the CX-100 wind turbine blade.

Table 11.2 Comparison of prior and posterior uncertainty of the FE model parameters

	Prior uncertainty			Posterior uncertainty		
FE model parameter	Lower	Upper	Range	Mean	Std. dev.	$\pm 2\sigma$ range
Trailing edge, density	274.60	823.80	549.20	335.62	49.49	197.95
Leading edge, density	858.20	2,574.60	1,716.40	1,165.30	248.76	995.03
Spar, modulus	29.92	53.56	23.64	43.40	5.51	22.05
Z-spring	1.00	100.00	99.00	71.91	15.98	63.91
Spar, density	1,267.00	3,801.00	2,534.00	1,673.57	335.74	1,342.94

11.2.3 Fixed-Free Model of the CX-100 Wind Turbine Blade

Calibration of the fixed-free model is re-considered due to the more rigid structure used in the NREL experiments. Five statistically significant parameters of the fixed-free FE model of the wind turbine blade are identified using a forward propagation of uncertainty and sensitivity analysis. The influential parameters are: density of the trailing edge, density of the leading edge, modulus of elasticity of the spar cap, translational springs used to model the boundary condition perpendicular to the base fixity, and density of the trailing edge. The uncertainty bounds of the trailing edge density, leading edge density, and spar cap density are increased from previous parametric studies [15], from ±25% bounds to ±50% bounds in the current study after an exploratory design-of-experiments suggests that larger bounds are needed for the FE model to envelope the experimental data. It is chosen to limit the increase of the parameter ranges to ±50% due to mode swapping when the parameters are allowed to vary past these values. The upper and lower uncertainty bounds chosen for the spar cap modulus reflect the posterior uncertainty obtained from previous free-free calibration. The uncertainty bounds of boundary springs are determined from parametric studies of the fixed-free boundary condition. Measurements of the natural frequencies obtained during the NREL testing are utilized to calibrate the FE model. Instead of performing calibration as an optimization of model parameters to best-fit the experimental data, inference uncertainty quantification is performed to explore the posterior probability distribution of these three parameters. To efficiently perform the inference uncertainty quantification, the FE model is executed using a three-level, full-factorial design-of-experiments to provide the simulation data required to train a fast-running Gaussian Process Model (GPM). A Markov Chain Monte Carlo (MCMC) algorithm is used to explore the posterior distribution of parameters of the GPM emulator, where a statistical T-test retains only parameters that provide frequency predictions sufficiently "close" to those measured experimentally. This methodology relies on the theory proposed by Kennedy and O'Hagan [39] and the computational framework developed by Higdon et al. [40].

The results of the inference are summarized in Table 11.2. Columns 2–4 summarize the prior uncertainty, which represents the range of values used in the full-factorial design to train the GPMs. Columns 5–7 provide the posterior uncertainty, which are the statistics inferred from the MCMC search algorithm. The inference is successful at "learning" the value of the parameters, as indicated by the reduction of the posterior uncertainty relative to the prior range.

A graphical representation of the posterior distribution is provided in Fig. 11.3, which plots the marginal distribution of each parameter on the diagonal and a probability contour of the pairs of parameters in the off-diagonal boxes. The inference is successful in reducing the uncertainty of the spar cap modulus of elasticity and the boundary spring constants. The clustering of samples drawn towards the lower bound in Fig. 11.3 suggests that the algorithm may be attracted to values outside of the initial range provided to the GPM. However, it is chosen to keep the current analysis due to the introduction of mode swapping when the parameter values are allowed to vary outside of the prior range.

Table 11.3 compares the experimentally obtained natural frequencies to those predicted by the FE model with the parameters set to the mean values of the posterior distribution. The simulation consistently under-predicts the experimental results by 0.5–3.4%. Due to the relatively small error in the predictions of resonant frequencies, and the fact that the predicted mode shape deflections correlate well with those measured experimentally, the accuracy of the calibrated fixed-free model is deemed acceptable. The same model is used next to explore the mass-added configuration.

11.3 Model Development for the Mass-Added Configuration of the CX-100 Wind Turbine Blade

This section discusses the development of two competing FE models used to simulate the NREL mass-added configuration. The two models pursued in this section are developed using different assumptions and simplifications, using either (i) a system of fictitious point masses and stiffening springs, or (ii) three-dimensional solid elements that represent the geometry of the added masses with high fidelity.

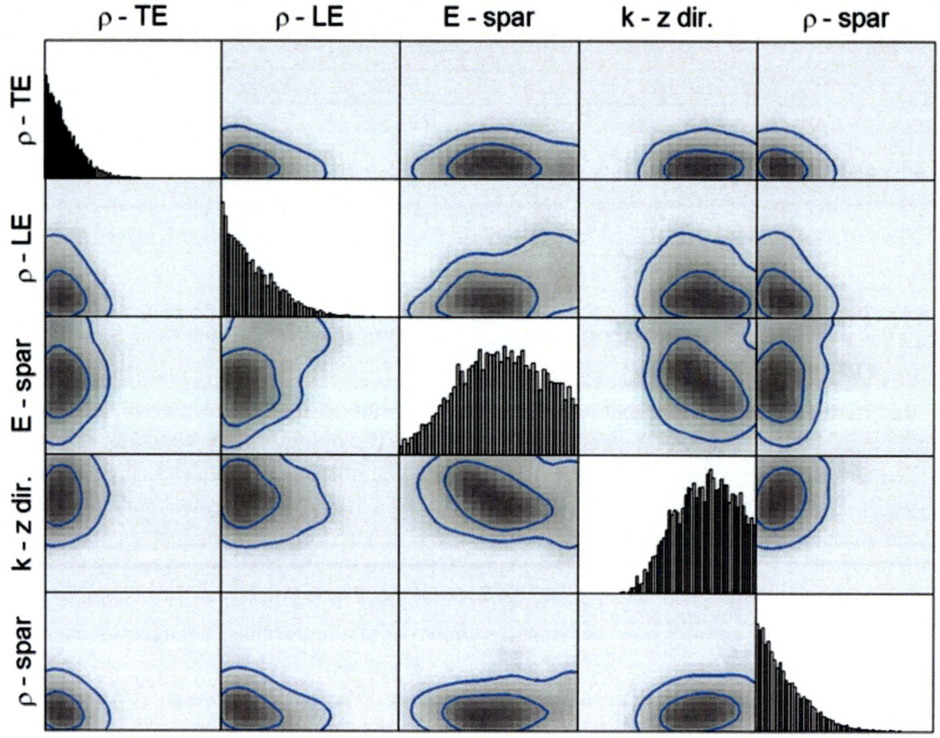

Fig. 11.3 Marginal distributions and correlation functions corresponding to Table 11.2

Table 11.3 Comparison of experimental and simulated results for the fixed-free model

Mode	Experimental (Hz)	Simulated (Hz)	Difference (%)
First flap	4.35	4.26	−2.1
Second flap	11.51	11.45	−0.5
Third flap	20.54	19.85	−3.4

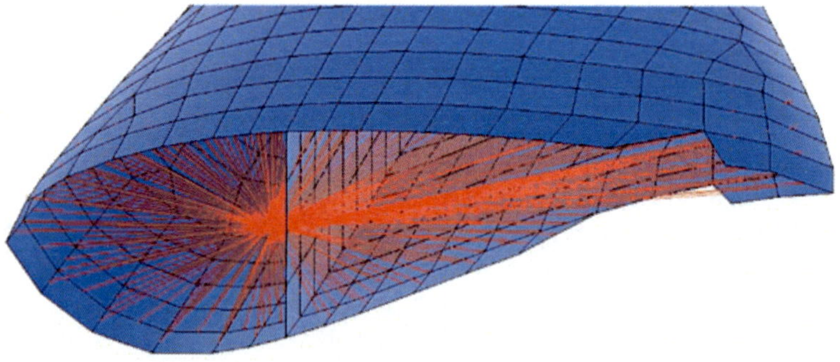

Fig. 11.4 Illustration of the blade cross-section with added point masses and springs

11.3.1 Development of the Point Mass Model

A central point mass is added to approximate the added masses at each of the two locations. The point masses are then connected by fictitious springs to the nodes of the blade to reflect the interaction between the blade and added masses. This configuration is shown in Fig. 11.4. The use of springs in this modeling strategy is adopted because exploratory FE simulations indicate that adding the point masses directly to the shell model of the blade introduces lower-order mode shapes with local deformations at the cross-sections with masses added. Fictitious springs are therefore introduced to maintain the rigidity of the blade cross-section at the locations where masses are added. This modeling strategy offers a compromise between low computational times-to-solution and a reasonable representation of the vibration mechanics.

Table 11.4 lists the parameters used to develop the point mass representation added to the three-dimensional blade model discretized with shell elements. The parameterization of this simplified modeling strategy results in the use of only six new model parameters.

Table 11.4 Parameters used to develop the point mass representation

Parameter	Description
(1; 2)	(Translation; rotation) springs at the 1.60-m section
3	Point mass at 1.60-m section
(4; 5)	(Translation; rotation) springs at 6.75-m section
6	Point mass at 6.75-m section

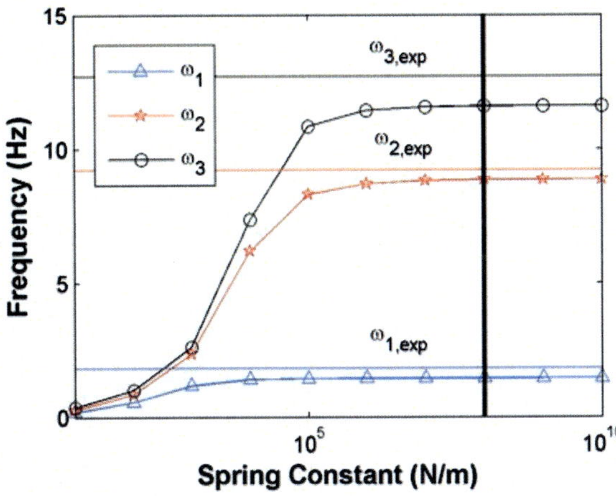

Fig. 11.5 Effect of spring stiffness coefficients on the first three bending frequencies

Table 11.5 Comparison of experimental and simulated frequencies for the mass-added model

Mode	Experimental (Hz)	Simulated (Hz)	Difference (%)
First flap	1.82	1.45	−20.3
Second flap	9.23	8.85	−4.1
Third flap	12.72	11.59	−8.9

To preserve the calibrated, fixed-free model discussed earlier in Sect. 11.2.3, only the spring stiffness coefficients are calibrated in the mass-added configuration (parameters 1, 2, 4, and 5 listed in Table 11.4). A parametric study is performed to evaluate the effect of the spring stiffness on frequency predictions.

Figure 11.5 shows the frequency predictions as the spring stiffness values are varied from 10 to 10^{+10} N/m. As the spring stiffness is increased, the natural frequencies also increase, due to the change in interaction between the blade and springs. Around a value of 10^{+6} N/m, the natural frequencies plateau to values that consistently under-predict the experimental natural frequencies. A calibration of the model parameters would likely converge to the upper bound of the spring stiffness values because the parametric study in Fig. 11.5 is unable to form an envelope around the experimentally obtained natural frequencies. Therefore, the value of 10^{+8} N/m is chosen for the spring stiffness, indicated by the vertical black line in Fig. 11.5.

Table 11.5 compares the natural frequencies measured experimentally to those predicted by the mass-added FE model. Again, the frequencies are consistently under-predicted by the model; however, the absolute differences for the first three modes, 0.37, 0.38, and 1.13 Hz, demonstrate an acceptable fidelity-to-data despite the minimal calibration activities performed after the model was modified to include the added masses.

11.3.2 Development of the Solid Mass Model

The second modeling strategy is to represent the added-mass configuration of the blade with the highest possible degree of geometrical fidelity. Three-dimensional, solid elements are utilized to represent the geometry of the experimental setup, implementing Solid-186 elements in ANSYS. Four sections are used to define the added masses. The sections are labeled the 6.75-m mass, 1.60-m mass, and two 1.60-m offset masses in Fig. 11.6.

Four sections are used to model the added masses onto the blade, as shown in Fig. 11.6: one section for the wooden form attached to the blade at the 1.60-m station, two symmetric sections for the hydraulic actuator system at the 1.60-m station, and one section for the wooden form attached to the blade at the 6.75-m station. The wooden forms are modeled using the geometry of the masses obtained from design specifications. The geometry of offset masses of the hydraulic actuator system at the 1.60-m station is simplified into rectangular solids. Figure 11.7 compares the actual geometry of the experimental setup

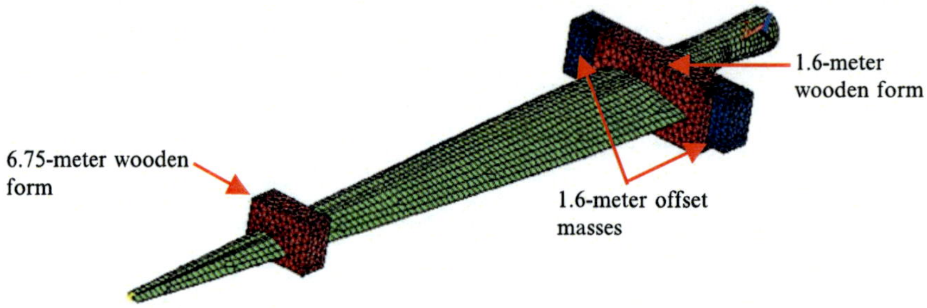

Fig. 11.6 Second modeling strategy that includes solid elements to represent the added masses

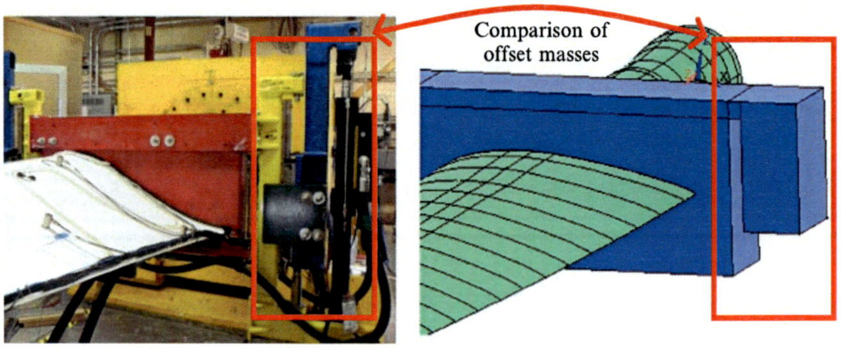

Fig. 11.7 Close-up of the offset mass modeled at the 1.60-m station

Table 11.6 Parameters used to develop the solid-mass representation

Parameter	Description
(1; 2)	(Elastic modulus; density) of 1.60-m section
(3; 4)	Center of gravity (X; Y) coordinates of 1.60-m offset mass
5	Density of 1.60-m offset section
(6; 7)	(Elastic modulus; density) of 6.75-m section

Table 11.7 Comparison of experimental and simulated frequencies for the solid-mass model

Mode	Experimental frequency (Hz)	Simulated frequency (Hz)	Difference (%)
First flap	1.82	1.44	−20.9
Second flap	9.23	9.29	0.7
Third flap	12.72	13.22	3.9

to the simplification implemented in the FE model. Due to its complicated geometry, expert opinion is used to identify the center of gravity of the offset mass, which introduces uncertainty in the analysis.

It is emphasized that, while this modeling strategy provides a better representation of the masses than the point-mass and spring system, the fundamental form of the model is still incorrect due to the assumption that the added masses can be represented using four homogenized sections. Another important factor is the time-to-solution: the vibration analysis of this model is significantly more expensive than the previous one. One objective of the investigation is to understand the extent to which the point-mass model is able to deliver a similar accuracy and robustness as the solid-mass model, at lower computational cost.

The parameters used to develop the solid-mass representation are listed in Table 11.6. The parameterization includes the geometry of the outset masses, represented by the center of gravity coordinates that define the masses, which influences the ability of the blade to bend in torsion. While the imperfect knowledge of the material parameters introduces parametric uncertainty in the prediction of vibration response, the imperfect knowledge of the center of gravity coordinates introduces numerical uncertainty due to the fact that the mesh changes each time that a different location of the center of gravity is implemented. This re-meshing step also contributes to significantly increasing the computational cost of the analysis. A close-up of the offset masses at the 1.60-m station is provided in Fig. 11.7, to highlight the good qualitative agreement between geometries of the experimental setup and FE model representation.

With the masses represented using solid elements, the density is back-calculated such that the weights of masses implemented in the FE model correspond to the weights of masses obtained experimentally. The elastic modulus of the wooden forms is assumed based on the documented value for balsa wood. The agreement between predictions of the FE model and experimental measurements is listed in Table 11.7.

11.4 Analysis of Robustness to Uncertainty Applied to Models of the CX-100 Wind Turbine Blade

This section discusses IGDT and how it is employed to evaluate the robustness to uncertainty of model predictions. By establishing robustness, one demonstrates the extent to which the predictions remain sufficiently accurate, even if some of the modeling assumptions and parameter values used in the simulation are incorrect. Lack-of-robustness, on the other hand, indicates that the expected level of accuracy obtained, for example, through calibration, may not be reached if some of the assumptions and parameter values happen to be incorrect. Analyzing the robustness of the FE model does not necessarily translate into a reduction of prediction uncertainty. It helps, instead, to identify a potentially dangerous situation whereby the predictions, and their accuracy, are sensitive to aspects of the modeling that may be unknown and/or uncontrolled.

11.4.1 Conceptual Demonstration of Robustness Analysis

For the info-gap analysis presented, the allowable range of variation of model parameters is controlled using an uncertainty parameter, α. The definition of α is not essential; what matters is that increasing its value defines a nested family of models. For simplicity, the definition of α is kept unit-less: the same level of uncertainty, α, can be applied simultaneously to multiple parameters from the two modeling strategies because it has no physical unit. The second attribute of the info-gap analysis is the performance metric of the model, herein quantified using the root mean squared difference between simulation predictions and experimental observations. A conceptual illustration is provided next to describe the evaluation of the robustness of the competing FE models.

Consider a model defined using two uncertain parameters, u_1 and u_2. Figure 11.8a describes the nominal performance of the model, where u_1 and u_2 are defined using initial, best guesses or nominal values. As uncertainty, α, is increased, the parameters are allowed to vary within a range of permissible values (see Sect. 11.4.2 for how the bounds are chosen for the parameters in our application). As a result, parameters are varied from their nominal settings to become u_1 and u_2. Herein, the allowable range of variation of u_1 and u_2 is referred to as the uncertainty space. It is represented as a two-dimensional rectangle of size $(\alpha_1)^2$ in Fig. 11.8b, c. With such changes in input parameters, the model performance either improves or degrades. We would like, therefore, to explore the best and worst achievable performances as $\tilde{u}_1$ and $\tilde{u}_2$ are allowed to venture away from their nominal values but remain within the uncertainty space defined by the parameter α_1. The improvement of the performance obtained from the model is described as the opportuneness, and the degradation of performance is the robustness. At any level of uncertainty α, the opportuneness and robustness points are obtained by solving two global optimization problems that search for the best and worst performances, respectively, given the space of allowable values for u_1 and u_2. Figure 11.8b, c illustrate the development of the robustness and opportuneness functions.

Figure 11.8d illustrates that as the uncertainty parameter, α, increases, the uncertainty space, also increases. If the uncertainty space is defined to have nested intervals as α increases, as suggested in Fig.11.8d, then the opportuneness and robustness curves will be monotonic functions because the global optimizations that they represent are performed within ever-growing spaces. Figure 11.8e then shows the resulting opportuneness and robustness curves, developed from the evaluation of best and worst performances at three levels of uncertainty. A particular focus is placed on the robustness curve, and its shape, that are useful to evaluate the worst-case performance of the model under increasing uncertainty bounds. A "steep," or nearly vertical, robustness curve indicates that the predictions are insensitive to increasing levels of uncertainty, α. Such an observation would be welcome as it would reinforce our conviction that the model can be applied with confidence even if some of the assumptions used for its development are questionable. Observing, on the other hand, a robustness curve with small slope, "$\Delta\alpha/\Delta R$," which denotes a small improvement of robustness "$\Delta\alpha$" relative to a change in performance "ΔR," indicates that the model predictions are sensitive to the values of u_1 and u_2 used in the simulation. Such a lack of robustness would decrease the level of trust placed in the assumptions upon which the model relies.

11.4.2 Rationale for the Definition of Uncertainty

To promote a fair comparison of the two modeling strategies, the parameters are varied in such as way that the effect on bending frequency predictions of the maximum parameter variation is consistent with the difference between the competing models at their nominal setting. Doing so ensures that the effect on predictions of the allowable range of parameter variation is consistent with the effect on predictions of varying the model forms of the competing modeling strategies. Because model

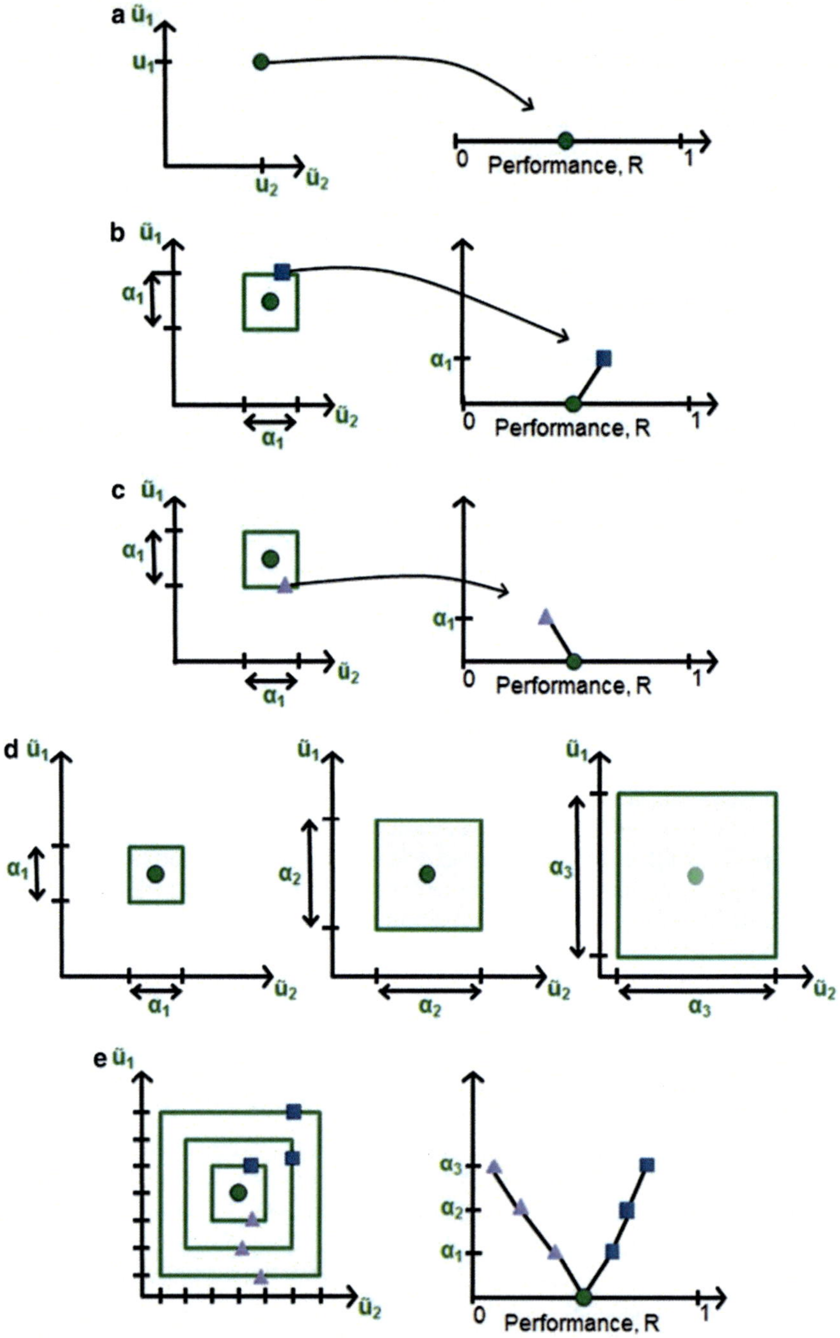

Fig. 11.8 Illustration of the successive steps of an info-gap analysis of robustness. (**a**) Analysis of nominal performance. (**b**) Development of the robustness function. (**c**) Development of the opportuneness function. (**d**) Increased uncertainty space for $\alpha_3 \geq \alpha_2 \geq \alpha_1$. (**e**) Robustness and opportuneness curves

selection is only applied to how the masses are modeled onto the existing shell representation, the info-gap analysis is restricted to the model parameters used to define the added masses. The two models at their nominal configuration exhibit a 20% average percent variation in the first three flapwise frequencies. The models are held at their nominal configuration while the weights of the masses are allowed to vary using the mass parameter for the point mass model, and the density parameter for the solid mass model. Table 11.8 summarizes the results of the mass-only variation, where 0% variation indicates the models held at their nominal setting.

The results presented in Table 11.8 are plotted in Fig. 11.9 for clarity. It is emphasized that the behaviors of the two curves are slightly different, even though only the mass parameters are allowed to vary in the two models. The observed difference is attributed to the combined effect of parameter variation and model form on frequency predictions. Figure 11.9 demonstrates that an 18% variation in masses is necessary to achieve the 20% variation observed between the two models at nominal configuration. Thus, the lower and upper bounds of the variation corresponding to $\alpha = 1$ are defined by allowing the

Table 11.8 Effect on natural frequency predictions of the mass-only variation

	Point mass model			Solid mass model		
Variation (%)	ω_1	ω_2	ω_3	ω_1	ω_2	ω_3
−50	1.94	10.42	13.43	1.94	10.59	15.38
−25	1.64	9.64	12.38	1.63	10.00	14.04
0	1.45	8.85	11.59	1.44	9.29	13.22
+25	1.30	8.17	10.78	1.30	8.59	12.66
+50	1.20	7.60	10.06	1.19	7.96	12.24

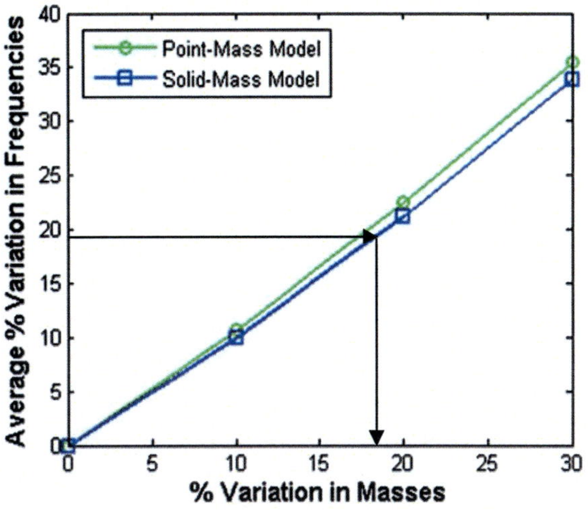

Fig. 11.9 Comparison of frequency prediction variation due to mass-only variation

mass parameters to vary up to ±18%. Having defined the parameter variations corresponding to any value of α, the info-gap analysis can be used to address the question of model selection.

11.4.3 Selection of the Mass Added Models

The info-gap analysis is performed for the competing FE models to explore the robustness of predictions obtained by each one of them. Upper and lower bounds of model parameters are defined corresponding to the level of uncertainty being evaluated, as suggested in Fig. 11.8 in the case of only two parameters. The uncertainty space is a hyper-cube defined from the lower and upper bounds for the vector of parameters u. Its size, or volume, increases monotonically depending on the level of uncertainty considered, α:

$$\tilde{u}(\alpha) = (1 \pm 0.2 \times \alpha)\, u. \tag{11.1}$$

The level of uncertainty, α, is multiplied by 0.2 such that when $\alpha = 1$ the parameters are varied by ±20%, consistent with the mass-added variation pursued in Sect. 11.4.2. The robustness and opportuneness functions are evaluated in increments of $\alpha = 0.5$. For each level of uncertainty evaluated, the "fmincon" optimization solver of Matlab™ is used to search for the combination of parameters within the family of models that produces the worst-case (for robustness) or best-case (for opportuneness) performance. It is emphasized that a new input deck, that includes re-meshing in the case of the solid element model, is generated and submitted to ANSYS each time that a combination of model parameters is evaluated during the optimization. Results of the ANSYS analysis must then be uploaded back in Matlab™ memory. This strategy requires significant scripting to automate the procedure but avoids the development of statistical emulators that may introduce unwanted approximations.

To ensure that the optimization is initiated with a high-quality guess, all combinations of the upper and lower bounds, or "corners" of the hyper-cube space, are evaluated using a two-level full-factorial design-of-experiments. The optimization is then initiated using the combination of model parameters that yields the maximum or minimum performance of the full-factorial design. Because the objective is to compare the prediction accuracies of the two modeling strategies, model performance is defined as the root mean squared error of natural frequencies for the first three flapwise bending modes:

Table 11.9 Range of variation for the parameters used in the point-mass model

Parameter	Description	Nominal value	± Variation
1	Translation springs at 1.60-m section	10^{+8}	$10^{+6.4}$–$10^{+9.6}$
2	Rotation springs at 1.60-m section	10^{+8}	$10^{+6.4}$–$10^{+9.6}$
3	Point mass at 1.60-m section	582.46	465.97–698.95
4	Translation springs at 6.75-m section	10^{+8}	$10^{+6.4}$–$10^{+9.6}$
5	Rotation springs at 6.75-m section	10^{+8}	$10^{+6.4}$–$10^{+9.6}$
6	Point mass at 6.75-m section	144.7	115.76–173.64

Table 11.10 Range of variation for the parameters used in the solid-mass model

Parameter	Description	Nominal value	± Variation
1	Elastic modulus of 1.60-m section	$8 \times 10^{+9}$	$6.4 \times 10^{+9}$–$9.6 \times 10^{+9}$
2	Density of 1.60-m section	636.1	508.88–763.32
3	Center of gravity X coordinates of 1.60-m offset mass	0.224	0.179–0.269
4	Center of gravity Y coordinates of 1.60-m offset mass	0.480	0.384–0.576
5	Density of 1.60-m offset section.	229.0	183.2–274.8
6	Elastic modulus of 6.75-m section	$8 \times 10^{+9}$	$6.4 \times 10^{+9}$–$9.6 \times 10^{+9}$
7	Density of 6.75-m section	1,644.5	1,315.6–1973.4

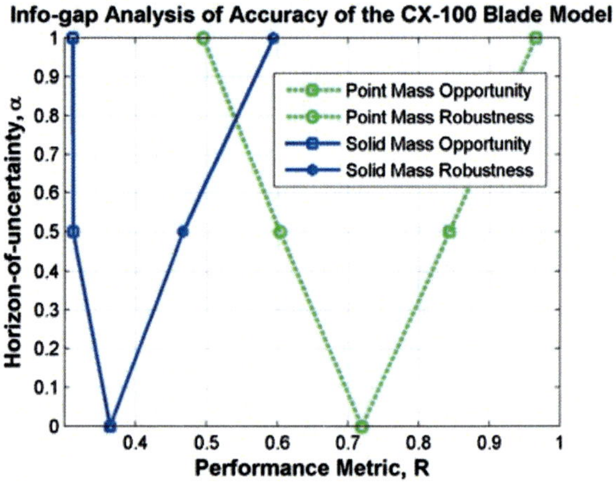

Fig. 11.10 Info-gap robustness and opportuneness curves of the two modeling strategies

$$R = \sqrt{\sum_{i=1}^{3} \left(\omega_{sim,i} - \omega_{\exp,i}\right)^2}, \tag{11.2}$$

where R denotes the model performance metric, ω_{sim} is the numerical prediction of natural frequency, and $\omega_{\exp}$ is the experimental measurement of the same frequency.

Tables 11.9 and 11.10 define the parameters associated with the competing models, along with the ranges of variation specified for the info-gap analysis at the level of uncertainty of $\alpha = 1$. Note that the center of gravity parameter in Table 11.10 affects the definition of meshes used in the solid mass representation. It means that the uncertainty parameter, α, influences both the material behavior (density, elastic modulus) and numerical uncertainty of FE predictions due to the changes that it brings to the mesh discretization.

Figure 11.10 presents the results of the info-gap analysis performed on the competing FE models. The nominal performance, associated with a level of uncertainty of $\alpha = 0$, clearly demonstrates that the solid mass model better reproduces the experimental data compared to the point mass model. Further, as the uncertainty parameter increases, the solid mass model remains the preferable modeling strategy. Conversely, it can be stated that the solid mass model provides a higher degree of accuracy at any level of modeling uncertainty, α. In fact, the robustness slopes of the competing models are relatively consistent despite the different representations of reality. The result of this analysis demonstrates unambiguously that the solid mass model is the preferable modeling strategy to utilize, despite the lack-of-knowledge associated with the modeling assumptions and parameters used in the simulation.

11.5 Conclusion

This manuscript discusses a decision analysis framework for model selection that considers the trade-offs in the ability of a numerical simulation to, first, replicate the experimental data and, second, provide predictions that are robust to the modeling assumptions identified in the model development process. Modeling assumptions are typically formulated when developing numerical simulations, such as the use of fictitious boundary springs or implementing smeared properties for composite materials instead of attempting to define the individual layers. Although such assumptions have become commonplace, their effect on model predictions often remains unknown. Another common practice is to consider that a model achieves sufficient "predictability" as long as its predictions reproduce the experimental measurements. Our contention is that assessing models based only on their fidelity-to-data while ignoring the effect that the modeling assumptions may exercise on predictions is not a sound strategy for model selection.

The framework discussed in this study is applied to competing models used to simulate an experimental configuration of the CX-100 wind turbine blade in which masses are added to the blade. Experimental data obtained from a fixed-free modal analysis performed at the National Renewable Energy Laboratory, with and without added masses, are utilized. The wind turbine blade is bolted to a 6,300-kg steel frame to define the fixed-free configuration. Masses are added at the 1.60- and 6.75-m sections to define the mass added configuration that enhances the flapwise bending vibrations. The FE model of the blade, developed from a previous verification and validation study, is first calibrated to measurements of the fixed-free configuration. Calibration results show that the FE model is able to replicate the experimental frequencies within an average 2% error. Two modeling strategies are then considered for implementing the masses onto the existing FE model, using (i) point masses and stiffening springs and (ii) high-fidelity solid elements. To examine the predictive capability of the mass-added FE models, limited calibration exercises are performed past the initial calibration to the fixed-free configuration. At their nominal configurations, the point mass model reproduces the experimental data to within 11.1% average error, and the solid mass model is within 8.5% average error for the first three flapwise bending natural frequencies.

An info-gap analysis is performed to address the question of model selection. An advantage of info-gap is that the formulation of prior probability distributions can be avoided because the analysis substitutes numerical optimization to statistical sampling. Further, the robustness to our lack-of-knowledge about the modeling assumptions and parameter values is accounted for when evaluating the model performance. The info-gap analysis is performed through parameter variation, where the maximum range of variation is chosen such that the change in model predictions is consistent with the change induced by the differing modeling strategies. It is observed that the solid mass model is not only more accurate, but also provides better behavior in robustness to modeling assumptions and unknown parameter values. Even though the solid mass model is a more complex representation of reality, and comes with higher computational cost, the analysis concludes unambiguously that it is the preferable modeling strategy for this application.

Acknowledgements This work is performed under the auspices of the Laboratory Directed Research and Development project "Intelligent Wind Turbines" at the Los Alamos National Laboratory (LANL). The authors are grateful to Dr. Curtt Ammerman, project leader, for his continued support and technical leadership. LANL is operated by the Los Alamos National Security, LLC for the National Nuclear Security Administration of the U.S. Department of Energy under contract DE-AC52-06NA25396.

References

1. Griffith DT, Ashwill TD (2011) The Sandia 100-meter all-glass baseline wind turbine blade: SNL100-00, Sandia technical report SAND2011-3779, Sandia National Laboratories, Albuquerque
2. Quarton DC (1998) The evolution of wind turbine design analysis: a twenty year progress review. Wind Energy 1(S1):5–24
3. Veers PS, Laird DL, Carne TG, Sagartz MJ (1998) Estimation of uncertain material parameters using modal test data. In: 36th AIAA aerospace sciences meeting, Reno
4. Det Norske Veritas (2010) Design and manufacture of wind turbine blades, offshore and onshore Wind Turbines. Det Norske Veritas, Høvik, http://link.springer.com/chapter/10.1007%2F978-1-4614-2431-4_2#
5. Jensen FM, Falzon BG, Ankersen J, Stang H (2006) Structural testing and numerical simulation of a 34-m composite wind turbine blade. Compos Struct 76(1–2):52–61
6. Leishman JG (2002) Challenges in modeling the unsteady aerodynamics of wind turbines. Wind Energy 5(2–3):85–132
7. Freebury G, Musial W (2000) Determining equivalent damage loading for full-scale wind turbine blade fatigue tests. In: 19th ASME wind energy symposium, Reno
8. Dalton S, Monahan L, Stevenson I, Luscher DJ, Park G, Farinholt K (2012) Towards the experimental assessment of NLBeam for modeling large deformation structural dynamics. In: Mayes R, Rixen D, Griffith DT, DeKlerk D, Chauhan S, Voormeeren SN, Allen MS (eds), Topics in experimental dynamics substructuring and wind turbine dynamics, vol 2. Springer, New York, pp 177–192
9. Myung J (2000) The importance of complexity in model selection. J Math Psychol 44(1):190–204

10. Atamturktur S, Hegenderfer J, Williams B, Egeberg M, Unal C (2012) A resource allocation framework for experiment-based validation of numerical models. J Mech Adv Mater Struct (conditionally accepted)
11. Martins M, Perdana A, Ledesma P, Agneholm E, Carlson O (2007) Validation of fixed speed wind turbine dynamic models with measured data. Renew Energy 32(8):1301–1316
12. Ben-Haim Y, Hemez FM (2012) Robustness, fidelity and prediction-looseness of models. Proc R Soc A 468(2137):227–244
13. Ben-Haim Y (2006) Info-gap decision theory: decisions under severe uncertainty, 2nd edn. Academic, Oxford
14. Mollineaux MG, Van Buren KL, Hemez FM, Atamturktur S (2012) Simulating the dynamics of wind turbine blades: part I, model development and verification. Wind Energy. doi:10.1002/we.1521
15. Van Buren KL, Mollineaux MG, Hemez FM, Atamturktur S (2012) Simulating the dynamics of wind turbine blades: Part II, model validation and uncertainty quantification. Wind Energy. doi:10.100/we1522
16. Draper D (1995) Assessment and propagation of model uncertainty. J R Stat Soc B 57(1):45–97
17. Larson SC (1931) The shrinkage of the coefficient of multiple correlation. J Educ Psychol 22(1):45–55
18. Kadane JB, Dickey JM (1980) Bayesian decision theory and the simplification of models. In: Kmenta J, Ramsey J (eds) Evaluation of econometric models. Academic, New York, pp 245–268
19. Arlot S, Celisse A (2010) A survey of cross-validation procedures for model selection. Stat Surv 4:40–79
20. Berger J, Pericchi L (1996) The intrinsic Bayes factor for model selection and prediction. J Am Stat Assoc 91(433):109–122
21. Graybill FA (1976) Theory and application of the linear model. Wadsworth, Belmont
22. O'Hagan A (1995) Fractional Bayes factor for model comparison. J R Stat Soc Ser B 57(1):99–138
23. Spiegelhalter DJ, Best NG, Carlin BP, van der Linde A (2002) Bayesian measure of model complexity and fit (with discussion). J R Stat Soc Ser B 64(4):583–649
24. Wasserman L (2000) Bayesian model selection and model averaging. J Math Psychol 44(1):92–107
25. Berger JO, Pericchi LR (2001) Objective Bayesian methods for model selection: introduction and comparison (with discussion). In: Lahiri P (ed) Model selection. IMS, Beachwood, pp 135–207
26. Kadane JB, Lazar NA (2004) Methods and criteria for model selection. J Am Stat Assoc 99(465):279–290
27. Terejanu G, Oliver T, Simmons C (2011) Application of predictive model selection to coupled models. In: Proceedings of the world congress on engineering and computer science, San Francisco
28. Beck JL, Yuen KV (2004) Model selection using response measurements: Bayesian probabilistic approach. J Eng Mech 130(2):192–203
29. Bozdogan H (2000) Akaike information criterion and recent developments in information complexity. J Math Psychol 44(1):62–91
30. Grunwald P (2000) Model selection based on minimum description length. J Math Psychol 44(1):133–152
31. Posada D, Buckley TR (2004) Model selection and model averaging in phylogenetics: advantages of Akaike information criterion and Bayesian approaches over likelihood ratio tests. Syst Biol 53(5):793–808
32. Bozdogan H, Bearse PM (1997) Model selection using informational complexity with applications to vector autoregressive (VAR) models. In: Dowe D (ed) Information, statistics, and induction in sciences (ISIS) anthology. Springer, Berlin/New York
33. Müller S, Welsh AH (2005) Outlier robust model selection in linear regression. J Am Stat Assoc 100(472):1297–1310
34. Johnson JB, Omland KS (2004) Model selection in ecology and evolution. Trends Ecol Evol 19(2):101–108
35. Vinot P, Cogan S, Ben-Haim Y (2002) Reliability of structural dynamic models based on info-gap models. In: 20th international modal analysis conference, Los Angeles
36. Hot A, Cogan S, Foltête A, Kerschen G, Buffe F, Buffe J, Behar S (2012) Design of uncertain pre-stressed space structures: an info-gap approach. In: Simmermacher T, Cogan S., Horta LG, Barthorpe R (eds) Topics in model validation and uncertainty quantification, vol 4. Springer, New York, pp 13–20
37. Deines K, Marinone T, Schultz R, Farinholt K, Park G (2011) Modal analysis and structural health monitoring investigation of CX-100 wind turbine blade. In: Proulx T (ed.) Rotating machinery, structural health monitoring, shock and vibration, vol 5. Springer, New York, pp 413–438
38. Farinholt K, Taylor SG, Park G, Ammerman CM (2012) Full-scale fatigue tests of CX-100 wind turbine blades: part I – testing. SPIE Proc
39. Kennedy M, O'Hagan A (2000) Predicting the output from a complex computer code when fast approximations are available. Biom 87(1):1–13
40. Higdon D, Gattiker J, Williams B, Rightley M (2008) Computer model calibration using high-dimensional output. J Am Stat Assoc 103(482):570–583

Chapter 12
Defining Coverage of a Domain Using a Modified Nearest-Neighbor Metric

Matthew C. Egeberg, Sez Atamturktur, and François M. Hemez

Abstract Validation experiments are conducted at discrete settings within the domain of interest to assess the predictive maturity of a model over the entire domain. Satisfactory model performance merely at these discrete tested settings is insufficient to ensure that the model will perform well throughout the domain, particularly at settings far from validation experiments. The goal of coverage metrics is to reveal how well a set of validation experiments represents the entire operational domain. The authors identify the criteria of an exemplary coverage metric, evaluate the ability of existing coverage metrics to fulfill each criterion, and propose a new, improved coverage metric. The proposed metric favors interpolation over extrapolation through a penalty function, causing the metric to prefer a design of validation experiments near the boundaries of the domain, while simultaneously exploring inside the domain. Furthermore, the proposed metric allows the coverage to account for uncertainty associated with validation experiments. Application of the proposed coverage metric on a practical, non-trivial problem is demonstrated on the Viscoplastic Self-Consistent material plasticity code for 5182 aluminum alloy.

Keywords Model validation • Design of experiments • Predictive maturity • Coverage • Predictive maturity index

12.1 Introduction

Numerical models are often used to make predictions at a range of settings, known as the operational domain. The inability of the model to match observations within this domain, known as discrepancy bias, is typically used to bias correct the model [1, 2]. This discrepancy bias is empirically trained from validation experiments, and is therefore dependent on the settings of the validation experiments until sufficient information is available to retrieve the true discrepancy bias. Conducting validation experiments in a limited region of the domain could result in a poorly trained discrepancy bias (Fig. 12.1), which in turn, can result in overconfidence in model predictions, represented by the dashed line in Fig. 12.1, as well as model prediction errors after bias correction, known as unforeseen forecasting errors [3]. The concept of forecasting errors is a well-known phenomenon and tends to increase when extrapolating the predictions of a model far away from where the model has been trained. To mitigate this problem, it is essential to conduct validation experiments at settings that provide a representation of the entire operational domain. A quantitative measure of the ability of a set of validation experiments to represent the entire domain is referred to as coverage.

Due to an increasing reliance on numerical models in decision-making, predictive capability metrics are needed to quantify the degree of confidence that decision-makers can place in supporting numerical models [4]. While predictive capability

M.C. Egeberg
Graduate Student, Clemson University, Clemson, SC 29634, USA
e-mail: megeber@clemson.edu

S. Atamturktur (✉)
Assistant Professor, Clemson University, Clemson, SC 29634, USA
e-mail: sez@clemson.edu

F.M. Hemez
Technical Staff Member, Los Alamos National Laboratory, Los Alamos, NM 87545, USA
e-mail: hemez@lanl.gov

T. Simmermacher et al. (eds.), *Topics in Model Validation and Uncertainty Quantification, Volume 5*,
Conference Proceedings of the Society for Experimental Mechanics Series 41,
DOI 10.1007/978-1-4614-6564-5_12,

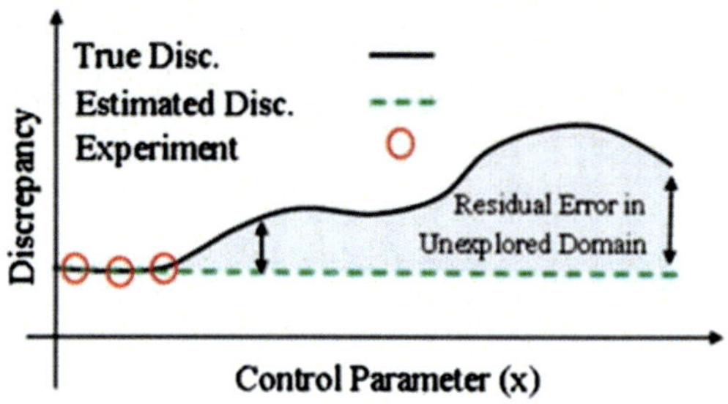

Fig. 12.1 Potential error in discrepancy estimation (Reprinted with permission from Hegenderfer [20])

is usually defined as the ability of predictions to reproduce experimental measurements, with little-to-no consideration for quality of validation experiments, the Predictive Maturity Index (PMI) developed in [4] incorporates the concept of coverage. It is the inclusion of coverage in the PMI that creates the importance of coverage metrics.

The PMI has recently been applied to the Preston-Tonks-Wallace model of plastic deformation [4], the Viscoplastic Self-Consistent (VPSC) material model [5], and the nuclear fuel performance code, LIFEIV [6]. Recently modified by Stull et al. [7], the PMI includes four attributes: coverage of the domain, discrepancy bias, robustness of predictions to uncertain model parameters, and complexity of the numerical model. As one of four components in the PMI, coverage has a major role in quantifying the predictive maturity of a model. Existing coverage metrics developed in Atamturktur et al. [8], Hemez et al. [4], and Stull et al. [7] typically provide a satisfactory definition of coverage, but each metric is subjected to limitations (discussed in detail in Sect. 12.3). With such importance placed upon coverage, it is important for coverage to be determined using the most refined definition available.

In Sect. 12.2 of this manuscript, we identify four essential criteria for coverage. In Sect. 12.3, we present three coverage metrics from the literature and investigate the ability of each metric to meet the identified criteria from Sect. 12.2. A new coverage metric is discussed in Sect. 12.4 based on the metric developed in Atamturktur et al. [8]. The proposed metric is applied to a non-trivial problem of polycrystal plasticity in Sect. 12.5. Section 12.6, provides an overview of the manuscript and suggests alternative uses of the proposed coverage metric.

12.2 Characteristics of Ideal Coverage Definition

Four criteria can be identified as essential characteristics for any coverage metric:

1. Coverage should improve if a new validation experiment is conducted at new, untested settings within the domain;
2. Poorer coverage should result from a clustered arrangement of validation experiments that limits exploration to certain regions of the domain, than an equal number of validation experiments spread more evenly throughout the domain;
3. Coverage should distinguish between interpolation and extrapolation, due to the lack of clear bounds for extrapolation (see discussion below);
4. Coverage should be objective, not subjective.

The first criterion is based on the assumption that conducting new validation experiments at untested settings provides additional information for calibration and validation, leading to a greater predictive maturity. If a validation experiment has already been conducted at that setting, then a repeated validation experiment should provide no additional coverage, provided that the experimental uncertainty is not reduced.

The second criterion is focused on even distribution of validation experiments as suggested by maximin, distance-based experimental designs [9]. Williams et al. [10] find that design strategies that spread points throughout the domain, particularly in input dimensions which have significant influence on the output of the model, result in lower average prediction errors. This criterion therefore attempts to incorporate the benefits of space-filling designs into the coverage metric.

Although evenly distributed experimental designs result in predictions with lower *average* prediction errors, Williams et al. [10] also find that experimental design strategies that concentrate runs near the boundaries of the domain tend to control the *maximum* prediction error better. This placement of validation experiments at the boundaries of the domain, similar to an entropy-based experimental design as described in [11], is the focus of the third criterion. While in the presence of systematic error, a good experimental design tends to distribute experiments throughout the domain, experiments located near the boundary of the domain are favorable in the presence of random error [11].

In the context of this work, "interpolation" refers to all predictions made within the region of validation experiments defined by a convex hull, and "extrapolation" refers only to predictions made outside the corresponding range of validation experiments. Under this definition, it is assumed that the mechanics or physics do not change within the region of validation experiments relative to those captured by the model. While interpolation is bounded, predictions made using extrapolation lack clear bounds. Montgomery [12] warns that it is possible for a model to provide poor predictions outside of the region of the available data even though the model may fit the observations well. Furthermore, experts assert that empirical models should not be used outside the range of calibration experiments [13–15]. The Challenger space shuttle disaster (which was caused by an O-ring failure) on January 28th, 1986, discussed in [16], exposes the dangers of extrapolation. Although there was concern about the effect of low temperatures on O-ring performance and historical data from previous shuttle flights was only available at temperature above 53°F, experts failed to recognize the true effect of temperature and decided to launch the shuttle as scheduled, at a temperature of 31°F.

The fourth criterion is obvious; a coverage based upon hard evidence should be more credible and reliable than one based on an individual's opinion. Herein, the quality of any coverage metric is measured by the ability to satisfy these four criteria.

12.3 Earlier Definitions of Coverage

This section reviews and compares coverage metrics defined earlier in published literature.

12.3.1 Atamturktur et al. [8]

Coverage is determined in Atamturktur et al. [8] using a sensitivity adjusted nearest-neighbor metric, in which control parameter ranges that define the domain of interest are normalized between 0 and 1. Next, each control parameter dimension is scaled according to the sensitivity of the model output. The degree of influence of each control parameter on the model output is determined through a sensitivity analysis. Atamturktur et al. [8] exploits the correlation length of the Gaussian Process Model (GPM), an emulator trained to replace the numerical model. The correlation length, which assumes a value between 0 and 1, is used to approximate the sensitivity of each control parameter, where a greater sensitivity causes the control parameter dimension to dilate, placing focus on more sensitive model inputs. This concept is analogous to the V-optimal design of experiments [17].

After scaling the axes, the domain is covered by uniformly distributed grid points and each grid point is appointed to the nearest validation experiment. Figure 12.2a shows the partitioning of the domain into nearest-neighbor regions. The distance between each grid point and the associated nearest validation experiment is summed for all grid points and normalized by the total number of grid points, as shown in Eq. (12.1).

$$\eta_c = \frac{1}{g} \sum_{i=1}^{g} \min(d_{E,i}) \tag{12.1}$$

Parameter η_c represents the coverage, g represents the total number of grid points, and $min(d_{E,i})$ is the minimum distance of the ith grid point to the nearest validation experiment. The result is a value that represents the average normalized and sensitivity-scaled distance between each point in the domain to the corresponding nearest validation experiment. Decreasing this value improves the coverage.

The nearest-neighbor metric of Atamturktur et al. [8] is sensitive to the addition of a new validation experiment as well as the clustering of validation experiments. The metric is also objective. However, the metric is incapable of showing preference to interpolation over extrapolation, as shown in Fig. 12.2b. Both grid point A and B are an equal distance from the nearest neighboring validation experiment; while point A involves an interpolative prediction, point B involves extrapolative prediction as it is located outside the region bounded by the dotted line. Furthermore, the nearest-neighbor metric supplies a value with unclear physical meaning, whereas the coverage defined using the methods presented in Hemez et al. [4] and Stull et al. [7], as discussed in the following sections, are physically meaningful and intuitive.

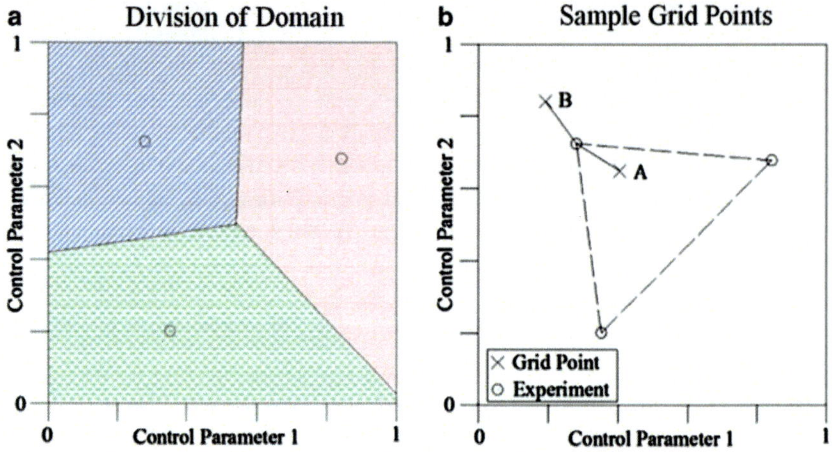

Fig. 12.2 Division of domain into nearest-neighbor regions

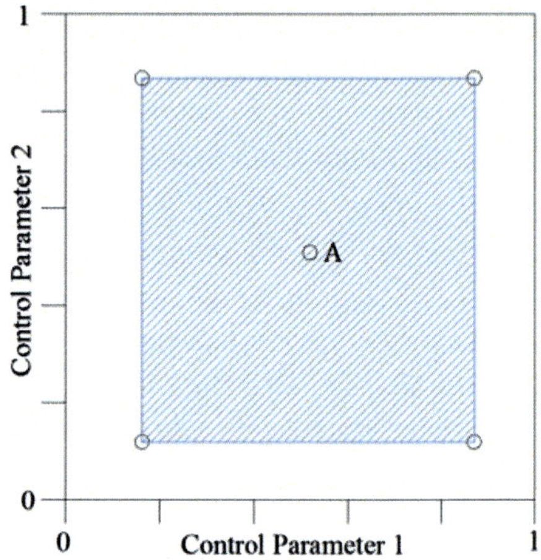

Fig. 12.3 Convex hull encompassing validation experiments

12.3.2 Hemez et al. [4]

In Hemez et al. [4], coverage is defined as the ratio of 2 multidimensional volumes defined by convex hulls, multi-dimensional domains with the smallest convex volume including all the points of interest. One convex hull encompasses the validation experiments while the other embodies the entire domain of interest. This is shown as:

$$\eta_c = \frac{Volume(\Omega_{CH})}{Volume(\Omega_V)} \tag{12.2}$$

where *Volume*($\cdot$) calculates the volume of a convex hull. Ω_{CH} is the convex hull that surrounds the validation experiments while Ω_V is the convex hull that includes the entire domain. The metric proposed by Hemez et al. [4] has a profound ability to show the distinction between interpolation and extrapolation. Moreover, the metric is objective. This metric however, is controlled by the positioning of the experiments at the boundaries of the domain, where the addition of experiments within the convex hull fails to reflect improvement in the coverage, as shown with experiment A in Fig. 12.3.

12.3.3 Stull et al. [7]

The metric defined in Stull et al. [7] creates a convex hull around each individual validation experiment rather than a single convex hull containing every experiment. The coverage is then defined as the ratio of the summation of the convex hulls surrounding the validation experiments to the convex hull defining the domain. This is defined as:

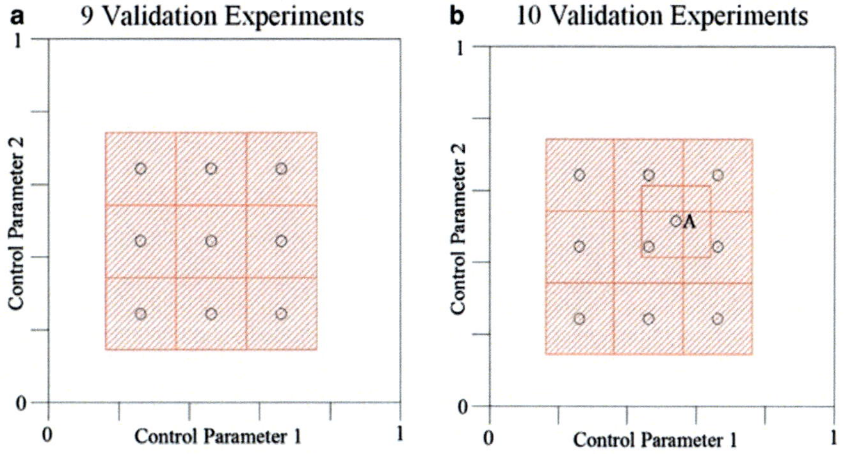

Fig. 12.4 Possible effect of adding validation experiments on coverage metric proposed by Stull et al. [7]

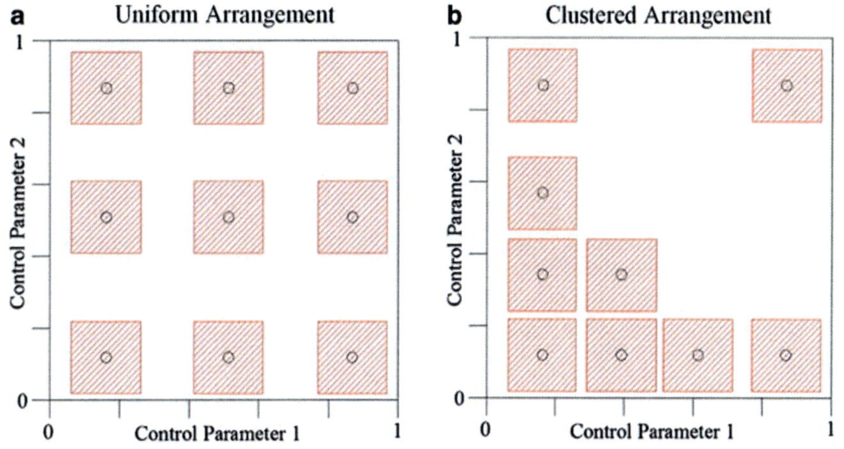

Fig. 12.5 Coverage of clustered versus uniform arrangement of validation experiments

$$\eta_c = \frac{\sum_{i=1}^{M} Volume(\Omega_{E,i})}{Volume(\Omega_V)} \tag{12.3}$$

where $\Omega_{E,i}$is the convex hull surrounding the *i*th validation experiment and M is the total number of validation experiments. If the convex hulls from more than one validation experiment occupy the same area, the area is only counted once.

The metric proposed by Stull et al. [7] is subjective as the size of the convex hull surrounding each validation experiment is based on expert opinion.[1] Furthermore, with this metric, a validation experiment could be added without improving the coverage, if the existing convex hulls completely engulf the convex hull of an additional validation experiment, as shown in Fig. 12.4.

Stull et al.'s [7] metric neither recognizes large unexplored regions in the domain nor differentiates between interpolation and extrapolation as the validation experiments could be clustered in one region of the domain and achieve the same coverage as a more distributed arrangement if there is no overlap of the convex hulls, as shown in Figs. 12.5 and 12.6.

The metric proposed by Stull et al. [7] should be given more credit than Figs. 12.4, 12.5, and 12.6 suggest, which present carefully-chosen, problematic situations for this metric. Using expert opinion to vary the size of the convex hull associated with each individual validation experiment may alleviate some limitations and provide an improved quantification of coverage. However, doing so forces the metric to rely heavily on experimental opinion and increases subjectivity.

The discussion presented in this section is summarized in Table 12.1. Note that each metric fails at least one criterion but each criterion is passed by at least one metric.

[1] A more objective criterion could also be used, where the size of each convex hull surrounding a validation experiment is based on a sensitivity analysis of the model; therefore, if the model predictions change rapidly around a point, then the size is made smaller.

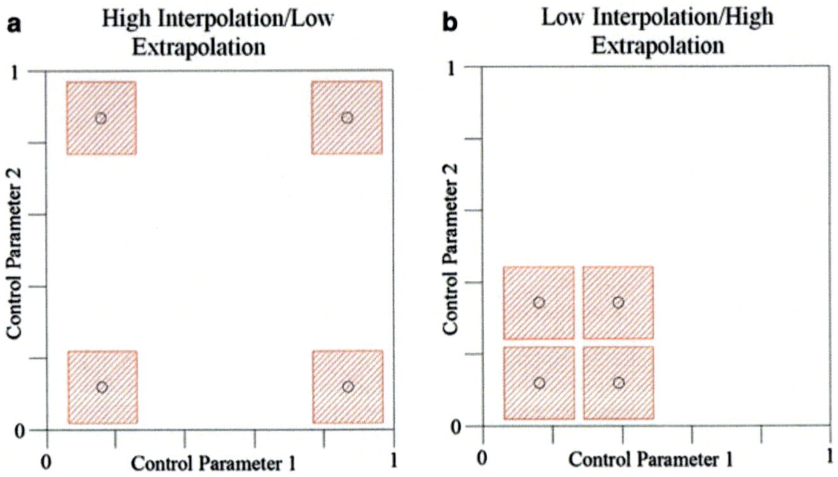

Fig. 12.6 Effect of interpolation/extrapolation ratio on coverage

Table 12.1 Criterion satisfaction for Atamturktur et al. [8], Hemez et al. [4], and Stull et al. [7]

Criterion	Atamturktur et al. [8]	Hemez et al. [4]	Stull et al. [7]
1	Pass	Fail	Improved but imperfect
2	Pass	Fail	Improved but imperfect
3	Fail	Pass	Fail
4	Pass	Pass	Fail

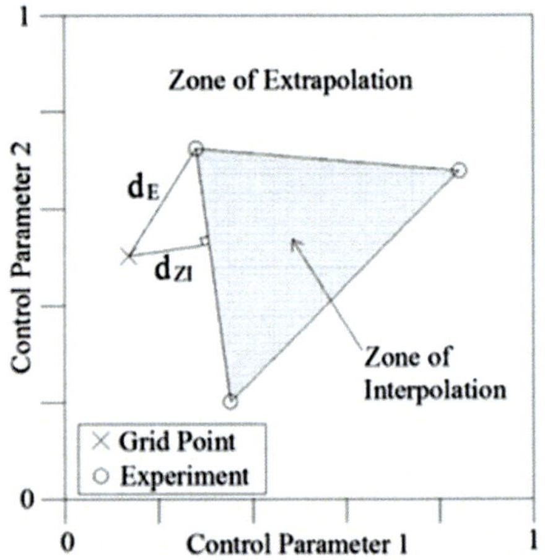

Fig. 12.7 Example zone of interpolation and extrapolation for a two dimensional domain

12.4 Proposed Coverage Definition

Hemez et al. [18] suggests that better definitions of coverage could be developed and applied to the PMI to account for the number, overall spread, and resolution of validation experiments. We attempt to account for the number and overall spread of the validation experiments by incorporating the existing sensitivity-adjusted nearest-neighbor metric developed in Atamturktur et al. [8] on the PMI and adding an extrapolation penalty. Moreover, we account for experimental uncertainty; validation experiments with a low level of variability would provide a more favorable coverage than validation experiments with a high level of variability.

12.4.1 Penalizing Extrapolation

The nearest-neighbor metric is modified to account for the dangers of extrapolation discussed in Sect. 12.2. A convex hull encompassing the validation experiments is defined as in Hemez et al. [4], creating a zone of interpolation as well as a zone of extrapolation, as shown in Fig. 12.7, where grid points (used to sample the operational domain) that lie outside the zone of interpolation are subject to an extrapolation penalty equal to the minimum distance between the grid point and the zone of

Table 12.2 PMI term definitions [7]

Term	Definition
Ψ_1	$\left\{\begin{matrix} \tanh(\gamma_1 \times \eta_c) \\ 1 \end{matrix}\right\} \begin{matrix} \eta_c < 1 \\ \eta_c \geq 1 \end{matrix}$
Ψ_2	$\tanh\left(\left(\frac{N_R}{N_K}\right)^{\gamma_2}\right)$
Ψ_3	$(1-\delta_S)^{\gamma_3}$
Ψ_4	$\left[1-\tanh\left(\frac{\gamma_4}{\alpha_S}\right)\right]$
Ψ_5	$e^{-\left[e^{-\eta_c} \times \delta_S \times e^{-\alpha_S}\right]}$

interpolation. This penalty is added to the distance between the grid point and the nearest validation experiment, as shown in Eq. 12.4:

$$\eta_c = \frac{1}{g}\sum_{i=1}^{g} \min(d_{E,i}) + d_{ZI,i} \tag{12.4}$$

where, $d_{ZI,i}$ is minimum distance between the *i*th grid point and the zone of interpolation. Distances d_E and d_{ZI} are shown in Fig. 12.7. Applying this extrapolation penalty increases η_c and thus, reduces coverage. Through this penalty, validation experiments are encouraged to be positioned nearer the boundaries of the domain, reducing the zone of extrapolation.

12.4.2 Accounting for Experimental Uncertainty

It is intuitive that a precise validation experiment with a low level of experimental uncertainty provides greater information than a validation experiment with a high level of experimental uncertainty. Therefore, a coverage metric that attempts to quantify the quality of a set of validation experiments should incorporate experimental variability. We propose a simple method to account for experimental uncertainty in the nearest-neighbor metric. The experimental coefficient of variation (standard deviation normalized with mean) is determined for each validation experiment and expressed as a decimal percentage. Next, the distance between a grid point and a validation experiment is multiplied by the coefficient of variation and added to the corresponding distance calculation. This is expressed as:

$$\eta_c = \frac{1}{g}\sum_{i=1}^{g} \min((1+U_E) \times d_{E,i}) + d_{ZI,i} \tag{12.5}$$

where, U_E is the coefficient of variation of validation experiment, E. Note that higher experimental variability increases η_c reducing coverage.

Under this definition, each grid point is classified with the validation experiment that minimizes the sum of the distance between the grid point and validation experiment with the additional uncertainty penalty. Therefore, a grid point may be classified with a validation experiment, which is distance-wise further away than another validation experiment, but has a lower level of experimental uncertainty.

12.4.3 Application to Predictive Maturity Index (PMI)

The PMI incorporates coverage, η_c, robustness, α_S, scaled discrepancy bias, δ_S, and complexity, N_K, under the following definition defined in [7]:

$$PMI(\eta_C; N_K; \delta_S; \alpha_S) = \prod_{i=1}^{5} \Psi_i \tag{12.6}$$

where ψ_i are shown in Table 12.2, with positive, user-defined coefficients γ_1, γ_2, γ_3, and γ_4. The purpose of the gamma values is to weigh the effect of each attribute on the PMI. Each attribute is bounded between 0 and 1 so the PMI is naturally bounded between 0 and 1.

In the current form of the PMI, improved coverage is indicated by a higher metric value. However, in the proposed coverage metric, a lower value indicates improved coverage. Therefore, the functional form of coverage in the PMI must be modified. From Table 12.2, coverage is included in two of the five terms in the PMI calculation: $\Psi_1$and Ψ_5. We propose the following definition for the Ψ_1 term to closely match the behavior of the existing definition:

$$\Psi_1 = e^{-\gamma_1(\eta_c)} \tag{12.7}$$

Using this definition, the Ψ_1 term is bounded between 0 and 1 with a higher value corresponding to a better coverage of the domain and therefore, an improved predictive maturity. The term asymptotically approaches a value of 1 as the coverage improves, preserving the asymptotic properties of the PMI.

The proposed coverage metric is implemented in the Ψ_5 term exactly as the discrepancy attribute, where a lower value indicates greater predictive maturity (Eq. 12.8). The Ψ_5 term ensures that predictive maturity asymptotes to 1, or "perfect" maturity, only if discrepancy and coverage both tend to zero. Hence, merely achieving "perfect" coverage (as $\eta_c \to 0$) or "perfect" predictions (as $\delta \to 0$) does not yield a PMI value equal to 1, both must occur simultaneously.

$$\Psi_5 = e^{-[\eta_c \times \delta_S \times e^{-\alpha_S}]} \tag{12.8}$$

This definition of the Ψ_5 term aligns well with the current definition and produces a greater value when the coverage is improved. All other terms of the PMI are unchanged.

12.5 Application to Viscoplastic Self-Consistent (VPSC) Code

The proposed coverage metric is implemented on the Viscoplastic Self-Consistent (VPSC) code for modeling stress-strain response and textural evolution of 5182 aluminum alloy. The VPSC code developed in [19] predicts plastic deformations considering both climb and glide dislocation at the single-crystal level. The multivariate C&G model has 2 control parameters (temperature and strain rate), 12 calibration parameters, and 3 outputs (stress-strain response, and textural evolution at two faces of the material). In [5], the 12 calibration parameters are calibrated against physical validation experiments measuring stress at a strain equal to 0.6, textural evolution of the 001 pole, and textural evolution of the 101 pole. The 001 and 101 poles represent orientations and are used to describe the structure of a crystal. In [20], these calibrated values are considered to be "true" values to allow for a simulated Batch Sequential Design (BSD) study.

12.5.1 Batch Sequential Design (BSD)

In BSD, information from available experiments is used to select the optimum settings of future experiments in batches of user-selected sizes. The proposed coverage metric is deployed on the experiments selected in [20] using the Expected Improvement for Predictive Stability (EIPS) criterion. The EIPS criterion is an index-based criterion that stabilizes the discrepancy bias using the Kullback-Leibler distance between the current and future distributions of discrepancy. The initial experimental settings (batch 0) and BSD selected settings are shown in Fig 12.8. A GPM for the discrepancy bias is trained using the data available from the initial experimental settings. Next, BSD is deployed to determine the optimal locations of experiments in the first batch. These experiments are simulated by running the VPSC code using the settings of control parameters (temperature and strain-rate) selected by BSD and the exact values of the calibration parameters determined in [5]. With the addition of new experimental data, the model is recalibrated and the process is repeated until completion of the tenth batch.

12.5.2 Results and Discussion

The operational domain is defined by temperatures between 200 and 550°C and strain-rates between 0.001 and 1 s^{-1}. The experimental variation is set to 5% for all experiments. Since the VPSC code is multivariate, the metric value is determined for stress-strain response, texture 001 evolution, and texture 101 evolution separately and also for the overall model by averaging the coverage of each output. Each output has a unique sensitivity to the inputs, so the axis scaling is non-uniform between outputs. The sensitivity of each control parameter to the model output is determined by the spatial dependence parameter, β, of the model output. The β parameter, trained using a GPM, describes the dependence of the output on each particular input;

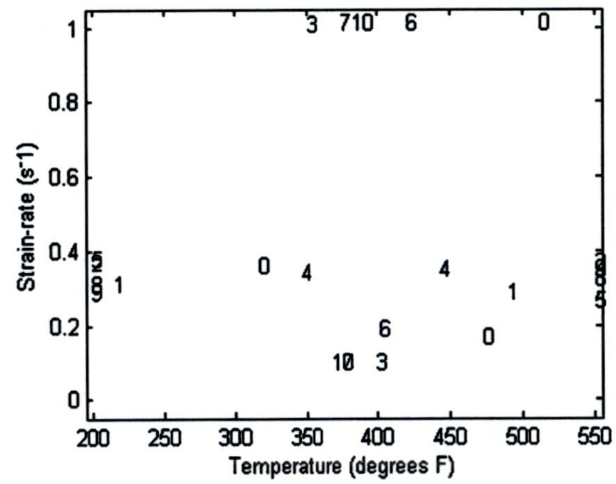

Fig. 12.8 Experimental settings selected through BSD (marker number denotes batch number)

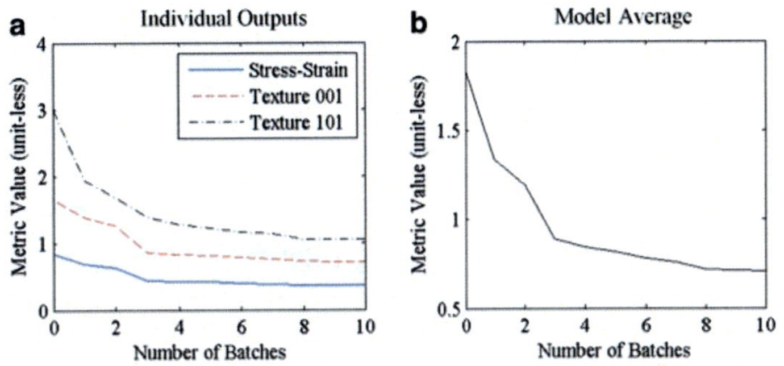

Fig. 12.9 Coverage versus number of batches

therefore, a control parameter with greater influence on the output will yield a larger β value than a control parameter with less influence. This sensitivity is determined after the tenth batch, when the maximum information is available. The metric value after each batch is shown in Fig. 12.9.

The coverage of each individual output as well as the overall model monotonically improves. Additionally, it becomes apparent that the overall metric value converges to a value around 0.7, and the experiments added in batches 4–10 insignificantly improve the coverage as compared to batches 1–3. This is due to the fact that most experiments added after the third batch are located near existing experiments (see clustering in Fig. 12.8). By doing this analysis, one may better understand the extent to which adding experiments beyond the first three batches would not significantly improve the coverage. This analysis may, therefore, avoid wasting valuable resources.

12.6 Conclusion

The proposed coverage metric is designed around four criteria established in Sect. 12.2: (i) coverage should improve if a new validation experiment is conducted at new, untested settings within the domain, (ii) poorer coverage should result from a clustered arrangement of validation experiments that limits exploration to certain regions of the domain, than an equal number of validation experiments spread more evenly throughout the domain, (iii) coverage should distinguish between interpolation and extrapolation, due to the lack of clear bounds for extrapolation, and (iv) coverage should be objective, not subjective. This manuscript modifies the sensitivity adjusted nearest neighbor metric developed in Atamturktur et al. [8] to encourage experimental designs with validation experiments nearer the boundaries of the domain, thus reducing extrapolation. The authors propose a method to incorporate experimental uncertainty, thereby accounting for the quality of each validation experiment. The authors also propose a new functional form of the Predictive Maturity Index (PMI) to implement the proposed coverage metric. The proposed coverage metric is demonstrated on the multivariate Viscoplastic Self-Consistent (VPSC) code.

The usefulness of the proposed coverage metric goes beyond implementation in the PMI. The metric can be used to directly compare multiple designs of experiments with varying experimental uncertainties. Furthermore, the metric could be implemented as a Batch Sequential Design (BSD) selection criterion to select the future settings of validation experiments.

As a distance-based criterion, the metric could be combined with an index-based criterion to create a selection criterion, similar to the Coverage Augmented Expected Improvement for Predictive Stability (C-EIPS) criterion developed in [20], which simultaneously explores the entire domain and exploits regions with high variance in the discrepancy bias.

Future work includes applying the proposed coverage metric to additional numerical models. Also, the metric must be applied to high-dimension models with an investigation of the computational demands of evaluating the metric at a high dimension and sufficient grid size.

Acknowledgements The authors would like to thank Godfrey Kimball for his editorial review of this manuscript.

References

1. Higdon D, Gattiker J, Williams B, Rightley M (2008) Computer model calibration using high-dimensional output. J Am Stat Assoc 103(482):570–83
2. Higdon D, Nakhleh C, Gattiker J, Williams B (2008) A Bayesian calibration approach to the thermal problem. Comput Method Appl Mech Eng 197(29–32):2431–2441
3. Atamturktur S, Hemez F, Williams B, Tome C, Unal C (2011) A forecasting metric for predictive modeling. Comput Struct 89(23, 24):2377–2387
4. Hemez F, Atamturktur S, Unal C (2010) Defining predictive maturity for validated numerical simulations. Comput Struct J 88:497–505
5. Atamturktur S, Hegenderfer J, Williams B, Egeberg M, Lebensohn R, Unal C (2013) A resource allocation framework or experiment-based validation of numerical models. Mech Adv Mater Struc (in print)
6. Stull C, Williams B, Unal C (2012) Assessing the predictive capability of the LIFEIV nuclear fuel performance code using sequential calibration. Los Alamos National Laboratory technical report, LA-UR-12-22712
7. Stull CJ, Hemez F, Williams BJ, Unal C, Rogers ML (2011) An improved description of predictive maturity for verification and validation activities. Los Alamos National Laboratory technical report, LA-UR-11-05659
8. Atamturktur S, Hemez F, Unal C, William B (2009) Predictive maturity of computer models using functional and multivariate output. In: Proceedings of the 27th SEM international modal analysis conference, Orlando
9. Johnson ME, Moore LM, Ylvisaker D (1990) Minimax and maximin distance designs. J Stat Plan Inference 26(2):131–148
10. Williams BJ, Loeppky JK, Moore LM, Macklem MS (2011) Batch sequential design to achieve predictive maturity with calibrated computer models. Reliab Eng Syst Saf 96:1208–1219
11. Sacks J, Welch W, Mitchell T, Wynn H (1989) Designs and analysis of computer experiments. Stat Sci 4:409–435
12. Montgomery DC (1997) Design and analysis of experiments, 5th edn. Wiley, New York, p 416
13. Fryer RJ, Shepherd JG (1996) Models of codend size selection. J Northwest Atl Fish Sci 19:51–58
14. Logan RW, Nitta CK, Chidester SK (2003) Risk reduction as the product of model assessed reliability, confidence, and consequence. Technical report UCRL-AR-200703, Lawrence Livermore National Laboratory
15. Oberkampf WL, Pilch M, Trucano TG (2007) Predictive capability maturity model for computational modeling and simulation. Sandia National Laboratory technical report, SAND-2007-5948
16. Draper D (1995) Assessment and propagation of model uncertainty. J R Stat Soc B 57(1):45–97
17. Thompson D, McAuley K, McLellan P (2010) Design of optimal sequential experiments to improve model predictions from a polyethylene molecular weight distribution model. Macromol React Eng 4(1):73–85
18. Hemez F, Atamturktur S, Unal C (2009) Defining predictive maturity for validated numerical simulations. In: Proceedings of the IMAC-XXVII, Orlando, 9–12 Feb 2009
19. Lebensohn RA, Hartley CS, Tomé CN, Castelnau O (2010) Modeling the mechanical response of polycrystals deforming by climb and glide. Philos Mag 90(5):567–83
20. Hegenderfer J (2012) Resource allocation framework: validation of numerical models of complex engineering systems against physical experiments. PhD dissertation, Clemson University

Chapter 13
Orthogonality for Modal Vector Correlation: The Effects of Removing Degrees-of-Freedom

Michael L. Mains

Abstract This paper reviews the weighted orthogonality property of modal vectors, the many Test-Analysis Model (TAM) transforms that have been developed to reduce Finite Element Model (FEM) based mass matrices and the Modal Assurance Criterion (MAC). These associated technologies have all been developed to try and correlate the FEM and experimentally obtained mode shapes. A case study is presented where the Effective Independence method for Degree-of-Freedom (DOF) selection was used to systematically reduce the DOF's of the FEM and understand the effects of DOF reduction on MAC, the Guyan TAM and the System Equivalent Reduction/Expansion Process (SEREP) TAM.

Keywords Modal correlation orthogonality TAM DOF

Nomenclature

$[Q]$	Orthogonal matrix
$[I]$	Identity matrix
λ_r	Natural frequency of mode r
$\{\phi_r\}$	Mode shape r
$[M]$	Mass matrix
[K]	Stiffness matrix
$[M_r]$	Diagonal matrix modal mass
$[K_r]$	Diagonal matrix modal stiffness
$\begin{Bmatrix} x_a \\ x_o \end{Bmatrix}$	Displacement vector at the full FEM DOF's
$\{x_a\}$	Displacement vector at the a-set DOF's
m	Dimension of the full FEM DOF's
a	Dimension of the matching a-set FEM and test DOF's

M.L. Mains (✉)
Brüel and Kjær, North America, 8180 Corporate Park Drive, Suite 310, Cincinnati, OH 45242, USA
e-mail: mike.mains@bksv.com

T. Simmermacher et al. (eds.), *Topics in Model Validation and Uncertainty Quantification, Volume 5*,
Conference Proceedings of the Society for Experimental Mechanics Series 41,
DOI 10.1007/978-1-4614-6564-5_13,

o Dimension of the omitted FEM DOF's (o-set)

N Number of target modes

$[T]$ mxa TAM transformation matrix

$[M_{red}]$ Reduced mass matrix

$[AOC]$ Auto-orthogonality check

$[XOC]$ Cross-orthogonality check

$[E]$ Matrix of experimentally obtained mode shapes

$[\Phi]$ Matrix of analytically obtained mode shapes

$[\Phi_a]$ a-set partition of the matrix of analytically obtained mode shapes

$[\Phi_o]$ o-set partition of the matrix of analytically obtained mode shapes

$\{\hat{e}\}$ Vector of an expanded test mode shape

$\{e\}$ Vector of a test mode shape

$[\widehat{E}]$ Matrix of expanded test mode shapes

$[E]$ Matrix of test mode shapes

$[K_{oo}]$ o-set matrix partition of the stiffness matrix

K_{aa} a-set matrix partition of the stiffness matrix

K_{oa} oxa-set matrix partition of the stiffness matrix

K_{ao} axo-set matrix partition of the stiffness matrix

M_{oo} o-set matrix partition of the mass matrix

M_{aa} a-set matrix partition of the mass matrix

M_{oa} oxa-set matrix partition of the mass matrix

$\{q\}$ Vector of modal coordinates

$[P]$ Projector matrix

13.1 Introduction

The concept of using the weighted orthogonality property to determine correlation between FEM and experimentally obtained modes shapes began in the 1960's when Guyan [1] and Irons [2] applied the static reduction to the dynamics problem. While this method has no error when used for developing a reduced stiffness matrix for linear and non-linear statics problems it has been shown to introduce error when used to reduce mass and stiffness matrices for computing mode shapes and natural frequencies. O'Callahan [3] introduced the Improved Reduced System (IRS) in 1989 to try and improve on the accuracy of Guyan reduction. Freed [4] showed that both Guyan and IRS introduce error in the frequencies of the calculated mode shapes and that IRS was indeed an improvement over Guyan.

Allemang and Brown [5] approached the problem of modal vector correlation by applying the ordinary coherence calculation associated with Frequency Response Function measurements to Modal Vectors. Their development of the Modal Assurance Criterion (MAC) has many uses in modal analysis and many similar criteria have been developed to better understand and compare the many vectors that are associated with structural dynamics [6] The primary advantage of their technique for FEM and Test Modal Vector Correlation is that a reduced mass matrix from the FEM is not required.

Once computer horsepower and algorithms such as Lanczos [7] and Simultaneous Vector Iteration(SVI) [8] allowed accurate and fast extraction of Mode Shapes from Larger FEM's, Kammer [9] and O'Callahan [10] developed similar methods that used a subset of modes to perform a model reduction that created minimal error in the mode shapes and natural frequencies of the reduced model. The main difference between Kammer's Modal Test-Analysis Model (TAM) and O'Callahan's System Equivalent Reduction/Expansion Process (SEREP) TAM can be viewed in the partition of the a-set rows of the TAM transformation matrix. In Kammer's Modal TAM the a-set partition is enforced to be identity while this may not the case with SEREP. A simple way to think about this is that when the Modal TAM is used for modal vector expansion, the a-set Degrees-of-Freedom (DOF's) of the expanded modes are unmodified, whereas with the SEREP TAM the a-set DOF's of the expanded modes are smoothed. One issue with both of these methods is that the resulting reduced mass and stiffness matrices are rank deficient and only the modes included in the analysis were extracted. Since the a-set DOF's are often larger than the number of modes that were extracted, this was seen as a deficiency. Kammer [11] developed the Hybrid TAM which combined the Modal and Guyan TAM to overcome this obstacle. Mains [12, 13] observed the different combinations of Guyan, IRS, Modal, SEREP and projector matrices could be combined to form eight different Hybrid TAM's each resulting in slightly different results for orthogonality comparisons. Most recently a new TAM was developed by Marinone [14] that provides reduced matrices of full rank similar to what a Hybrid TAM using SEREP and Guyan or IRS might provide.

Freed [3] introduced the concept of the TAM which is the full FEM DOF's (denoted by m) by the Test DOF's (denoted by the subscript a or commonly referred to as the "a-set") transform matrix used by all the aforementioned methodologies (except MAC) to reduce the mass matrix of the FEM to the a-set dimensions of the test. This transform can also be used to expand experimentally derived mode shapes or operating data to the full FEM shapes. The TAM concept is a simple way express and to see conceptually the similarities and differences amongst the various methods.

Because all of the above methods for vector correlation are so sensitive a-set DOF selection or sensor placement, much research and development has been focused on pre-test Analysis sensor selection [15–17]. Much of pre-test Analysis focuses on using the FEM target modes for developing the best a-set DOF's and evaluating orthogonality using the various TAM's with respect to the reduced FEM modes [3,18–20]. Also, the Coordinate Orthogonality check (CORTHOG) was developed in order to try and understand which DOF's were contributing to the poor diagonal and off-diagonal terms or the orthogonality checks [19].

Much current research work, known as VIKING [21–23], still uses the SEREP TAM but avoids any attempt at correlation of FEM and Test altogether. Instead the TAM, using carefully selected test DOF's (avoiding non-nodal or noisy DOF's) and Test data are used augment the FEM. The test augmented FEM is then used to try and calculate full field stress and strain distributions, damage detection, Frequency Response Functions (FRF's) at rotational or unmeasured DOF's, etc.

13.2 Weighted Orthogonality of Modal Vectors

The word orthogonal comes from the Greek (orthos), meaning "straight", and (gonia), meaning "angle". In elementary geometry, two vectors are orthogonal if they are perpendicular; their dot product is zero; if they form a right angle [24]. In some parameter estimation algorithms we use "orthogonal polynomials" to improve the conditioning of the numerics. In linear algebra we have, an orthogonal matrix, which is a square matrix with real entries whose columns and rows are orthogonal unit vectors and the transpose of the matrix times itself, is an identity matrix:

$$[Q]^T[Q] = [Q][Q]^T = [I] \tag{13.1}$$

Thus the transpose of the matrix is also the inverse:

$$[Q]^T = [Q]^{-1} \tag{13.2}$$

In Modal Analysis we have the undamped equations of motion described in an eigenvalue problem of the form:

$$-\lambda_r^2[M]\{\phi_r\} = [K]\{\phi_r\} \tag{13.3}$$

The eigenvectors or mode shapes resulting from the eigenvalue solution to this problem posses a weighted orthogonality of the form:

$$[\Phi]^T[M][\Phi] = [M_r] = \text{Modal Mass} \tag{13.4}$$

$$[\Phi]^T[K][\Phi] = \left[\lambda_r^2 M_r\right] = [K_r] = \text{Modal Stiffness} \tag{13.4a}$$

When the modal vectors are scaled to unity modal mass Eqs. (13.4) and (13.4a) become:

$$[\Phi]^T[M][\Phi] = [I] \tag{13.5}$$

$$[\Phi]^T[K][\Phi] = \left[\lambda_r^2\right] \tag{13.5a}$$

While the weighted orthogonality of modal vectors can be determined by Eqs. (13.4), (13.4a), (13.5) or (13.5a), it is Eq. (13.5) that is most typically used for orthogonality checks. This is done since the diagonals are conveniently scaled to one and the fact that most FEM codes scale their mode shapes to Eq. (13.5). For experimentally obtained modal vectors it is not possible to obtain a mass matrix but it would be convenient to somehow use the FEM mass matrix to validate the FEM and the experimentally obtained modal vectors, thus motivating the idea of the TAM.

13.3 Test Analysis Models (TAM)

The basic idea behind the TAM is to develop a transform between the full FEM m-set and the Experimental Modal Model (EMM) a-set space. This can be written as:

$$\left\{\begin{matrix} x_a \\ x_o \end{matrix}\right\}_{mx1} = [T]_{mxa}\{x_a\}_{ax1} \tag{13.6}$$

where $[T]_{mxa}$ is a TAM transformation matrix which maps the EMM a-set space to the full FEM m-set space.

Mass matrix reduction is accomplished by pre-multiplying the FEM mass matrix by the transpose of the TAM transformation matrix and post-multiplying by the TAM matrix as follows:

$$[M_{red}]_{axa} = [T]_{axm}^T[M]_{mxm}[T]_{mxa} \tag{13.7}$$

The Auto-Orthogonality check (AOC) often referred to as the Pseudo-Orthogonality (POC) is the triple product of the experimentally obtained mode shapes transposed, multiplied by the reduced mass matrix, multiplied by the experimentally obtained mode shapes:

$$[AOC]_{NxN} = [E]_{Nxa}^T[M_{red}]_{axa}[E]_{axN} \tag{13.8}$$

To check the error introduced by the TAM reduction, the AOC can use the a-set partition of the analytically obtained mode shapes:

$$[AOC]_{NxN} = [\phi_a]_{Nxa}^T[M_{red}]_{axa}[\phi_a]_{axN} \tag{13.8a}$$

The Cross-Orthogonality (XOC) is the triple product of the experimentally obtained mode shapes transposed, multiplied by the reduced mass matrix, multiplied by the a-set partition of the analytically obtained mode shapes:

$$[XOC]_{NxN} = [E]_{Nxa}^T[M_{red}]_{axa}[E]_{axN} \tag{13.9}$$

Both AOC and XOC assume that the FEM and Test modes are scaled to unity modal mass. For perfect correlation between FEM and test models one would hope that the AOC and XOC matrices would be identity, in reality that is never the case. The industry standard for acceptable correlation is less the 0.10 on the off-diagonals and between 0.9 and 1.1 on the diagonal or correlated modes.

Alternatively, modal vector expansion (typically expanding experimental mode shapes or operating shapes) to the FEM space is accomplished by simply applying Eq. (13.6) to the shapes of interest:

$$\{\hat{e}\}_{mx1} = [T]_{mxa}\{e\}_{ax1} \tag{13.10}$$

$$\left[\hat{E}\right]_{mxn} = [T]_{mxa}[E]_{axn} \tag{13.11}$$

While AOC and XOC could be computed using the expanded test modes $[\hat{E}]$ with the full FEM modes the computation is more expensive and the results can be shown to be equivalent.

13.3.1 Guyan

The Guyan Reduction is the static superelement analysis applied to the dynamics problem. The Guyan Reduction was historically used to reduce to the size of the FEM eigensolution. There are now better methods to perform large eigensolutions (Lanczos [7], SVI [8], etc.). Guyan Reduction is still probably the most common way to reduce the FEM mass matrix. It is typically required for Airplane and Satellite Ground Vibration Tests (GVT's) and it is easy to perform in FEA software. To see how Guyan Reduction is derived let's start with the linear statics problem partitioned into our a-set DOF's and the omitted or o-set DOF's:

$$\begin{bmatrix} K_{oo} & K_{oa} \\ K_{ao} & K_{aa} \end{bmatrix}_{mxm} \begin{Bmatrix} x_o \\ x_a \end{Bmatrix}_{mx1} = \begin{Bmatrix} f_o \\ f_a \end{Bmatrix}_{mx1} \tag{13.12}$$

Since the o-set are internal, it is assumed that no external forces are applied, since we can set f_o to zero we can develop an expression for x_o in terms of x_a:

$$[K_{oo}]\{x_o\} + [K_{oa}]\{x_a\} = \{0\} \tag{13.13}$$

$$\{x_o\} = [K_{oo}]^{-1}[K_{oa}]\{x_a\} \tag{13.14}$$

Referring back to Eq. (13.6), the Guyan TAM can be written as:

$$\begin{Bmatrix} x_a \\ x_o \end{Bmatrix}_{mx1} = \begin{bmatrix} [I] \\ [K_{oo}]^{-1}[K_{oa}] \end{bmatrix}_{mxa} \{x_a\}_{ax1} \tag{13.15}$$

where:

$$[T_S] = \begin{bmatrix} [I] \\ [K_{oo}]^{-1}[K_{oa}] \end{bmatrix}_{mxa} = \begin{bmatrix} [I] \\ [t_s] \end{bmatrix} \tag{13.16}$$

Note that the transform only contains stiffness terms, making the accuracy of this transform very sensitive to a-set quantity and selection.

13.3.2 Improved Reduced System (IRS)

The IRS TAM was developed to try and improve on the Guyan reduction. Refer to O'Callahan's [3] paper for the complete derivation.

$$[T_{IRS}] = \begin{bmatrix} [I] \\ [t_S] + [t_D] \end{bmatrix} \tag{13.17}$$

where:

$$[t_D] = -[K_{oo}]^{-1}[[M_{oa}] + [M_{oo}][t_s]][M_{aa}][K_{aa}] \tag{13.18}$$

Note that the transform, while, more computationally expensive, does contain mass and stiffness terms, making this transform more accurate and less sensitive to a-set quantity and selection.

13.3.3 System Equivalent Reduction/Expansion Process (SEREP)

The derivation of SEREP [10] begins with recalling the modal transform for the FEM:

$$\{x\}_{mx1} = [\Phi]_{mxN}\{q\}_{Nx1} \tag{13.19}$$

Partitioning the equation into a-sets and o-sets yields:

$$\begin{Bmatrix} x_a \\ x_o \end{Bmatrix}_{mx1} = \begin{bmatrix} \Phi_a \\ \Phi_o \end{bmatrix}_{mxN} \{q\}_{Nx1} \tag{13.20}$$

$$\{x_a\}_{ax1} = [\Phi_a]_{axN} \{q\}_{Nx1} \tag{13.20a}$$

$$\{x_o\}_{ox1} = [\Phi_o]_{oxN} \{q\}_{Nx1} \tag{13.20b}$$

Solving for $\{q\}$ in Eq. (13.20a) yields:

$$\{q\}_{Nx1} = [\Phi_a]^{+}_{Nxa} \{x_a\}_{ax1} \tag{13.21}$$

where $(^{+})$ denotes the pseudo-inverse of a matrix. Substituting (13.21) into Eq. (13.20) provides the relationship between the a-set and full m-set (a-set and o-set) DOF's as:

$$\begin{Bmatrix} x_a \\ x_o \end{Bmatrix}_{mx1} = \begin{bmatrix} \Phi_a \\ \Phi_o \end{bmatrix}_{mxN} [\Phi_a]^{+}_{Nxa} \{x_a\}_{ax1} \tag{13.22}$$

$$[T_{SEREP}] = \left[\begin{bmatrix} \Phi_a \\ \Phi_o \end{bmatrix}_{mxN} [\Phi_a]^{+}_{Nxa} \right]_{mxa} = [\Phi]_{mxN} [\Phi_a]^{+}_{Nxa} \tag{13.23}$$

When using this transform to reduce mass and stiffness matrices, N is typically less than a, resulting in rank deficient matrices, but retaining the full accuracy of the original N FEM modes in frequency and full mode shape accuracy. When this transform is used for test shape expansion the modes at the a-set are smoothed.

13.3.4 Modal

The Modal TAM [9] is very similar to SEREP, except that the a-set modes are unmodified (not smoothed). This is accomplished by realizing there is a direct representation of $\{X_a\}$:

$$\{x_a\} = [I] \{x_a\} \tag{13.24}$$

and substituting Eq. (13.21) into Eq. (13.20b) providing the relationship between the a-set and o-set DOF's as:

$$\{x_o\} = [\Phi_o]_{oxN} [\Phi_a]^{+}_{Nxa} \{x_a\}_{ax1} \tag{13.25}$$

Referring back to Eq. (13.6), the Modal TAM can be written as:

$$\begin{Bmatrix} x_a \\ x_o \end{Bmatrix}_{mx1} = \begin{bmatrix} [I] \\ [\Phi_o]_{oxN} [\Phi_a]^{+}_{Nxa} \end{bmatrix}_{mxa} \{x_a\}_{ax1} \tag{13.26}$$

defining the Modal TAM as:

$$[T_{Modal}] = \begin{bmatrix} [I] \\ [\Phi_o] [\Phi_a]^{+} \end{bmatrix}_{mxa} \tag{13.27}$$

When using this transform to reduce mass and stiffness matrices, N is typically less than a, resulting in rank deficient matrices, but retaining the full accuracy of the original N FEM modes in frequency and full mode shape accuracy. When this transform is used for test shape expansion the modes at the a-set are unmodified.

13.3.5 Hybrid

The Hybrid TAM was developed by Kammer [11] to provide the exact modal representation provided by the Modal TAM and the residual information provided by the Guyan reduction. The reduced matrices of the Hybrid TAM are not rank deficient with the eigenvalue decomposition giving the mode shapes and natural frequencies that the Modal TAM would provide for

Table 13.1 Summary of hybrid TAM combinations

Hybrid no.	Projector	TAM combinations	
1	Oblique	Modal	Guyan
2	Orthogonal	Modal	Guyan
3	Oblique	SEREP	Guyan
4	Orthogonal	SEREP	Guyan
5	Oblique	Modal	IRS
6	Orthogonal	Modal	IRS
7	Oblique	SEREP	IRS
8	Orthogonal	SEREP	IRS

the N target modes included in the aforementioned Modal TAM plus the additional information that the Guyan Reduction would provide for the remaining model order (a-set minus the target modes (a – N)). The Hybrid TAM is given as:

$$[T_{Hybrid}] = [T_s] + [[T_{Modal}] - [T_S]]\,[P] \tag{13.28}$$

The projector matrix $[P]$ is given in both its oblique and orthogonal forms as:

$$[P_{oblique}] = [\Phi_a] + [\Phi_a]^T\,[T_{Modal}]^T\,[M]\,[T_{Modal}] \tag{13.29}$$

$$[P_{ortho}] = [\Phi_a]\,[\Phi_a]^+ \tag{13.30}$$

Mains [13] observed that in addition to the oblique and orthogonal forms of the Hybrid TAM, that SEREP could be substituted for the Modal TAM and IRS could be used instead of the Guyan TAM. Table 13.1 summarizes the eight different Hybrid TAM's combinations that are possible.

13.4 Modal Assurance Criteria (MAC)

In the absence of a reducible FEM mass matrix, correlation between modal vectors can be determined thru the use of the MAC. One advantage of MAC is that because it doesn't use a mass matrix, errors created by reducing a mass matrix are not incorporated into the calculation. MAC was modeled after the development of the ordinary coherence calculation associated with interpreting the quality of measured FRF's. MAC is defined as the scalar constant relating the degree of consistency (or linearity) between one modal vector and another:

$$MAC_{ij} = \frac{\{\phi_j\}^H\{\phi_i\}\{\phi_i\}^H\{\phi_j\}}{\{\phi_j\}^H\{\phi_j\}\{\phi_i\}^H\{\phi_i\}} \tag{13.31}$$

Refer to Allemang [6] for all the various forms the MAC can be expressed in.

13.5 Orthogonality Using SEREP

One interesting characteristic of the SEREP TAM is that only the information needed to develop a reduced mass matrix is an accurate calculation of the FEM mode shapes and a specification of the a-set DOF's.

Substituting the SEREP TAM of Eq. (13.23) into the general Eq. (13.7) for reducing mass matrices to the a-set results in:

$$[M_{red}]_{axa} = \left[[\Phi][\Phi_a]^+\right]^T_{axm}\,[M]_{mxm}\,\left[[\Phi]\,[\Phi_a]^+\right]_{mxa} \tag{13.32}$$

$$[M_{red}]_{axa} = [\Phi_a]^{+T}_{axN}[\Phi]^T_{Nxm}\,[M]_{mxm}\,[\Phi]_{mxN}[\Phi_a]^+_{Nxa} \tag{13.33}$$

Since the property of weighted Orthogonality holds true for the FEM modal vectors:

$$[I]_{NxN} = [\Phi]^T_{Nxm}[M]_{mxm}[\Phi]_{mxN} \tag{13.34}$$

Equation (13.33) then reduces to:

$$[M_{red}]_{axa} = [\Phi_a]^{+T}_{axN}[\Phi_a]^{+}_{Nxa} \tag{13.35}$$

Resulting in a SEREP reduced mass matrix that only used the FEM mode shapes at the a-set DOF's. Note that this same simplification does not work for the other TAM's including the Modal TAM. This makes SEREP orthogonality easy and convenient to compute since only the FEM modes are required. The other TAMS typically must be computed in FEM software, sometimes with special scripts (referred to DMAP alters in Nastran) and exported to Correlation software for orthogonality computations.

13.6 Previous Work

Freed [3] and Mains [12, 13] looked at the performance of the various TAM's by investigating several of the different TAM methods with several sets of sensors. From the analysis performed it was very difficult to determine if a given TAM performed better than another. Mains [20] then looked at the effect of sensor cross-sensitivity and mis-alignment on modal vector correlation. He found that it was very easy to find significant variations in the orthogonality calculations (values greater than 0.1) for the normal amounts of error typically found in today's transducers. This previous work strongly suggests that it is very easy to see off-diagonal terms greater than 0.1 just from sensor selection error, poor choice of TAM or common measurement errors. In other words off-diagonal terms greater than 0.1 can occur even if the FEM and test are perfectly correlated.

13.7 Optimal DOF Selection and Decreasing DOF'S

13.7.1 Expected Results

Given an initial set of Orthogonal Modes from a FEM, one would expect that as we reduce the DOF's the MAC and Orthogonality will degrade. Given that MAC is missing the mass matrix it is expected that it will degrade quicker as a function of decreased DOF, than orthogonality checks. Since Guyan is a static reduction ignoring the effects of the mass matrix, we also expect it to degrade rather quickly as DOF's are reduced. The IRS TAM tries to account for missing mass yet, is will still degrade, but perhaps at a slower rate. Because Modal and SEREP include the target modes of interest they are much less sensitive to sensor location and in fact may not degrade severely until the pseudo-inverse involved in both methods becomes ill-conditioned. Because we are starting with perfectly correlated modes with no noise, we expect that the Hybrid TAM's will produce the same results as the as their Modal and SEREP cousins. As the studies grow to include modes that are not exactly correlated, it will then be interesting to see how the Hybrid TAM's perform. Figure 13.1 illustrates what result we might expect from any FEM, with any set of target modes as the number of DOF's are decreased in an optimal manner. For each TAM it's also expected that a different DOF selection method must be used to reach the optimal or minimum error.

13.7.2 Removing DOF's Using Effective Independence

A FEM was built for an Aluminum plate of dimensions $290 \times 250 \times 8$ mm using 1116 CQUAD4 elements resulting in 1,184 nodes and 7,104 DOF's. The first six flexible modes were used as target modes. For this study MAC, SEREP and the Guyan TAM were used to reduce the mass matrix. DOF's were first removed by removing all rotational DOF's, then removing DOF's in the X and Y directions. DOF's then were then removed one-by-one using the Effective Independence [15] as the method of selection.

Because Effective Independence tries to maximize the amount of modal response, thus avoiding nodes of modes and picking locations at the edges, instead of picking areas of high mass to stiffness ratio we see that MAC and Guyan degrade

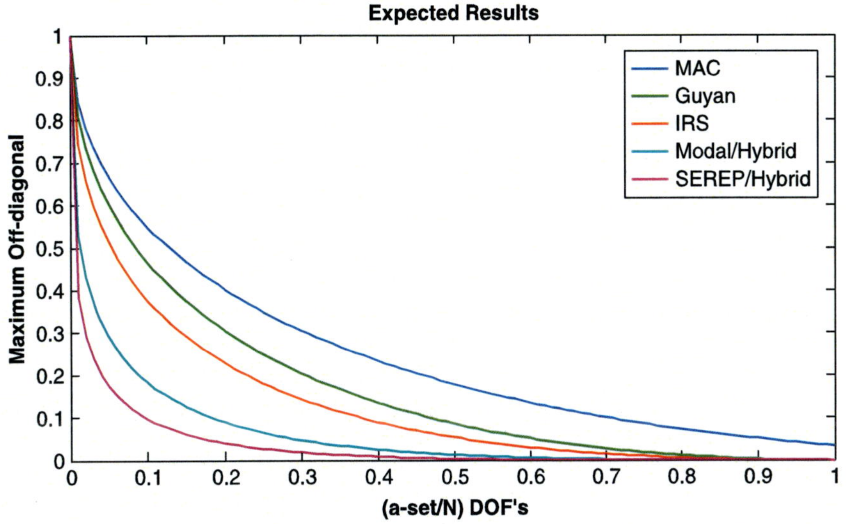

Fig. 13.1 Expected results of optimally reducing DOF for each TAM

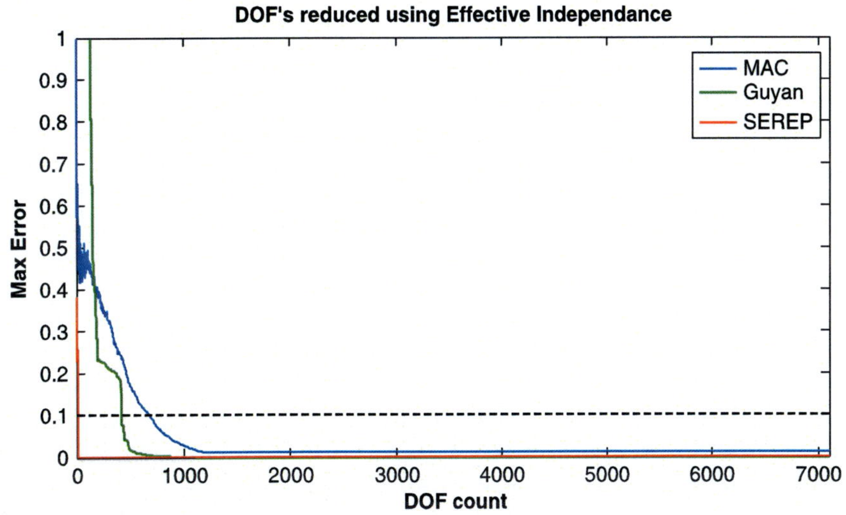

Fig. 13.2 Results of using effective independence for MAC, Guyan and SEREP

rather quickly. We see MAC rise above 0.1 on the off-diagonal as soon as 650 DOF's and Guyan is above 0.1 at 410 DOF's. It's expected that if a Genetic Algorithm was used with the proper fitness equation for MAC and for Guyan results could be improved substantially (Figs. 13.2 and 13.3).

Much work has been done on pre-test and sensor placement algorithms [16,17] especially in the area of genetic algorithms to select a near-optimal set of sensors for a fitness equation that is appropriate for the MAC or TAM orthogonality condition that is to be met. It is expected that if a similar study was performed as above only using the appropriate fitness equation for each TAM, then curves would be calculated that more closely meet the expected results presented in Fig. 13.1. Future work would not only include using genetic algorithms to find optimal sensor sets, but also the introduction of various forms of noise or modifications to see if the TAM methods blur or distort the expected changes in orthogonality.

13.8 Future Work

There are two areas where this study of DOF reduction can be expanded. Certainly the TAM's that where omitted should be included. It's expected that IRS will outperform Guyan and that Modal and Hybrid TAM's will give results very similar to SEREP. This belief about the Modal and Hybrid TAM's is held because we are using the original analytical vectors. It is only when we introduce noise or modes that are not correlated with the target modes that we would start to see differences. Also, any future study should include the use of the genetic algorithm for sensor selection since its optimization criteria are better tailored to select sensors that minimize the error of each respective TAM and the MAC.

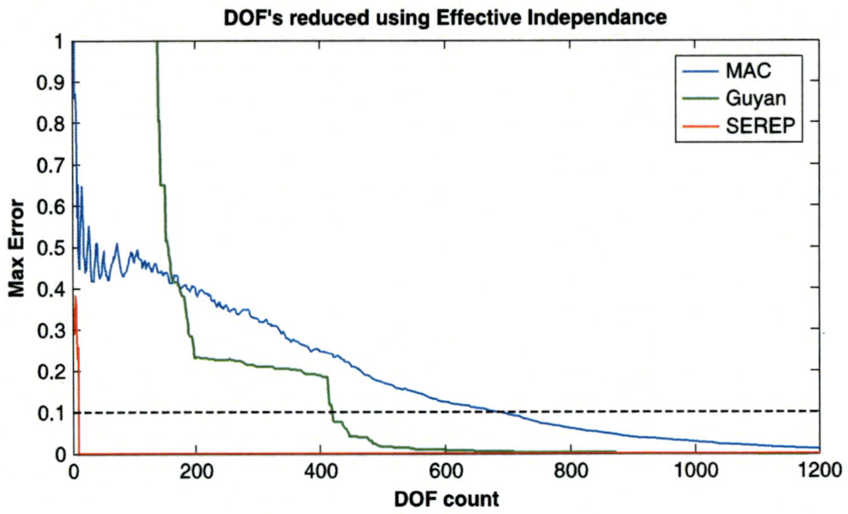

Fig. 13.3 Zoomed into where z DOF's are being removed via effective independence

Further, once performance of all TAM methods have been analyzed, different structures should be analyzed confirming the author's hypothesis, that the basic shapes of the curves shown in Fig. 13.1 don't change substantially regardless of the structure or the target modes.

Next, error needs to be introduced into the model. Perhaps that needs to be an error that creates a value of 0.1 in at least one off-diagonal term of the orthogonality matrix. Each method then, as a function of optimal DOF, needs to be evaluated on its accuracy of identifying the correct orthogonality check matrix.

13.9 Conclusion

While much work is yet to be done, it is believed that an understanding of which TAM to use for modal vector correlation under different test and model validation circumstances can be determined. While Guyan reduction is the industry standard for modal vector orthogonality determination, the author believes other methods should prove to be more robust given the fact that Guyan reduction makes such a gross assumption about the effects of the mass at the o-set or unmeasured degrees of freedom, while other methods are much more forgiving. It is also believed that the rather arbitrary requirement that the diagonal terms of the modal orthogonality must between 0.9 and 1.1 and that off-diagonal terms must be less than 0.1 can more justifiably be set based on the number of DOF's or sensors used to measure the test structure and the types target modes (local or global) to be used, thus more reasonable requirements can be set based on a rigorous pre-test analysis.

References

1. Guyan RJ (1965) Reduction of stiffness and mass matrices. AIAA J 3(2):380
2. Irons B (1965) Structural eigenvalue problems: elimination of unwanted variables. AIAA J 3(5):961–962
3. O'Callahan JC (1989) A procedure for an improved reduced system (IRS) model. In: Seventh international modal analysis conference, Las Vegas, pp 17–21
4. Freed AM, Flannigan CC (1990) A comparison of test-analysis model reduction methods. In: Eighth international modal analysis conference, Kissimmee, pp 1344–1351
5. Allemang RJ, Brown DL (1982) A correlation coefficient for modal vector analysis. In: First international modal analysis conference, Orlando, pp 110–116
6. Allemang RJ (2003) The modal assurance criterion (MAC): twenty years of use and abuse. In: Twentieth international modal analysis conference, Los Angeles, pp 397–405, 2002. Sound Vib Mag 37(8):14–23
7. Weingarten VI, Ramanathan RK, Chen CN (1983) Lanczos eigenvalue algorithm for large strucutres on a minicomputer. Comput Struct 16(1–4):253–257
8. Klahs JW (1985) Simultaneous vector iteration for the eigensolution of nonconservative system dynamics. In: Third international modal analysis conference, Orlando, pp 515–522
9. Kammer DC (1987) Test-analysis-model development using an exact modal reduction. J Anal Exp Modal Anal 2:174–179

10. O'Callahan JC, Avitabile P, Riemer R (1989) System equivalent reduction expansion process (SEREP). In: Seventh international modal analysis conference, Las Vegas, NV, pp 29–37
11. Kammer DC (1991) A hybrid approach to test-analysis-model development of large space structures. J Vib Acoust 133(3):325–332
12. Mains ML, Nicolas VT (1992) Investigation of the effects of sensor location and quantity for test-analysis models. In: Seventeenth international seminar on modal analysis, Leuven, Belgium
13. Mains M (1994) Investigation of methods for pre-testing, correlating and optimizing analytical and experimental modal models. Master's thesis, University of Cincinnati
14. Marinone T, Butland A, Avitable (2012) A reduced model approximation approach using model updating methodologies. In: Thirtieth international modal analysis conference, Jacksonville, vol 5. paper 118
15. Kammer DC (1990) Sensor placement for on-orbit modal identification and control of large space structures. In: Proceedings of the American control conference, San Diego, pp 2984–2990
16. Stabb M, Blelloch P (1995) A genetic algorithm for optimally selecting accelerometer locations. In: Thirteenth international modal analysis conference, Nashville
17. Linehan D, Napolitano K (2012) Accelerometer selection methods for modal pretest analysis. Sound Vib Mag 46(2):5–8
18. Avitabile P, Pechinsky F, O'Callahan XC (1992) Study of modal vector correlation using various techniques for model reduction. In: Tenth international modal analysis conference, San Diego, pp 572–583
19. Avitabile P, Pechinsky F (1994) Coordinate orthogonality check (CORTHOG). In: Twelfth international modal analysis conference, Honolulu, pp 753–760
20. Mains M, Vold H (1995) Investigation of the effects of transducer cross-sensitivity and misalignment error on modal vector correlation. In: Thirteenth international modal analysis conference, Nashville, pp 1048–1056
21. Nicgorski D (2008) Investigation on experimental issues related to frequency response function measurements for frequency based substructuring. Master's thesis, University of Massachusetts Lowell
22. Butland A, Avitabile P (2010) A reduced order, test verified component mode synthesis approach for system modeling applications. Mech Syst Signal Process 24(4):904–921
23. Thibault L, Butland A, Avitabile P Variability improvement of key inaccurate node groups – VIKING. In: Thirtieth international modal analysis conference, Jacksonville
24. Barile M Orthogonal. From MathWorld – a wolfram web resource, created by Eric W. Weisstein. http://mathworld.wolfram.com/Orthogonal.html

Chapter 14
CAE Model Correlation Metrics for Automotive Noise and Vibration Analysis

Qijun Zhang, Shawn Hui, and Kurt Schneider

Abstract CAE models have become more and more critical for decision making in various stages of the product development process with less hardware builds and shorten development time. To serve their purpose, it is important to understand the CAE capability of a model. A metric is essential to quantify this capability as well as track its history for improvement over the time. In particular, the noise and vibration sensitivity of a vehicle to a unit force or torque in a wide frequency range is of interest. In this paper, various correlation analysis techniques are applied to the Frequency Response Spectrum data from a CAE model and hardware test to develop a correlation metric. The comparison between Principal Component Analysis, Canonical Correlation Analysis and the developed metrics in this paper is presented. The correlation metrics developed in this paper also matches the subjective evaluation of our CAE capability for different measurement.

Keywords CAE • Correlation • Principal component analysis • Canonical correlation analysis • Noise and vibration

14.1 Introduction

In the production development process, CAE model has become more and more important, specifically in early stage when hardware is not available. Even in the late stage, less hardware and more CAE is used due to the high cost of the prototype. Quality CAE models capable to produce credible prediction are essential to support this process. To measure such quality a methodology needs to be developed with engineering metrics. This metrics not only indicate the CAE capability level but also identify the weakareas of CAE analysis. Over the time, it also provides a way to track the CAE capability improvement for both management and technology development purposes.

In Noise and Vibration (NV) analysis of the automotive industry, various methods are used to solve different problems. In high frequency range (400 Hz–10K Hz), Statistically Energy Analysis (SEA) is used for airborne noise analysis. In low frequency (< 150 Hz), Finite Element Analysis (FEA) is used for both vibration and noise analysis. In the mid-frequency range (150–500 Hz), a hybrid method of FEA and SEA combination can be used for structure-borne noise analysis.

Regardless the methods or the frequency range, at the end, a frequency spectrum of either noise or vibration is of interest, which will influence the customer perception ultimately.

Traditionally, a correlation between a CAE analysis and hardware test is a comparison of such frequency spectrum as shown in Fig. 14.1. A visual comparison of these two curves will provide certain evaluation of the correlation level.

In this comparison, both the level of the peak and the frequency of the peak are important. This plot is only one load case and one response. Even so, this single spectrum comparison already contains a wide frequency range with many peaks. A data plot like this provides only a way for an objective evaluation of the correlation level, which certainly depends on each engineer's experience and judgment.

In real life, multiple load cases (excitation at different locations, such as torque on engine block, force on spindle, etc.) and multiple responses (acceleration on seat track, steering wheel, noise in driver and passenger head space etc.) need to be considered. Such comparison quickly becomes a lot more complicated. A systematic approach to quantify the difference

Q. Zhang • S. Hui • K. Schneider (✉)
General Motors Company, 3300 General Motors Road, Milford, MI 48380, USA
e-mail: kurt.schneider@gm.com

T. Simmermacher et al. (eds.), *Topics in Model Validation and Uncertainty Quantification, Volume 5*,
Conference Proceedings of the Society for Experimental Mechanics Series 41,
DOI 10.1007/978-1-4614-6564-5_14, © The Society for Experimental Mechanics, Inc. 2013

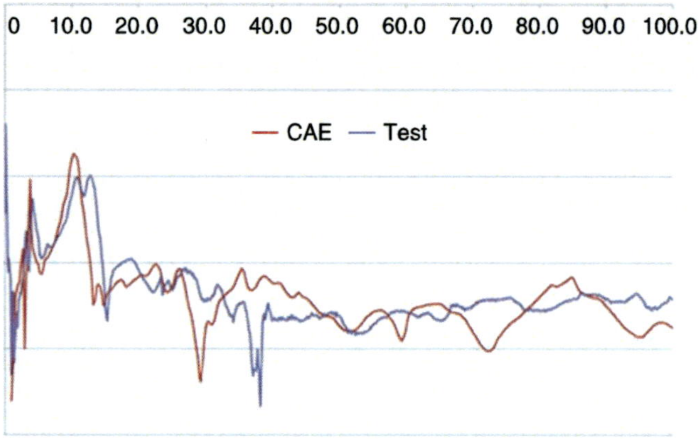

Fig. 14.1 FRF comparison between test and CAE

between the test results and analysis results will not only provide a consistent objective measure of the correlation level but also identify the area where the correlation level is low.

Another important aspect of the correlation is A-B comparison. Often time, it is desired to evaluate two or even more design proposals and the best design is chosen based on CAE analysis. In this case, whether the CAE analysis predicts the same difference between two designs as the hardware does is more important than the absolute correlation.

14.2 Statistical Analysis Methods [1, 2]

14.2.1 Principal Component Analysis

Consider a set of data (from either CAE or Test) $[X] = [\{X_1\}, \{X_2\}, \ldots \{X_q\}]$ with N observations (in rows, such as location/direction), each row consists of q variables (in columns such as frequency). The sample covariance matrix $[S_{\tilde{X}\tilde{X}}]$ is defined as:

$$\begin{aligned}
[S_{\tilde{X}\tilde{X}}] &= [S_{ij}] \\
S_{ij} &= \frac{1}{N-1}\sum_{n=1}^{N}\tilde{X}_i^{(n)}\tilde{X}_j^{(n)} \\
\tilde{X}_i^{(n)} &= X_i^{(n)} - \bar{X}_i \\
\bar{X}_i &= \frac{1}{N}\sum_{n=1}^{N}X_i^{(n)} \\
i,j &= 1\ldots q
\end{aligned} \tag{14.1}$$

where S_{ii} is the variance of $\tilde{X}_i$, S_{ij} the covariance between $\tilde{X}_i$ and $\tilde{X}_j$

The covariance matrix, $[S_{\tilde{X}\tilde{X}}]$, is a $q \times q$ symmetric matrix. It is reduced to a diagonal matrix $\Lambda_{\tilde{X}}$ by pre-multiplying and post-multiplying a unitary matrix $E_{\tilde{X}}$ such that (prime denotes transpose)

$$E'_{\tilde{X}}S_{\tilde{X}\tilde{X}}E_{\tilde{X}} = \Lambda_{\tilde{X}} \tag{14.2}$$

The diagonal elements of $[\Lambda_{\tilde{X}}]$, $\lambda_1 \geq \lambda_2 \geq \cdots \geq \lambda p$, $p \leq q$, are the eigenvalues (also referred to as latent) of $[S_{\tilde{X}\tilde{X}}]$. The columns $\{e_i\}$ of $[E_{\tilde{X}}]$ are the eigenvectors of $[S_{\tilde{X}\tilde{X}}]$, where $\|e_i\| = 1$.

The principal component score (PC) $\{P_i\}$ is the representation of $[\tilde{X}]$ in the Principal component space with its rows corresponding to observations,

$$\{P_i\} = [\tilde{X}]\{e_i\} \tag{14.3}$$

Each originally observed deviation (row vector) can be written as

$$\{\tilde{X}_{i.}\} = \sum_{j=1}^{p} \{e_j\} P_{ij} \tag{14.4}$$

Each originally observed variable (column vector) can be written as

$$\{\tilde{X}_{.j}\} = \sum_{i=1}^{p} \{P_i\} e_{ji} \tag{14.5}$$

Usually only first a few PCs will capture the majority of the information contained in the original observed deviation. Rather than requiring all q variables to describe the original observations, now we only need a few PCs. Also important is the original q variables are all correlated, now these reduced components are orthogonal and in the decreasing order of significance. The significance of each PC can be defined as the contribution using the latent from Eq. (14.2),

$$\delta_i = \lambda_i / \sum_{i=1}^{p} \lambda_i \tag{14.6}$$

For correlation purpose between CAE data and Test data, we now only compare a few PCs instead of all the variables.

PCA is applied to CAE data ([X]) and Test data ([Y]) separately to produce two sets of PCs $[P_X]$ and $[P_Y]$ as Eq. (14.3). Since each is defined by different projections $[E_{\tilde{X}}]$ and $[E_{\tilde{Y}}]$, the direct correlation between them does not make good sense. Also chances are the number of PCs required to make up 90 % accumulated contribution can be 5 or even more. Canonical Correlation Analysis comes to play for further data reduction.

An alternative concept to Canonical Correlation Analysis is that the linear combination given as $\{e_i\}$ in Eq. (14.3) for ith PC defines one dimension in the Principal component space. The PC is just a projection of the original data into this space. Once this space is developed for the first set of the data, say Test data, we can project the second set of the CAE data to this same space without conducting another PCA to the second data set. The correlation between the two sets of data becomes the correlation between their projections to the same space.

14.2.2 *Canonical Correlation Analysis*

In statistics, canonical correlation analysis (CCA) is used to understand the correlation between two sets of multivariate variables. That is to say we want to make sense of their cross-covariance matrices. We have two sets of variables, $X_1, X_2, X_3, \ldots, X_p$ and $Y_1, Y_2, Y_3, \ldots, Y_q$, where p and q are the dimensionality of the two observation vectors respectively. Suppose we found there are correlations among the variables, then canonical correlation analysis will enable us to find linear combinations of the X's and Y's which have maximum correlation with each other. Such pairs of linear combinations are the canonical variates and their correlation is called canonical correlation.

Typically, CCA is performed on the data matrices *X(CAE)* and *Y(Test)*, where each of these two matrices consist of the same N rows of observation vectors. Here the basic thrust is in the use of principal component scores. Suppose that the first m_x and m_y principal component scores are retained from P_x and P_y as Eq. (14.3),

Let

$$P = \lfloor P_y P_x \rfloor \tag{14.7}$$

The covariance matrix of P is

$$\begin{aligned} S_{PP} &= \frac{1}{N-1} P' P = \frac{1}{N-1} \begin{bmatrix} P_y' P_y & P_y' P_x \\ P_x' P_y & P_x' P_x \end{bmatrix} \\ &= \begin{bmatrix} S_{yy} & S_{yx} \\ S_{xy} & S_{xx} \end{bmatrix}_{(m_y+m_x)\times(m_y+m_x)} \end{aligned} \tag{14.8}$$

The total covariance matrix is a block matrix where S_{yy} and S_{xx} are the within-sets covariance matrices of P_y and P_x respectively (which are diagonal matrix $\Lambda_{\tilde{X}}$ as in Eq. (14.2)) and S_{xy} and S_{yx} are the between-sets covariance matrices.

The canonical correlation between P_x and P_y can be found by solving the eigenvalue equations

$$S_{xx}^{-1}S_{xy}S_{yy}^{-1}S_{yx}\hat{a} = \rho^2\hat{a} \tag{14.9}$$

$$S_{yy}^{-1}S_{yx}S_{xx}^{-1}S_{yx}\hat{b} = \rho^2\hat{b} \tag{14.9'}$$

where the eigenvalues ρ^2 are the squared canonical correlations and the eigenvectors $\hat{a}$ and $\hat{b}$ are the normalized canonical correlation basis vectors. The number of non-zero solutions to these equations is limited to the smallest dimensionality of P_x and P_y, i.e., the smaller of m_x and m_y.

The canonical correlations $\rho_1 > \rho_2 > \cdots > \rho_s$, where $s = \min(m_x, m_y)$, correspond to the s pairs of canonical variates,

$$u_i = P_y\hat{b}_i \quad \text{and} \quad v_i = P_x\hat{a}_i \quad i = 1,2,\ldots,\text{s} \tag{14.10}$$

Together, they provide the s dimensions of relationship between the two sets of data. The s dimensions of relationship (u_i, v_i) are nonredundant; the information each pair provides is totally unavailable in the other pairs.

Also noticed that the canonical correlations are in decreasing order so only the first a few are strongly correlated.

14.3 Noise and Vibration Engineering Metrics

In automotive NVtest and analysis, the response of the vehicle against multiple load conditions is of interest. The load conditions include operational loads, such as Hard Acceleration, Cruising, and Coarse Road Surface. The operational loads involve the variation of the excitation, which adds another uncertainty in the correlation. Another type of load condition is a unit load, or the vehicle sensitivity measurement, such as a unit torque on the engine block or a unit force on spindle. In this paper, we will focus on the sensitivity correlation.

Since most NV issues are due to either powertrain excitation or road input, following vehicle sensitivity will be considered,

Powertrain Vibration Sensitivity (A/T):
Acceleration at seat track, steering wheel and other body locations to a unit torque on engine block
Road Noise Sensitivity (p/F):
Noise at driver/passenger ear to a unit force on each spindle and each direction
Road Vibration Sensitivity (A/F):
Acceleration at the seat track, steering wheel and other body locations to a unit force on each spindle and each direction

Each one of these engineering metrics will have many spectra to form a large data set.

For statistical analysis as described above, we consider each frequency line as a variable, each response an observation. A matrix can be used to represent each data set, with each row representing one observation and each column one variable (frequency line). Various statistical analyses can be applied to this matrix.

In the traditional way, correlation is based on the comparison between CAE and test. When we compare a set of spectra, the line plot as in Fig. 14.1 will be very crowded and not easy to draw any conclusion. Instead a color map as in Fig. 14.2 will be much better, where the horizontal axis is still the frequency, the vertical axis is the observations. Each horizontal row represents one observation (one curve as in Fig. 14.1), with the color indicating the level.

This comparison does show the similarity between Test (Left) and CAE (Right). It also shows a similar change of the response after certain components are disconnected (Top: baseline, Bottom: after disconnection). But still it is a subjective evaluation with no metrics defined to measure how similar or how different the Test and CAE are.

14.4 Correlation Analysis and Metrics

When the data in Fig. 14.2 is arranged in a matrix form [X] with each row representing an observation and each column a variable (frequency line in this case), then the statistical analysis as described above can be applied. In this example, a matrix of 162 × 198 is comprised for the test data from 162 observations with each having 198 frequency points. The matrix of the CAE prediction has the same size. The correlation between them then becomes the comparison of these 162 observations in a 198 dimension space. The comparison is difficult due to the large number of variables. A Principal Component Analysis (PCA) will extract the most representative characteristics from the raw data so that the correlation will be on a much smaller

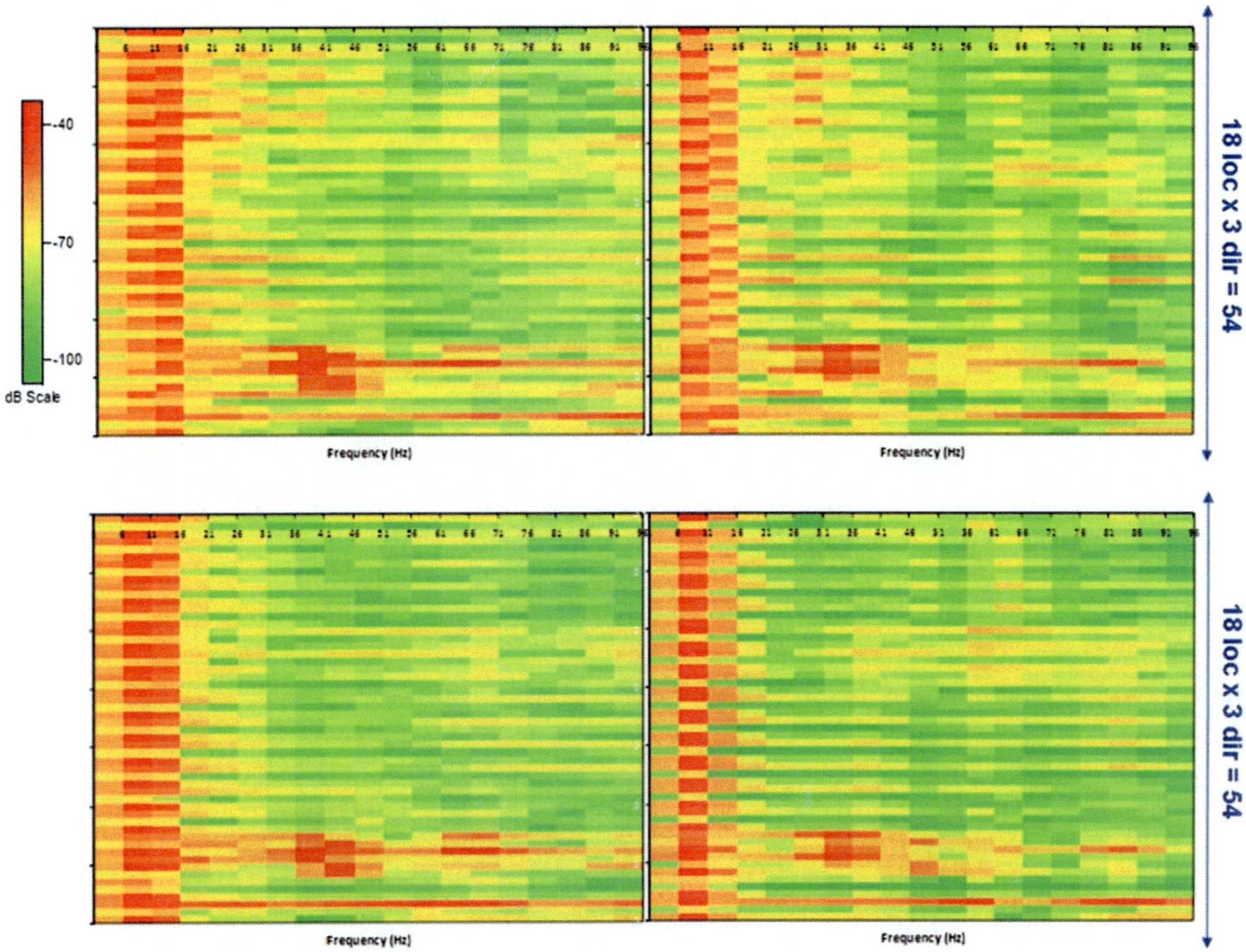

Fig. 14.2 Test vs. CAE with 54 observations. A/T with 18 acceleration locations and 3 directions. *Left*: test, ***Right***: CAE, *Top*: baseline, *Bottom*: with components disconnected

set of data. In other words, we want to compare the dispersion patterns of these two sets of 162 point in a 198 dimension space. By using principal component analysis, such comparison is simplified to a reduced, but most relevant subspace.

14.4.1 PCA Based Correlation Metrics

A PCA is applied to the test data matrix first. The latent (as defined above) and accumulated contribution of the first 12 Principal Components are shown in Fig. 14.3. Each Principal Component Score $\{P_i\}$ as Eq. (14.3) is a vector, which represents the characteristic of the data set across all the variables (frequency lines). This analysis shows that the variation between observations (response at different locations and directions) has different characteristics at different frequency (variable) such that no single Principal Component Score dominates the entire data population. To reach 90 % accumulated contribution as many as 12 PC's are required. Even so, the number of dimension is now reduced from 198 to 12.

Each PC is a vector (162 × 1), which is actually a projection of the original data $[\tilde{X}]$ to this Principal Component space as in Eq. (14.3), where $\{e_i\}$ is a unit length vector $\sum_{j=1}^{198} e_{ji}^2 = 1.0$. This also shows that the PC is a weighted average of all variables (frequency).

Apply the same projection of $\{e_i\}$ onto the CAE deviation matrix $\tilde{Y}$ to produce a new vector

$$\{P_i\} = \left[\tilde{X}\right]\{e_i\}, \{Q_i\} = \left[\tilde{Y}\right]\{e_i\} \tag{14.11}$$

where $\left[\tilde{Y}\right]$ is the CAE deviation matrix in the same observation order as in the test.

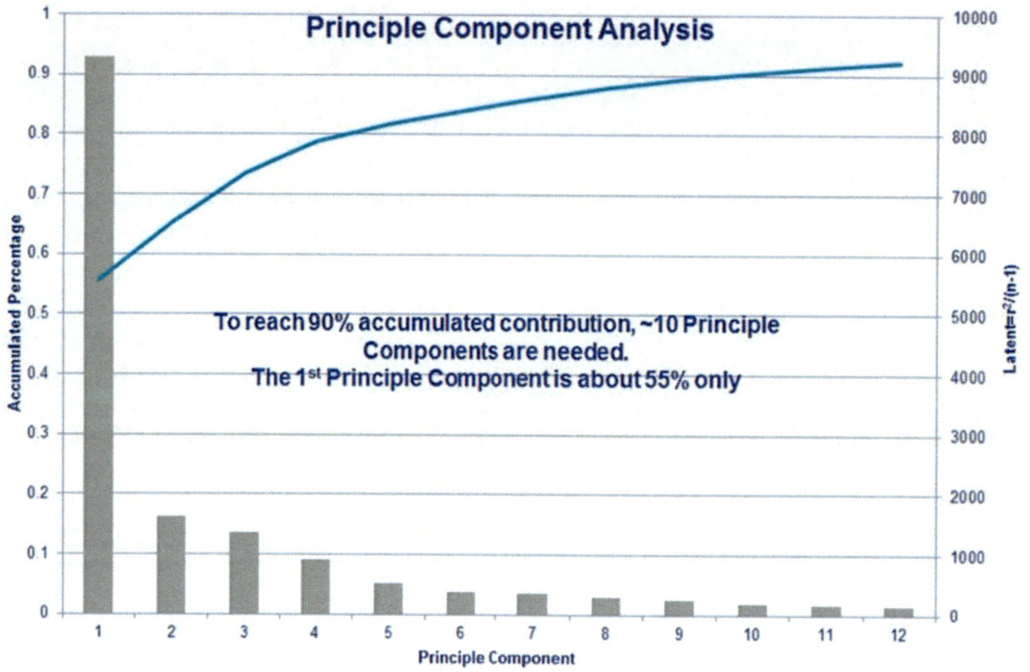

Fig. 14.3 PCA of the test data. *Bar*: latent of each PC, *Line*: accumulated contribution of each PC

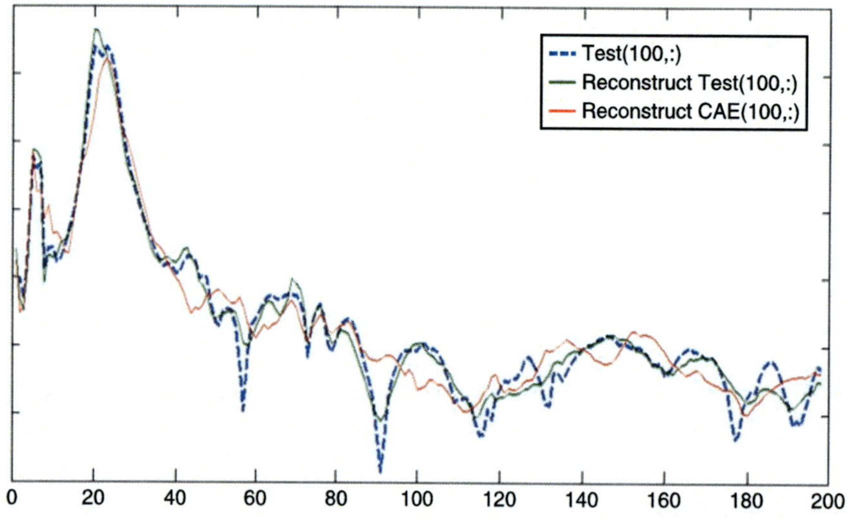

Fig. 14.4 Reconstructed observations compared to raw data. — test raw data, reconstructed test data with $\{P_i\}$, reconstructed CAE data with $\{Q_i\}$

Even with a small set of PC's, the majority information of the original data can be captured. As an example, the observation #100 can be reconstructed using only 12 PC's as shown in Fig. 14.4, which is very close to the original data,

Certainly if CAE is a perfect match to the test data, $\{P_i\} = \{Q_i\}$. In reality, these two vectors are different. A correlation coefficient is defined between these two vectors as below,

$$\rho_i = \frac{\sum_{j=1}^{n}(x_j - \bar{x})(y_j - \bar{y})}{\sqrt{\sum_{j=1}^{n}(x_j - \bar{x})^2}\sqrt{\sum_{j=1}^{n}(y_j - \bar{y})^2}} \tag{14.12}$$

where x_j and y_j are jth entry in $\{P_i\}$ and $\{Q_i\}$.

The correlation coefficient defined above has the range of -1.0 to 1.0, with 1.0 for a perfect match, 0 for completely uncorrelated and -1.0 indicating a negative correlation (negative correlation could mean that the model prediction indicates opposite direction as the test). In real cases, it is most often in the range of 0.0–1.0.

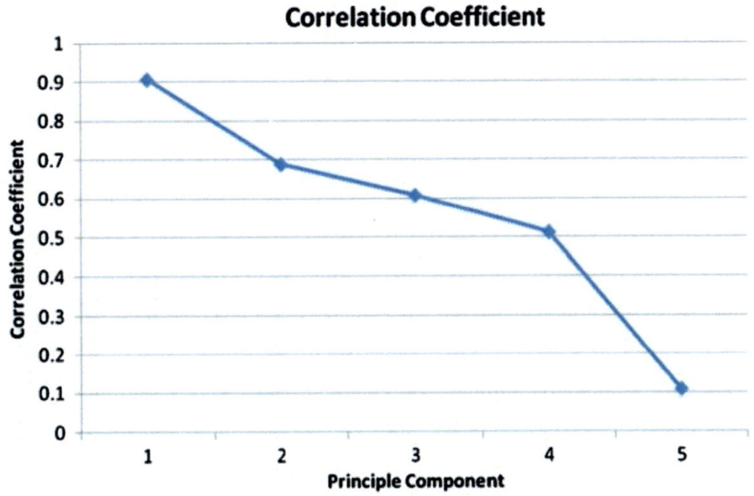

Fig. 14.5 Correlation coefficient of the first five principal components

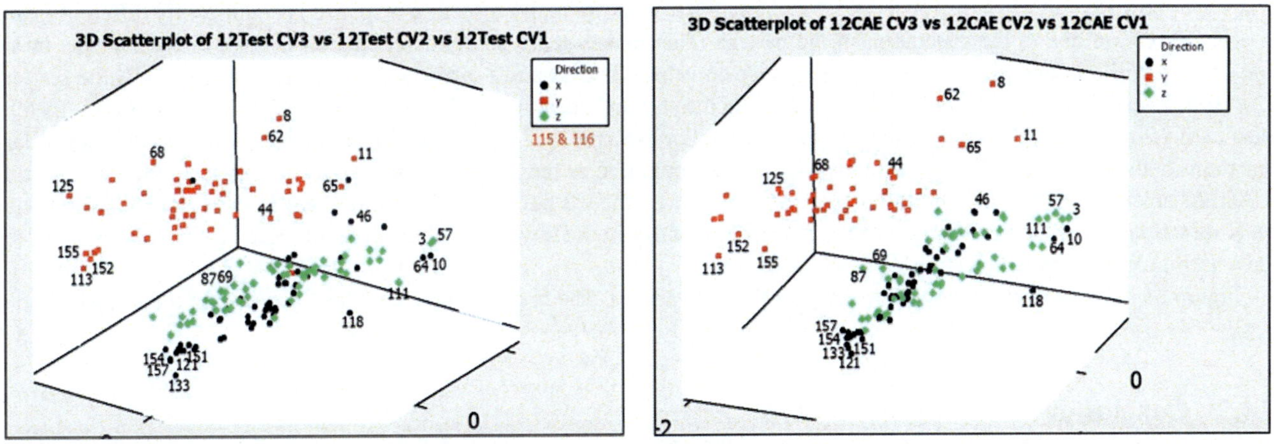

Fig. 14.6 The first 3 CV's of canonical correlation analysis (*Left*: test, *Right* CAE)

The correlation coefficients of the first 5 PC's are shown in Fig. 14.5.

The overall correlation coefficient is expressed as the weighted average of all ρ_i,

$$\rho = \sum_{i=1}^{12} w_i \rho_i \tag{14.13}$$

where w_i is the contribution of each PCs with $\sum w_i = 1$.

14.4.2 Canonical Correlation Analysis

When a PCA is applied to the CAE data in the same way as to the test data, another 12 PCs are calculated. Following analysis is between these two sets of PC's. Even though the number of variables is greatly reduced after PCA (from 198 to 12), but it is still too many for a correlation.

Canonical Correlation Analysis (CCA) is a method to measure the linear relationship between two multidimensional variables. It finds the maximum correlation coefficient between these two sets of 12 PC's from Test and CAE by finding and applying a transformation to each. It then finds the next maximum, which is orthogonal to the previous projection, and so on. This analysis will help to investigate the relationship between two large data sets.

As an example shown in Fig. 14.6, a 3D plot shows the top 3 CV's (Canonical Variates as in Eq. (14.10)) of CCA for both Test (left) and CAE (right) with x/y/z direction response separated. Each dot represents the top 3 CV's for one observation.

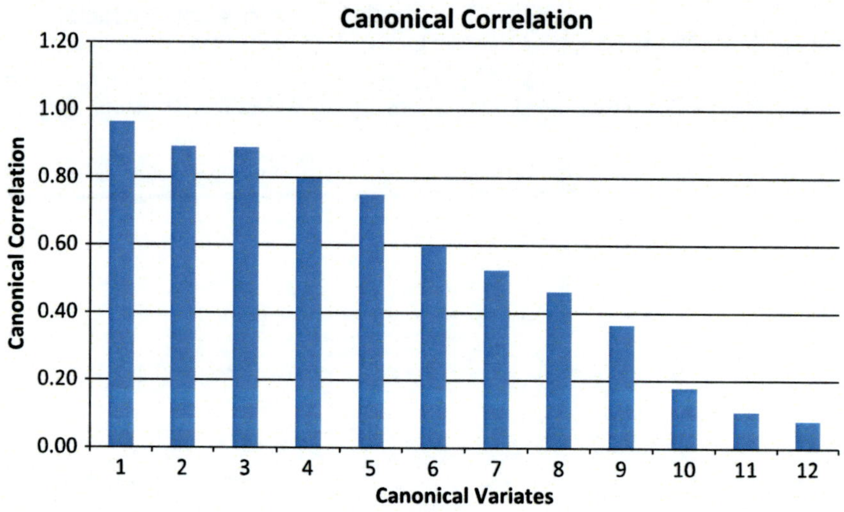

Fig. 14.7 The canonical correlation

They are shown in three groups for X/Y/Z directions. First it shows that y direction response is significantly different from x and z direction due to the separation in the pattern. This makes sense since the excitation is engine block torque in Y axis, which will produce more translational response on vehicle body x and z direction than y direction. Secondly the scatter pattern between CAE and Test is similar, which indicates a good correlation. Even further, the labeled dots closely match also in this 3-d space, which represent each observation. For example, #62 for Y direction, #118 for X direction are in similar locations between Test and CAE. This plot is a way to emphasize or magnify the similarity or difference between Test and CAE. As an example, there is a black dot mixed in red dots on the left but not on the right. The black dot (for an observation in X direction) is certainly an outlier, which is hard to identify in the raw data. In this case, though, this outlier turns out to be a user error by mislabeling X as Y in the test data.

Figure 14.7 shows the Canonical Correlation of all 12 variates. The first few have very high correlation (close to 1.0).

14.5 Conclusion

In an effort to understand the correlation of the CAE model to the test data taken, Principal Component Analysis and Canonical Correlation Analysis methods were used to evaluate how similar the model and the test data were. These methods looked across multiple load cases such that a simple and objective statement of correlation could be made, relative to the overall performance of the model as compared to test, rather than focusing on a single load case. This gives a view of the capability of the CAE model as whole, against many potential loads and boundary conditions.

Acknowledgements The authors would like to acknowledge Dr. Jim DeClereck of Michigan Technological University. During his time at General Motors, he conducted foundational work for this paper, in conjunction with the authors.

References

1. Rencher AC (1995) Methods of multivariate analysis. Wiley, New York
2. Johnson RA, Wichern DW (1992) Applied multivariate statistical analysis. Prentice-Hall, Englewood Cliffs

Chapter 15
Damage Localization Using a Statistical Test on Residuals from the SDDLV Approach

L. Marin, M. Döhler, D. Bernal, and L. Mevel

Abstract Mechanical systems under vibration excitation are prime candidate for being modeled by linear time invariant systems. Damage localization in such systems, when the excitation is not measurable, can be carried out using the Stochastic Dynamic Damage Locating Vector (SDDLV) approach, a method that interrogates changes in a matrix that has the same kernel as the change in the transfer matrix at the sensor locations. Damage location is related to some residual derived from the kernel. Deciding that this residual is zero is up to now done using empirically defined thresholds. In this paper, we describe how the uncertainty of the state space system can be used to derive uncertainty on the damage localization residuals to decide about the damage location. The results are illustrated in finite element models of a truss and of a plate.

Keywords Ambient vibration • Covariance analysis • Damage localization • Hypothesis testing • Load vectors.

15.1 Introduction

Alternative approaches to visual inspections of physical structures, such as bridges and buildings, have been provided by vibration-based monitoring techniques. Sensors installed in the structures collect data and statistical approaches originated from stochastic system realization theory for linear systems provide estimates for the parameters of interest. The time evolution of the data is characterized by system matrices of the underlying linear system. The eigenstructure of those matrices relate directly to some parameterization of interest for the monitoring of structures, usually the modal parameters (natural frequencies, damping ratio and mode shapes), and subsequently to the finite element model (FEM). Fault detection (damage detection for mechanical structures) and fault isolation (damage localization) can be inferred from changes in these parameters.

Assuming that damage occurs, [1] presents alternate damage localization techniques using both finite element information and modal parameters: the Stochastic Dynamic Damage Location Vector (SDDLV) approach. From estimates of the system matrices in both reference and damaged states, the null space of the difference between the respective transfer matrices is obtained. Then, damage is related to a residual derived from this null space and located where the residual is close to zero. Empirical thresholds are currently used for decision without considering the intrinsic uncertainty, which happens due to unknown noise excitation and limited data length in the identification of system matrices. The lack of uncertainty consideration is critical: no information is available on the choice of threshold for deciding whether the lowest residual is zero or not in practical situations. Nevertheless, sensitivity based methods, such as presented in [2] and [3], provide some guidelines to derive uncertainty estimates for modal parameters, and an efficient sensitivity computation of these quantities has been derived in [4–6].

This paper aims to replace empirical rules by sensitivity-based rules for applying some damage localization criterion, and is organized as follows. In Sect. 15.2 the SDDLV approach is introduced as a method for stochastic damage localization

L. Marin (✉) • L. Mevel
INRIA, Centre Rennes – Bretagne Atlantique, Campus de Beaulieu, F-35042 Rennes, France
e-mail: luciano.gallegos@inria.fr; laurent.mevel@inria.fr

M. Döhler • D. Bernal
Department of Civil & Environmental Engineering, Northeastern University, Boston, MA 02115, USA
e-mail: michael.doehler@bam.de; d.bernal@neu.edu

T. Simmermacher et al. (eds.), *Topics in Model Validation and Uncertainty Quantification, Volume 5*, Conference Proceedings of the Society for Experimental Mechanics Series 41,
DOI 10.1007/978-1-4614-6564-5_15, © The Society for Experimental Mechanics, Inc. 2013

of mechanical structures from output-only signals. In Sect. 15.3, the covariance of the system matrices is propagated to the damage localization residuals. In Sect. 15.4, numerical examples are provided. Finally, some conclusions of this work are presented in Sect. 15.5.

15.2 The SDDLV Approach

The SDDLV, derived in [1, 7], is an output-only damage localization method based on interrogating changes δG in the transfer matrix G of a system, which is related to a FEM to localize damage without using a detailed model. Typically, the SDDLV is performed in two data sets: one from undamaged (reference) state and another from damaged state. *Load vectors* of the null space of δG are then used for the computation of a *stress field* over the structure in order to indicate the damage location: *Stresses* are measures of internal reactions to external forces applied on a deformable body, where (in the method to be described) zero stress over elements of a structure indicates changes in the flexibility and consequently damage. The basic principles and underlying models of the SDDLV are introduced in this section.

15.2.1 Dynamical Equation and State-Space Model

The behavior of a mechanical structure is assumed to be described by a linear time-invariant (LTI) system and represented by the corresponding *continuous-time state-space model*

$$\begin{cases} \dot{x} = A_c x + B_c e \\ \eta = C_c x + D_c e \end{cases}, \tag{15.1}$$

where $x \in \mathbb{R}^n$ is the state, $\eta \in \mathbb{R}^r$ is the output, $A_c \in \mathbb{R}^{n \times n}$ is the state transition matrix, $B_c \in \mathbb{R}^{n \times r}$ is the input influence matrix, $C_c \in \mathbb{R}^{r \times n}$ is the output mapping matrix, $D_c \in \mathbb{R}^{r \times r}$ is the direct transmission matrix. The fictive force $e(t)$ acts only in the measured coordinates and that re-produce the measured output, n is the system order and r is the observed outputs coordinates. If all the modes of the LTI system were identified then $n = 2d$. In practice this is seldom the case, so what one gets from identification is a reduced model order $n \ll 2d$. Since SDDLV is an output-only method, the non-identified matrices B_c and D_c are used in order to derive properties of the transfer matrix [1]. Only the system matrices A_c and C_c are relevant for system identification in this paper.

15.2.2 Damage Localization Procedure

The damage localization in mechanical structures with output-only data can be determined with the null space vectors for the SDDLV technique. Like this, damage localization information from structural changes (stress over elements) is extracted with the underlying idea of detecting changes in the flexibility. Note that while the transfer matrix is defined at the coordinates defined by the sensors, damage can be localized at any point of the structure because the stress field generated from the sensor coordinate loads covers the full domain.

Consider now the transfer matrix of model (15.1), which is given by

$$G(s) \stackrel{\text{def}}{=} R(s) D_c, \tag{15.2}$$

where

$$R(s) \stackrel{\text{def}}{=} C_c A_c^{-b} [sI - A_c]^{-1} \begin{bmatrix} C_c A_c^{1-b} \\ C_c A_c^{-b} \end{bmatrix}^{\dagger} \begin{bmatrix} I \\ 0 \end{bmatrix} \tag{15.3}$$

with $G(s) \in \mathbb{C}^{r \times r}$, $b = 0, 1, 2$ the output measurements (displacements, velocities, or accelerations respectively) and I the identity matrix.

Using (15.2) for the damaged (variables with tilde) and reference states, respectively, and dropping the Laplace variables s for simplicity, gives the difference in the transfer matrices $\delta G = \tilde{G} - G$. Neglecting D_c in (15.2) in both damaged and reference states (see [1] for more details), the desired null space of δG has the same null space of $\delta R^T = \tilde{R}^T - R^T$. Then, the null space of δR^T is finally obtained from the Singular Value Decomposition (SVD)

$$\delta R^T = \begin{bmatrix} U_1 & U_2 \end{bmatrix} \begin{bmatrix} \Sigma_1 & 0 \\ 0 & \Sigma_2 \end{bmatrix} \begin{bmatrix} V_1 & V_2 \end{bmatrix}^H, \tag{15.4}$$

where $U, \Sigma, V \in \mathbb{C}^{r \times r}$, $\Sigma_2 \approx 0$ and $V = (v_1, \ldots, v_r) = [V_{(1)} \;\; V_{(2)}]$ the right singular vectors. Note that $V_{(1)}$: $(v_1, v_2, \ldots, v_t)$ is the nonzero singular vectors and $V_{(2)}$: $(v_{t+1}, v_{t+2}, \ldots, v_r)$ is the ideally zero singular vectors (in practice small), where a desired load vector v in the null space of δR^T is then any linear combination of the vectors in $V_{(2)}$, e.g. $v = v_r$. For any chosen value s, the load vector $v = v(s)$ in the null space of $\delta G(s)$ can be computed as described above, where only model (15.1) has been used without using information about the geometry of the structure.

The computation of the stress implies knowledge of the model of the structure (coming e.g. from a FEM) and is a linear function of displacement. The function that maps the displacement to the stress resultant is denoted as matrix $Q \in \mathbb{R}^{d \times d}$, the transfer matrix is $G_{\text{model}}(s) \stackrel{\text{def}}{=} (Ms^2 + Cs + K)^{-1}$ of model of LTI system in the reference state, and the sensors mapping matrix $P \in \mathbb{N}^{d \times r}$ with 1's where each line (position in the structure) and each column (sensor number) agree and zeros elsewhere. Let this function be given by $\mathcal{L}_{\text{model}}(s) = QG_{\text{model}}(s)P$, such that the stresses $S(s) \in \mathbb{C}^d$ for a chosen value s write as [1]

$$S(s) = \mathcal{L}_{\text{model}}(s) v(s). \tag{15.5}$$

If an element at some degree of freedom j is damaged, the resulting stress $S_j(s)$ at coordinate j from the load $v(s)$ is zero [1]. Thus, the stresses in $S(s)$ are considered as damage localization residuals, where the entries close to zero correspond to elements that are potentially (but not necessarily) damaged.

15.3 Uncertainties on Damage Localization Residuals

System matrices A_c and C_c estimated from a finite number of data samples (e.g. using Stochastic Subspace Identification (SSI) methods [8,9]) are used for the damage localization both in the reference and damaged state. Due to the reduced model order that represents the identified bandwidth, what are obtained is the estimated matrices $\hat{A}_c$ and $\hat{C}_c$ and not the "true" system matrices A_c and C_c. The input of system (15.1) is unmeasured noise, leading $\hat{A}_c$ and $\hat{C}_c$ to variance errors depending on the data and the estimation method. A variance analysis of the system matrices obtained from Stochastic Subspace Identification is made e.g. in [10] and expressions for their computation in the context of structural vibration analysis are given in [3,5,6].

When estimating the load vectors in the null space of δG and the related stress field, the uncertainty of the system matrices is propagated to the uncertainty in the damage localization results. In this section, the variances of damage localization results are evaluated in order to support the decision between undamaged and damaged elements.The decision if the stress $S_j(s)$ at element j is zero (potentially damaged) or not is facilitated when knowing the variance of the estimate.

Note in this section that first-order matrices and derivation were dropped for simplicity. For further details, see [11,12].

15.3.1 Covariance of R

In this section, the sensitivity of the matrix R in (15.3) with respect to the system matrices A_c and C_c is derived, which is needed for the damage localization in (15.4). First, assume that the data is given by acceleration sensors ($b = 2$). Derivations for displacement and velocity data ($b = 0, 1$) follow analogously. A perturbation of R is linked to a perturbation of A_c and C_c by the relation

$$\text{vec}(\Delta R) = \begin{bmatrix} \mathcal{J}_{A_c} & \mathcal{J}_{C_c} \end{bmatrix} \begin{bmatrix} \text{vec}(\Delta A_c) \\ \text{vec}(\Delta C_c) \end{bmatrix}, \tag{15.6}$$

and with notation $M_{\text{re}} \stackrel{\text{def}}{=} \begin{bmatrix} \text{Re}(M) & \text{Im}(M) \end{bmatrix}^T$ and (15.6), the relation

$$\text{cov}\left((\text{vec}(R^T))_{\text{re}}\right) = \mathcal{J}_R \, \text{cov}\left(\begin{bmatrix} \text{vec}(A_c) \\ \text{vec}(C_c) \end{bmatrix}\right) \mathcal{J}_R^T \tag{15.7}$$

holds for the asymptotic covariance of the real and imaginary parts of R^T, where $\mathcal{J}_R$ is defined as

$$\mathcal{J}_R = \begin{bmatrix} \mathcal{P}_{r,r} & 0_{r^2,r^2} \\ 0_{r^2,r^2} & \mathcal{P}_{r,r} \end{bmatrix} \begin{bmatrix} \mathrm{Re}(\mathcal{J}_{A_c}) & \mathrm{Re}(\mathcal{J}_{C_c}) \\ \mathrm{Im}(\mathcal{J}_{A_c}) & \mathrm{Im}(\mathcal{J}_{C_c}) \end{bmatrix}.$$

15.3.2 Covariance of Damage Localization Residuals

In order to compute the covariance of the damage localization residual – the stresses $S(s)$ from (15.5) for a chosen value s – , the covariance of the load vector v is needed, which is a singular vector in the null space of $\delta R^T = \tilde{R}^T - R^T$ in (15.4). In the following, the first-order perturbation of right singular vectors v in the null space is provided in order to obtain the covariance of the stresses $S(s)_{\mathrm{re}}$ in (15.8).

Let $\mathrm{cov}((\mathrm{vec}\, R^T)_{\mathrm{re}})$ and $\mathrm{cov}((\mathrm{vec}\, \tilde{R}^T)_{\mathrm{re}})$ from the reference and damaged state be given in (15.7) and the sensitivity in $\mathcal{J}_v$. Then,

$$\Sigma_S \stackrel{\mathrm{def}}{=} \mathrm{cov}(S(s)_{\mathrm{re}}) = \mathcal{J}_{S(s)} \left(\mathrm{cov}((\mathrm{vec}\, \tilde{R}^T)_{\mathrm{re}}) + \mathrm{cov}((\mathrm{vec}\, R^T)_{\mathrm{re}})\right) \mathcal{J}_{S(s)}^T, \tag{15.8}$$

where

$$\mathcal{J}_{S(s)} = (\mathcal{L}_{\mathrm{model}}(s))_{\mathrm{Re}}\, \mathcal{J}_v$$

with $\mathcal{L}_{\mathrm{model}}(s)$ defined in Sect. 15.2.2.

15.3.3 Hypothesis Testing for Damage Localization

If the stress $\hat{S}_j(s)$ over a finite element j is close to zero, the element is a candidate for being classified as damaged. The values in the stress vector $\hat{S}(s)$ are complex values, whose real and imaginary parts can have different signs. One could for example test if the real parts are close to 0 (neglecting the imaginary part if it is small), or, more general, if both the real and imaginary parts are close to 0 for an element. For each element j, this corresponds to the hypotheses

$$\begin{cases} \mathbf{H}_0: \ \hat{S}_j(s) \neq 0 & \text{(element is undamaged)} \\ \mathbf{H}_1: \ \hat{S}_j(s) = 0 & \text{(element is potentially damaged)} \end{cases} \tag{15.9}$$

The elements in the vector $\hat{S}(s)_{\mathrm{re}}$ are asymptotically Gaussian distributed with non-zero mean under $\mathbf{H}_0$ and zero mean under $\mathbf{H}_1$. A consistent estimate of the covariance $\hat{\Sigma}_S$ of $\hat{S}(s)_{\mathrm{re}}$ can be obtained from (15.8). Then, testing $\mathbf{H}_0$ against $\mathbf{H}_1$ can be done by computing the variables

$$\hat{\chi}_j^2(1) \stackrel{\mathrm{def}}{=} \frac{(\mathrm{Re}(\hat{S}_j(s)))^2}{\hat{\sigma}_j^2} \quad \text{or} \quad \hat{\chi}_j^2(2) \stackrel{\mathrm{def}}{=} \begin{bmatrix} \mathrm{Re}(\hat{S}_j(s)) \\ \mathrm{Im}(\hat{S}_j(s)) \end{bmatrix}^T \hat{\Sigma}_j^{-1} \begin{bmatrix} \mathrm{Re}(\hat{S}_j(s)) \\ \mathrm{Im}(\hat{S}_j(s)) \end{bmatrix}, \tag{15.10}$$

where $\hat{\sigma}_j^2 = \hat{\Sigma}_S(j,j)$ is the entry (j,j) of $\hat{\Sigma}_S$ and $\hat{\Sigma}_j$ is the covariance of $\left[\mathrm{Re}(\hat{S}_j(s))\ \mathrm{Im}(\hat{S}_j(s))\right]^T$.

Thresholds t_1 and t_2 are defined in $\int_0^{t_i} f_{\chi^2(i)}(x)dx = 1 - \beta$, where $f_{\chi^2(i)}(x)$ is the probability density function of the central χ^2 distribution with i degrees of freedom ($i = 1,2$), and β is the probability of deciding that an element is undamaged while it is potentially damaged (type II error of the hypothesis test (15.9)). Then, using the test $\hat{\chi}_j^2(1)$, $\mathbf{H}_0$ is rejected and $\mathbf{H}_1$ is accepted for an element j (damage occurred), if $\hat{\chi}_j^2(1) \leq t_1$. Using test $\hat{\chi}_j^2(2)$, $\mathbf{H}_0$ is rejected and $\mathbf{H}_1$ is accepted for an element j, if $\hat{\chi}_j^2(2) \leq t_2$.

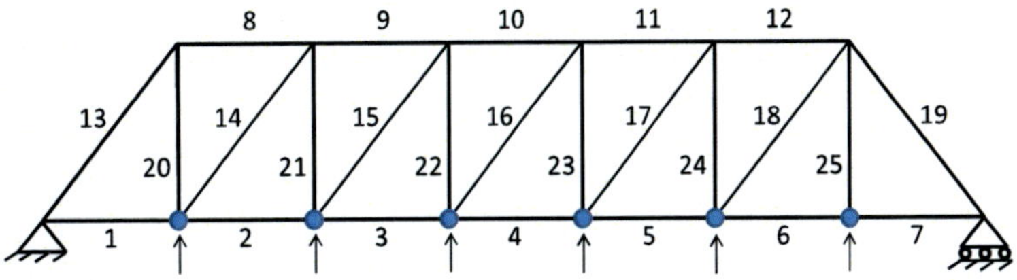

Fig. 15.1 Truss structure with six sensors

15.4 Numerical Application

Two numerical applications using simulated structures were used to validate the damage localization algorithm with hypothesis test from Sect. 15.3.3 where both applications have threshold $t_2 = 2.16$ computed at $\beta = 0.34$ to decide if an element is potentially damaged or not (the horizontal line in Figs. 15.2, 15.3, 15.5 and 15.7). Recall that the residual (the stress) is close to zero for damaged elements. Computational time in both applications was a few seconds after the uncertainty computation of the system identification results with elapsed time around 30 s in each case.

15.4.1 *Truss Structure*

The first numerical application uses a simulated 25 DOF truss structure (Fig. 15.1) to validate the damage localization algorithm. Damage was simulated by stiffness reduction on bars. For both the undamaged and the damaged state, a data sample of length $N = 25{,}000$ of acceleration data ($b = 2$) was generated with added output noise using Gaussian white noise excitation. From the output-only data, the system matrices and their covariances were estimated of the discrete-time state-space system corresponding to (15.1), using SSI and the uncertainty quantification in [3]. In order to obtain the matrices $\hat{A}_c$ and $\hat{C}_c$ of the continuous-time system and their respective covariances, a discrete to continuous transformation was made. The Laplace variable s was empirically chosen near a pole of $\hat{A}_c$ to compute the stress $\hat{S}(s)$ in (15.5). The covariance $\hat{\Sigma}_S$ of $\hat{S}(s)_{\text{re}}$ was computed from (15.8), and the corresponding χ^2 was computed from (15.10).

First, the output was generated at six sensor positions in vertical direction at the lower chord (see Fig. 15.1) with 5 % output noise added. Damage was simulated by decreasing the stiffness of element 16 by 20 %. From system identification, not all of the 25 theoretical modes could be identified at model order $n = 50$. Four well-estimated modes were chosen in both the undamaged and the damaged state using a stabilization diagram procedure [13], from where the matrices $\hat{A}_c$ and $\hat{C}_c$ and their covariances are obtained in both states. From these system matrices, the real and imaginary parts of the stress values and their covariance are computed for $s = 2\text{i}$. Once the identified modes are selected from the system identification for $\hat{A}_c$ and $\hat{C}_c$, the method is automated.

In Fig. 15.2, the values $\hat{\chi}_j^2(2)$ in (15.10) are computed on the real and imaginary parts of the stress and their covariance. Two elements are below this threshold: the damaged element 16 as well as the undamaged element 23 that is a neighboring element of 16 (see Fig. 15.1). It can be shown, in fact, that for the used sensor set element 23 is inseparable from element 16 at $s = 0$ [14], i.e. if the stress in 16 is zero, so it must be in 23. Although $s = 2\text{i}$ (used here) is not zero, it is small and the noted behavior is clearly manifested. The corresponding χ^2-values are $\hat{\chi}_{16}^2 = 0.30$ for element 16 and $\hat{\chi}_{23}^2 = 1.26$ for element 23. Thus, the lowest χ^2-value corresponds correctly to the damaged element. The elements 1–15, 17–22 and 24–25 are correctly classified as undamaged. In Fig. 15.2, the χ^2-values are only displayed until the value 300, while some of them were at more than 10^5.

The theory shows that as the number of damaged bars increases the dimension of the theoretical null space decreases and thus the estimation of vectors in the null space under noisy conditions becomes more difficult. In this case, a more precise estimation of δR^T is necessary, which requires more modes to be estimated from the system identification and thus more sensors due to the constraint $2r \geq n$. In Fig. 15.3, results are presented for two damages in bars 3 and 18, where 12 sensors (the present 6 sensors and more 6 sensors on the upper cord) were used. Twelve estimated mode pairs were chosen with both damaged and undamaged cases. It should be noted that the estimated stresses in the damaged elements are small but different from zero due to modal truncation and noise, which become more important when multiple damages are present. While the resulting χ^2 values of the damaged elements are the lowest and the value of bar 3 is correctly under the threshold, the values of bar 18 exceeds it slightly.

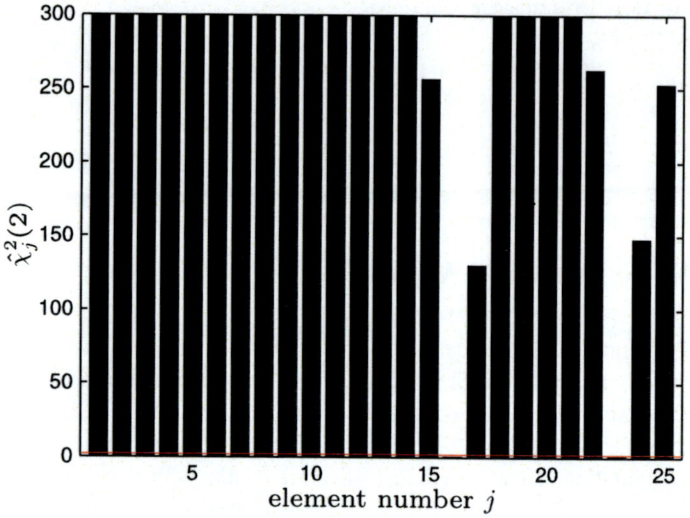

Fig. 15.2 χ^2-test values with one damage – 5 % output noise, 6 sensors, 20 % stiffness reduction in bars 16

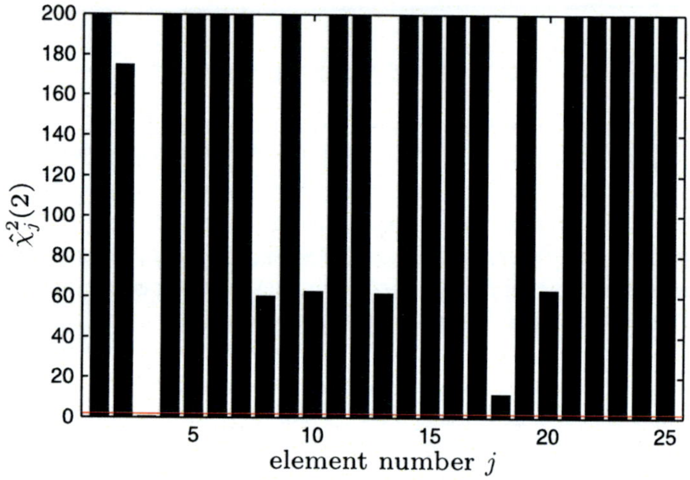

Fig. 15.3 χ^2-test values with multiple damages – 5 % output noise, 12 sensors, 20 % stiffness reduction in bars 3 and 18

15.4.2 Plate

The second numerical application was simulated to validate the damage localization algorithm. The plate (Fig. 15.4) nodes and positions "P" were spatially determined to locate undamaged and damaged regions for didactic reasons, unlike in reality. With 81 DOF, the plate has 150 cm width, 100 cm height and 1 cm of thickness. Edges are fixed and do not have DOF. The output was generated at ten sensor positions in nodes 16, 25, 29, 31, 47, 65, 70, 83, 95 and 98 with added 5 % output noise. Downsampling filter on system loads (inputs) limiting the output frequency in 150 Hz was applied to avoid undesirable folding frequencies (Nyquist frequencies) initially generated. From system identification at model order $n = 30$, matrices $\hat{A}_c$ and $\hat{C}_c$ and their covariances are obtained in both states. From these system matrices, the real and imaginary parts of the stress values and their covariance are computed for $s = 1 + 140i$. Once the identified modes are selected from the system identification for $\hat{A}_c$ and $\hat{C}_c$, the method is automated.

Two simulations were performed on the plate: the first is presented in Fig. 15.5 with damage in element 34 by decreasing the stiffness in 50 %. Nine well-estimated modes were chosen in both the undamaged and the damaged state. The values $\hat{\chi}^2_j(2)$ in (15.10) are computed on the real and imaginary parts of the stress and their covariance. Some elements are below the threshold: the damaged element 34 as well as the damage elements around the damaged region (see Fig. 15.5). The corresponding χ^2-values is $\hat{\chi}^2_{34} = 0.2$ for element 34 and is correctly classified as damaged. Neighboring elements with low results such as 25, 35–36, 45–47, 54–58 and 65–66 are near to the damaged region and also classified as damaged. In Fig. 15.5, the χ^2-values are only displayed until the value 6, while some of them were at more than 650.

Alternatively, damage position 34 and damaged region can also be visualized in Fig. 15.6. Each position "P" can be composed by vertical axis as the first number and horizontal axis as the second number (i.e. vertical axis position 4 and

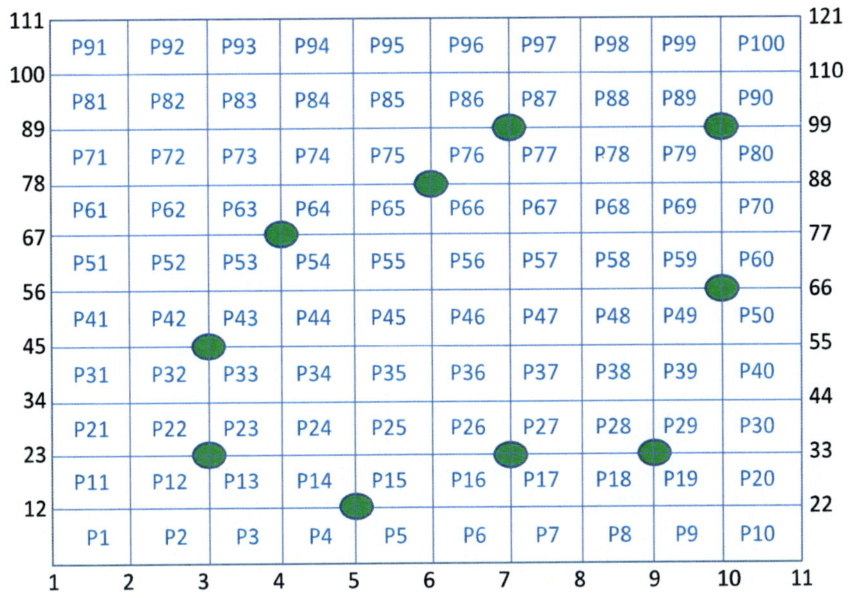

Fig. 15.4 Plate with ten sensors

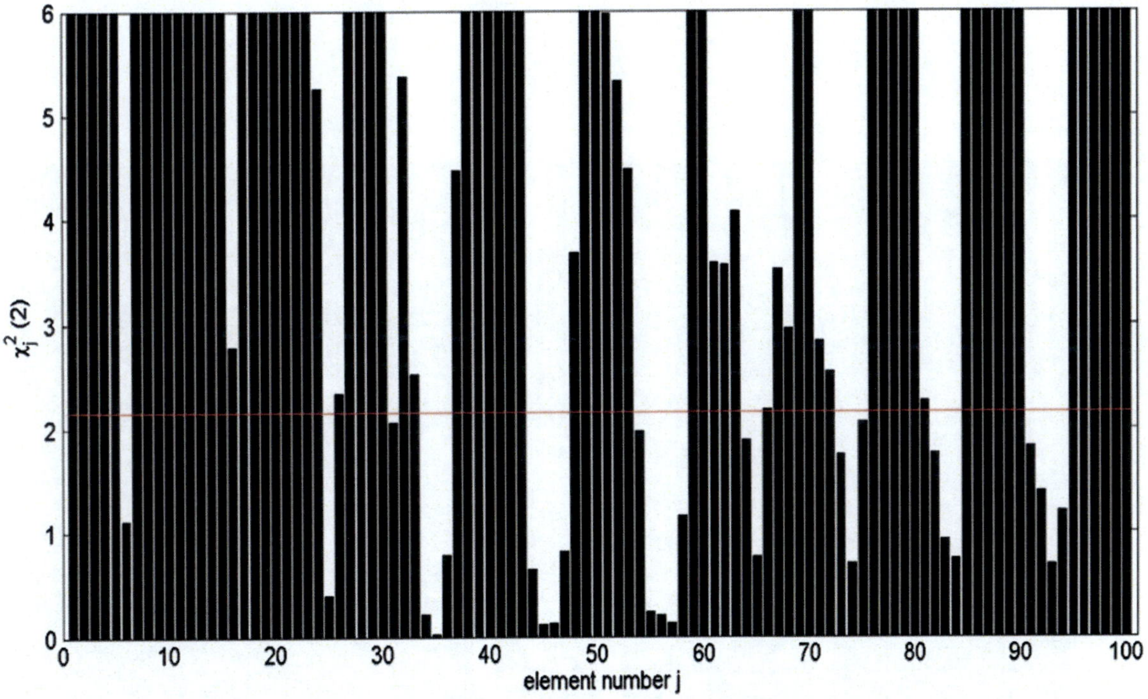

Fig. 15.5 χ^2-test values with one damage – 5 % output noise, 10 sensors, 50 % stiffness in position P34

horizontal axis position 9 lead to the position P49). The damaged/undamaged scale vary from damaged state value -3 (in red) to undamaged state value 6 (in blue). The damage in position P34 as well as the damaged region are presented in red as expected.

The second simulation on the plate has damage in position P66 by decreasing the stiffness in 50 %. Eight well-estimated modes were chosen in both the undamaged and the damaged state. The values $\hat{\chi}_j^2(2)$ in (15.10) are computed on the real and imaginary parts of the stress and their covariance. Some elements are below this threshold: the damaged element 66 as well as some elements around (see Fig. 15.7). The corresponding χ^2-values is $\hat{\chi}_{66}^2 = 0.60$ for element 66. The elements 44–46 and 56–57 are located in the damaged region and also classified as damaged. The elements 24–25 are not around the damaged position and in practical cases should be discarded after visual verification. In Fig. 15.7, the χ^2-values are only displayed until the value 6, while some of them were at more than 4,000.

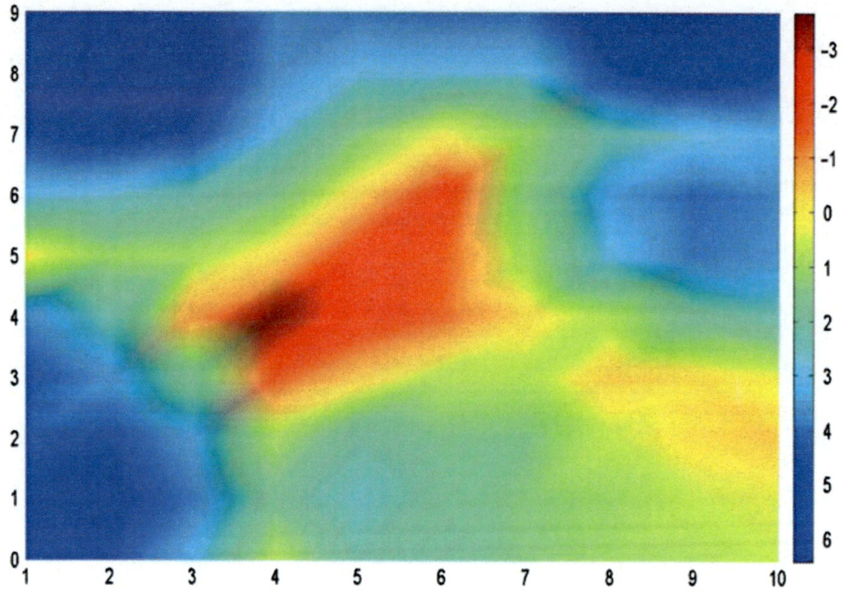

Fig. 15.6 χ^2-test from Fig. 15.5 on a plate representation – damage in P34 (*red region*)

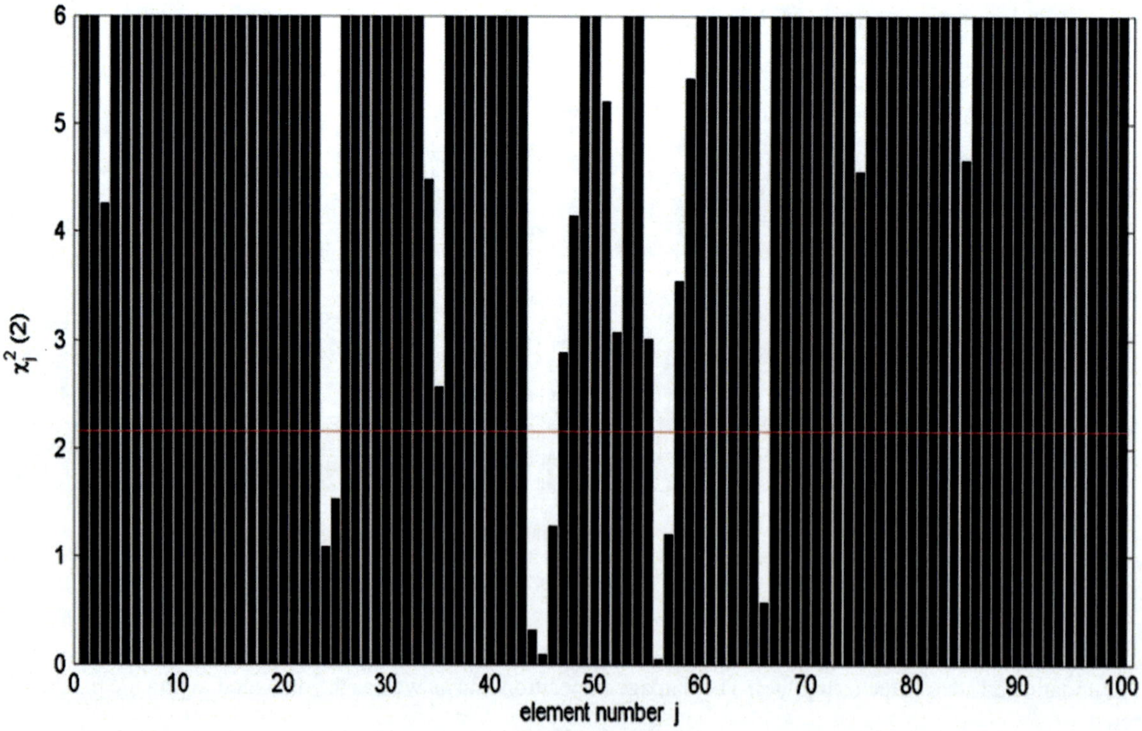

Fig. 15.7 χ^2-test values with one damage – 5 % output noise, 10 sensors, 50 % stiffness in position P66

The damage position 66 in the plate and the correspondent damaged region is also presented in Fig. 15.8. The damaged/undamaged scale values vary from damaged state value -2 (in red) to undamaged state value 8 (in blue). The damage in position P66 as well as the damaged region are presented in red as expected.

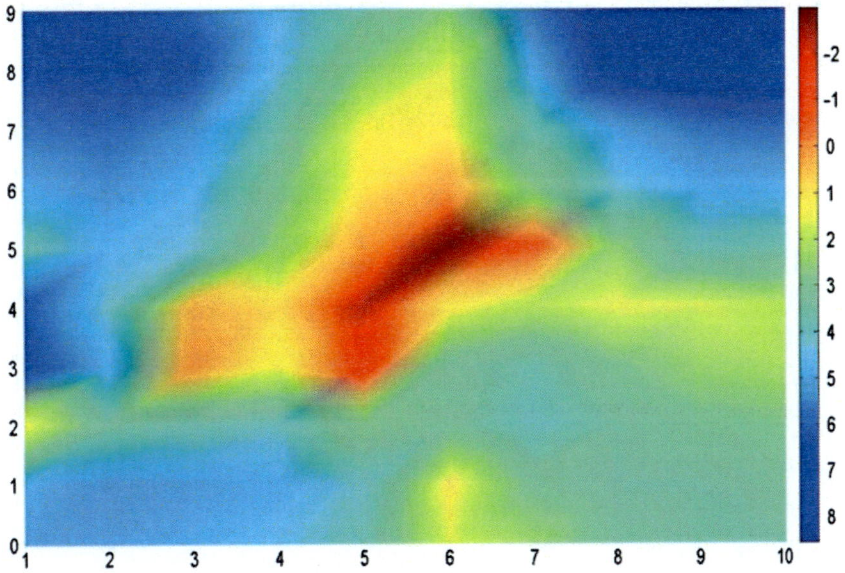

Fig. 15.8 χ^2-test from Fig. 15.7 on a plate representation – damage in P66 (*red region*)

15.5 Conclusion

Deciding whether a damage localization residual is zero or not is no more based on empirical thresholds, but on uncertainty bounds, which are now obtained for each element that is tested for damage separately, unlike in [1]. Thus, the intrinsic uncertainty from the data is propagated properly for each evaluated element in the damage localization residual $S(s)$. Then, it can be decided if an element is potentially damaged or undamaged by using hypothesis test that takes into account the uncertainties, which was successfully performed in two numerical applications. Choosing a different identification method could yield a different performance. Future work includes the aggregation of the damage localization residual at different values of the Laplace variable s using other statistical techniques and the validation of the method on a large-scale example under realistic noisy conditions.

Acknowledgements This work was partially supported by the European project FP7-PEOPLE-2009-IAPP 251515 ISMS.

References

1. Bernal D (2010) Load vectors for damage location in systems identified from operational loads. J Eng Mech 136(1):31–39
2. Pintelon R, Guillaume P, Schoukens J (2007) Uncertainty calculation in (operational) modal analysis. Mech Syst Signal Process 21(6):2359–2373
3. Reynders E, Pintelon R, De Roeck G (2008) Uncertainty bounds on modal parameters obtained from stochastic subspace identification. Mech Syst Signal Process 22(4):948–969
4. Döhler M, Mevel L (2011) Robust subspace based fault detection. In: Proceedings of 18th IFAC world congress, Milan
5. Döhler M, Lam X-B, Mevel L (2011) Uncertainty quantification for stochastic subspace identification on multi-setup measurements. In: Proceedings of 50th IEEE conference on decision and control, Orlando
6. Döhler M, Mevel L, Efficient multi-order uncertainty computation for stochastic subspace identification. Mech Syst Signal Process (2013). In press. doi:10.1016/j.ymssp.2013.01.012
7. Bernal D (2006) Flexibility-based damage localization from stochastic realization results. J Eng Mech 132(6):651–658
8. Van Overschee P, De Moor B (1996) Subspace identification for linear systems: theory, implementation, applications. Kluwer Academic Publishers, Kluwer, Boston
9. Peeters B, De Roeck G (1999) Reference-based stochastic subspace identification for output-only modal analysis. Mech Syst Signal Process 13(6):855–878
10. Chiuso A, Picci G (2004) The asymptotic variance of subspace estimates. J Econ 118(1–2):257–291
11. Marin L, Döhler M, Bernal D, Mevel L (2012) Uncertainty quantification for stochastic damage localization for mechanical system. In: Proceedings of 8th IFAC safeprocess, Mexico City

12. Döhler M, Marin L, Bernal D, Mevel L Statistical decision making for damage localization with stochastic load vectors. Mech Syst Signal Process (2013). In press. doi:10.1016/j.ymssp.2012.12.011
13. Döhler M, Mevel L (2012) Fast multi-order computation of system matrices in subspace-based system identification. Control Eng Pract 20(9):882–894
14. Bernal D (2002) Load vectors for damage localization. J Eng Mech 128(1):7–14

Chapter 16
Robust Tolerance Design in Structural Dynamics

Chaoping Zang, Jun Yang, and M.I. Friswell

Abstract Tolerance is a major source of uncertainty and contributes significantly to the variation of dynamic responses, worsening the repeatability of assembled products. A systematic tolerance design strategy is required to control this kind of variation, and an approach for tolerance design based on robust design theory is proposed in this paper with a focus on the optimization of the dynamic response. The approach is based on Taguchi's method, and performed by the following steps: (1) define the input and output parameters for the problem; (2) determine the effects of the control factors on the dynamic responses of interest; (3) identify factors to be adjusted and transform the problem into a multi-objective optimization. A benchmark tolerance design of a joint assembly of aero engine casings is used to verify the feasibility of the approach.

Keywords Dynamics • Robust design • Tolerance design • Taguchi's method • Multi-objective optimization

16.1 Introduction

In practical industrial cases, an assembled structure usually cannot exactly meet the nominal design due to uncertain inputs such as loads, material properties, and environmental factors. On the other hand, manufacturing and assembly tolerances, as a major source of uncertainties, significantly contribute to the variability of assembled products that are built to the same design, worsening the performance variation of the ultimate products. In the case of aero-engines, even though components are all manufactured within the tolerance limits and assembly is performed according to the given handbook, the whole-engine vibration of the final products can be quite different. In the worst case, disqualification may happen. Similar cases can be found in the automobile industry, robotics [1], and other fields. Effective and efficient tolerance design to assure the boundary of the system dynamic response has drawn considerable industrial attention and still requires investigation.

Conventional parameter optimization focuses on keeping the performance on target, ignoring the effects of input variations (Fig. 16.1a). Clearly, the main task is to determine *the optimal solution* that simply brings the objective of interest on target. But when uncertainties arise, deciding whether the performance of the solution remains on target is difficult. Large variations may exist due to the uncertainty.

Robust design is different to conventional parameter optimization by focusing on both "on target" and also "low variation" (Fig. 16.1b). The definition of robustness can be described as: *"A product or process is said to be robust when it is insensitive to the effects of sources of variability, even though the sources themselves have not been eliminated"* [2]. It should be noted that the optimization within robust design involves at least two targets – the mean and standard deviation of the response of interest – forming a multi-objective problem. Solving the multi-objective problem directly leads to the Pareto set – a set of *optimal solutions* – giving designers a choice to balance on target performance and low variation.

Robust design only attracted limited attention in dynamics applications until Zang et al. [3] optimized a tuned mass damper using this strategy. Since then, a number of attempts, both in research and practical industry, have been made [4–6].

C. Zang (✉) • J. Yang
College of Energy and Power Engineering, Nanjing University of Aeronautics and Astronautics, Nanjing, 210016 China
e-mail: c.zang@nuaa.edu.cn; j.yang.cherry@gmail.com

M.I. Friswell
College of Engineering, Swansea University, Swansea SA2 8PP, UK
e-mail: m.i.friswell@swansea.ac.uk

T. Simmermacher et al. (eds.), *Topics in Model Validation and Uncertainty Quantification, Volume 5*,
Conference Proceedings of the Society for Experimental Mechanics Series 41,
DOI 10.1007/978-1-4614-6564-5_16, © The Society for Experimental Mechanics, Inc. 2013

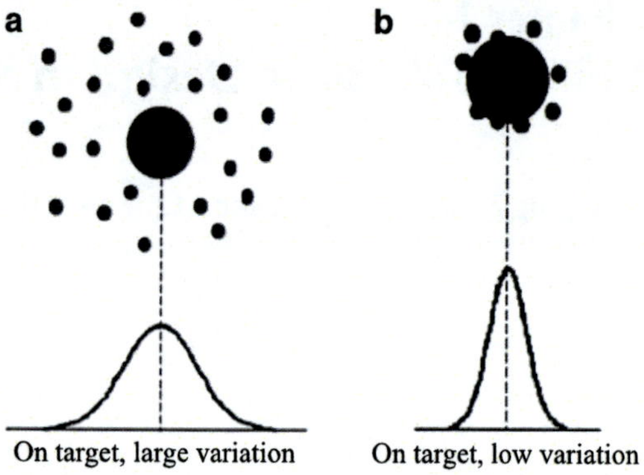

Fig. 16.1 Optimization result types [3]. (**a**) On target, large variation. (**b**) On target, low variation

The most practical and efficient robust design methodology today is still Taguchi's method, which has been widely accepted and applied in various fields, such as chemistry [7], manufacturing [8], and electronics [9].

In this paper, a robust tolerance design methodology in structural dynamics is proposed based on Taguchi's method. By considering the tolerances for component manufacturing and assembly, the strategy aims to limit the range of dynamic responses of the assembled products for the same nominal design, i.e. with a desired mean value. Generally, the method involves two goals – to bring the nominal design on target and to limit performance variation within a tolerable range. Three steps are required: define the input/output parameters for the problem; determine the effect of the control factors on the responses of interest by an orthogonal array experiment and analysis of variance (ANOVA), and determine possible nominal values; identify factors for adjustment and transform the problem into a multi-objective optimization problem.

16.2 Robust Tolerance Design Strategy

16.2.1 Robust Design Theory

In robust design theory, the input/output relation of a dynamic system should be investigated first, and this can be represented by a P-diagram (Parameter-diagram), as shown in Fig. 16.2 [10]. Input parameters can be classified as signal factors (s), control factors (x) and noise factors (z): signal factors are the trigger to generate responses; control factors are basically design parameters that designers can choose; and noise factors are elements beyond designer's control, but have an effect on the output of the system. For example, during a vibration test of a cantilever beam, the impulse from the hammer is the signal factor, the locations of the excitation and sensors are control factors, and the vibrations of the test table are noise factors.

By considering these three factors, the robust design problem is to determine the design variables with regard to uncertainties, or mathematically

$$\begin{cases} \text{find} x \in \Omega \\ \text{to minimize; } \{\mathrm{E}(f(s,x,z)), \sigma(f(s,x,z))\} \\ \text{subject to}: h_k = 0 \, (k = 1,2,\ldots,l) \\ g_j \leq 0 \, (j = 1,2,\ldots,m) \end{cases} \tag{16.1}$$

where $E(f(s,x,z))$and $\sigma(f(s,x,z))$ respectively reflect the mean and standard deviation of the desired output about control and noise factors. The former reflects whether the system response is on-target, and the latter reflects its variation. This may be compared to conventional parameter optimization, given by

$$\begin{cases} \text{find} x \in \Omega \\ \text{to minimize}: f(x) \\ \text{subject to}: h_k = 0 \, (k = 1,2,\ldots,l) \\ g_j \leq 0 \, (j = 1,2,\ldots,m) \end{cases} \tag{16.2}$$

where $x = [x_1, x_2, \ldots, x_n]$ denotes the design vector, Ω the design domain of x, $f(x)$ the objective function, $h_k = 0$ the equality constraints and $g_j \leq 0$ the inequality constraints. Clearly instead of simply considering the product performance target, robust

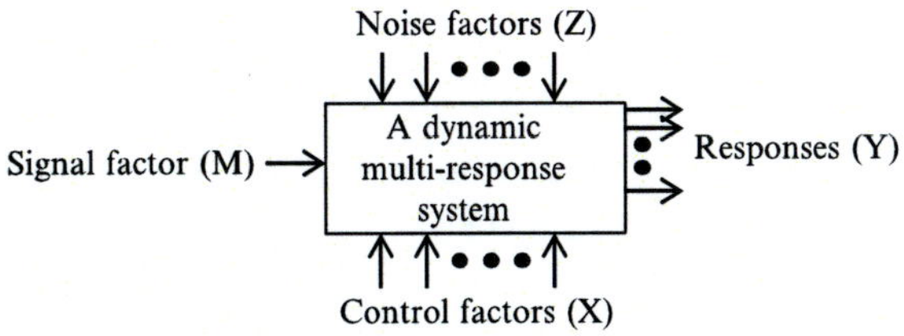

Fig. 16.2 The P-diagram for a dynamic system [10]

design also focuses on the control of variations due to the noise factors. Here, to examine the effect of uncertainties, methods of uncertainty propagation from the literature [11] are used, and uncertainty is represented in probabilistic form.

16.2.2 Taguchi's Method

Taguchi's method is a three-stage design procedure: system design, parameter design, and tolerance design. In system design, the target is to decide the requirements of the product, basically the functions and configurations of the product. The optimum set of control factors to give the best nominal design, in addition to low variation, is identified in the parameter design stage. Finally, tolerances are determined in the tolerance design stage, taking both costs and effects on the product response variation into consideration. More attention has been paid to the parameter design stage, while far fewer technical professionals have become proficient in the follow-on process known as tolerance design [12].

16.2.2.1 Orthogonal Array Experiment

Instead of a full factorial design of experiments, which requires large calculation effort, the orthogonal array experiment needs only a fraction of the full factorial combination, dramatically reducing time and resource consumption. Also, it allows an independent and efficient mathematical assessment of the effect on the objective of each variable. In this special matrix, the columns are mutually orthogonal. That is, for any pair of columns, all combinations of factor levels occur, and they occur an equal number of times. The number of combinations required for s parameters with r levels is $s(r-1)+1$, based on the assumption that there is no interaction among the design parameters. The procedure of constructing orthogonal arrays is given in [12].

16.2.2.2 Analysis of Variance (ANOVA)

The ANOVA process is used to decompose the main parameter effects and the random experimental effects (experimental error). ANOVA has the mathematical capability to process the experimental data into sums of squares that hold the key to the quantification of the main effect contribution for each design parameter. Furthermore, errors are represented by empty columns in orthogonal array, just like other parameters. Two forms of analysis are used in this paper to quantify the variation resident in the experimental parameters and the experimental error effects:

1. Sum of squares (SS): Indicates the differences associated with a parameter or error compared to the overall mean of the data set, i.e.

$$\text{SS parameter or error} = \sum_{i=1}^{n} (y_i - \bar{y})^2 \tag{16.3}$$

   For example, the calculation of SS for parameter A with r levels is

$$SS_A = (no.of\,exp.at\,A1)\,(\bar{y}_{A1} - \bar{y})^2 + (no.of\,exp.at\,A2)\,(\bar{y}_{A2} - \bar{y})^2 + (no.of\,exp.at\,A3)\,(\bar{y}_{A3} - \bar{y})^2 \tag{16.4}$$

   where $no.of\,exp.at\,Ai$ indicates the total number of experiments at level Ai, $\bar{y}_{Ai}$ is the average of the dynamic responses at level Ai, and $\bar{y}$ is the average of the dynamic responses of all experiments.

2. Mean of the sum of squares (MS): Quantifies the mean value of the sum of squares associated with a parameter or with an error, obtained by dividing by the DOF allocated to the parameter or set of errors. The number of DOF for a parameter with r levels is $(r-1)$, and the DOF for the error is (number of experiments $-\sum$DOF of each parameter $-$ 1). Thus

$$MS = \frac{SS\ parameter\ or\ error}{DOF\ for\ the\ parameter\ or\ error} \tag{16.5}$$

To test the significance of the parameter effects compared with the experimental error, the F-ratio is used, also referred to as the variance ratio, given by

$$F = \frac{MS}{MS_e} = \frac{mean\ square\ due\ to\ design\ parameter}{mean\ square\ due\ to\ experimental\ error} \tag{16.6}$$

Some suggested generalized values for F ratios are

- $F < 1$: the experimental error outweighs the parameter effect; the parameter is insignificant and indistinguishable from the experimental error.
- $F \approx 2$: the parameters have only a moderate effect compared to the experimental error.
- $F > 4$: the parameter effect is strong compared to the experimental error and is clearly significant.

In the proposed robust tolerance design strategy, the effectiveness of each parameter is determined by its F-ratio, and insignificant parameters are eliminated.

16.2.2.3 The Application of Taguchi's Method

To achieve the goal that dynamic performance should both be on target and possess low variation, the nominal design and the tolerance of the design parameters are determined separately. The orthogonal array experiment and ANOVA is applied to partially fulfill the first mission. More specifically, Taguchi's method is used first to localize the most likely solutions to bring the objectives on target, thus reducing the search space for further optimization.

Parameters are determined in Taguchi's method using the following steps:

1. Define upper and lower bounds for each critical parameter under evaluation and the objectives, and then select the appropriate orthogonal array;
2. Set up and run the experiments;
3. Conduct a formal analysis of variance (ANOVA) on the data, obtaining the sensitivity analysis of each parameter and decide on the nominal design;
4. If necessary, set up a quality-loss function, and obtain the tolerance adjustment for each parameter.

Due to its efficiency and relatively high accuracy in examining the effect of the parameters on the objectives over the whole design space, Taguchi's method has attracted a lot of attention, and the orthogonal array experiment and ANOVA has shown great practical use in applications.

16.2.3 Multi-objective Optimization

16.2.3.1 Calculation Strategy

Problems with only one target hardly exist in the real industrial world. Usually several (often competing) objectives are to be simultaneously optimized, forming a multi-objective optimization problem. Instead of a single optimal solution set, there tends to be a set of alternative solutions, each of which must be considered equivalent in the absence of further information regarding the relative importance of each of the objectives in the solution vectors. Such a solution set is called the Pareto-optimal set.

Traditionally, the core approach to solve a multi-objective problem is to turn it into a single-objective problem. One way is to keep the most essential target and turn others into constraints. This may lead to the problem that only the response corresponding to the kept target is optimum, while others just meet the upper or lower limit of the desired output. Another approach is to form a weighted sum, combining multiple objectives into a single objective given by

$$y = \sum_{i=0}^{n} w_i y_i \quad \text{where} \quad \sum_{i=0}^{n} w_i = 1, \tag{16.7}$$

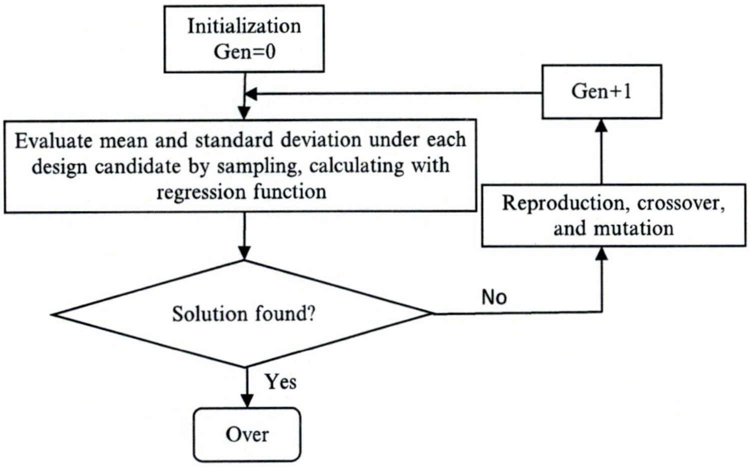

Fig. 16.3 Framework of proposed MOGA

n is the total number of objectives, y_i the ith objective, and w_i the ith weighted factor. Despite its easy implementation, the fatal defect in obtaining the Pareto set via this approach has been pointed out decades ago [13].

Several stochastic optimization techniques, such as simulated annealing and ant colony optimization, have also been used to generate the Pareto set. Unfortunately these methods often get stuck at a good approximation and do not guarantee to identify the optimal trade-offs [14]. The application of evolutionary algorithms (EA) to multi-objective optimization is computational efficient since it is easy to parallelize, enabling several members of the Pareto-optimal set to be found in a single run. Many attempts have been made to develop efficient evolutionary multi-objective algorithms. Particle Swarm Inspired Evolutionary Algorithm (PS-EA) [14] uses a synergistic combination of Particle Swarm Optimization (PSO) and Evolutionary Algorithm (EA) to overcome the limitations of both strategies. The Strength Pareto Evolutionary algorithm [15] is based on the theory of evolution strategies. The main feature of the algorithm is that it maintains diversity in the population so that a well distributed wide spread trade-off front is reached, preventing premature convergence to a part of the Pareto front. In [16] a Non-dominated Sorting Genetic Algorithm (NSGA) is proposed and stands out in its efficiency. The algorithm uses elitism and a crowded comparison operator that keeps diversity without specifying any additional parameters.

16.2.3.2 The Application of Multi-objective Optimization

To simultaneously control the mean and standard deviation, multi-objective optimization is applied in the second stage of the proposed strategy. Tolerances are considered as a major source of performance variation. Thus noise factors turn out to be the actual value of the design parameters and the control factors are the nominal values and tolerances of the design parameters.

Another key requirement for the multi-objective optimization in robust tolerance design in structural dynamics is to efficiently evaluate the mean and standard deviation of the system performance for each design candidate. The system dynamic response cannot be written in simple closed form equations; accurate responses require either extensive experimental data or simulation (if an FE model is available). To evaluate the mean and standard deviation in probabilistic form requires many samples. Both factors lead to high costs in time and resources for uncertainty problems concerning dynamics.

To obtain a relatively accurate function, instead of analyzing the whole design space, regression analysis is performed on a small range. A second orthogonal array experiment, around the possible best solution given at the first stage, is set up and the regression function of interest is formed. During the operation of the multi-objective genetic algorithm, the regression function helps to evaluate the mean and standard deviation for each set of design candidates.

The optimization procedure using multi-objective genetic algorithm is as follows (see Fig. 16.3):

1. Initialize the first generation within given lower and upper bounds;
2. Evaluate the mean and standard deviation response for each design candidate;
3. Rank the candidates using a non-dominated sorting procedure;
4. Check whether the stopping criteria are satisfied (the stopping criteria can be either a certain number of generations, or a certain desired target for the problem). If not, repeat procedure starting from 2).

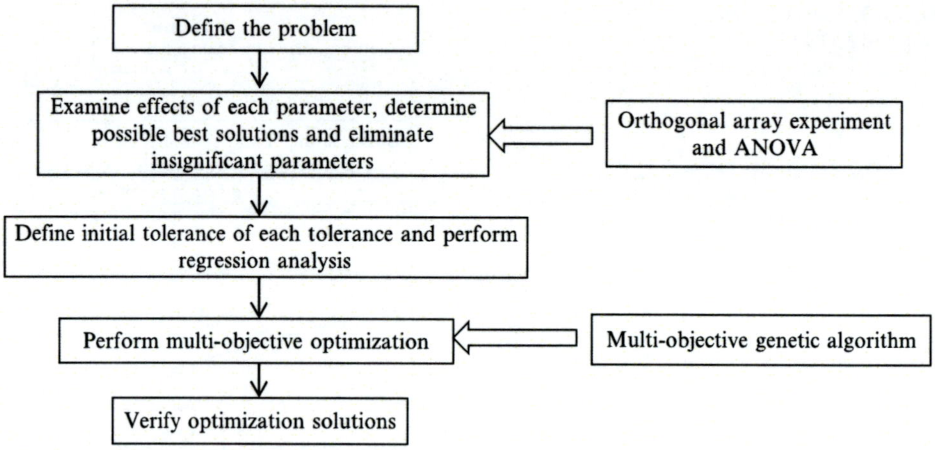

Fig. 16.4 Proposed tolerance design strategy

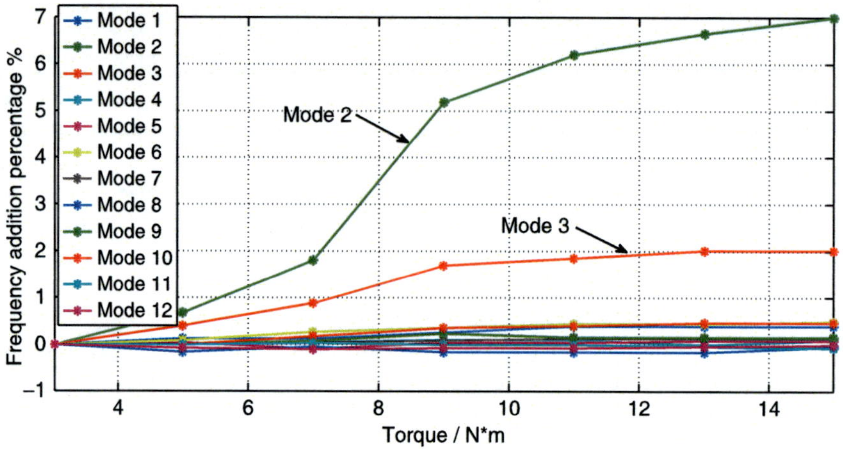

Fig. 16.5 The effect of torque on modes 1–12

16.2.4 Proposed Tolerance Design Strategy

In summary, the detailed steps of proposed tolerance design strategy are as follows (see Fig. 16.4):

1. Define the problem – the three factors, design space and objectives;
2. Select the appropriate orthogonal array, and set up and run experiments;
3. Conduct a formal analysis of variance (ANOVA) on the data, obtaining the sensitivity of each parameter, determine the nominal design for each parameter that brings the objective on target, and eliminate insignificant parameters;
4. Define the initial tolerance of each parameter and run experiments for regression analysis;
5. Perform the multi-objective optimization by the proposed strategy in Sect. 16.3.
6. Verify the optimal solutions.

16.3 Case Study: Joint Assembly of Two Aero-Engine Casings

The dynamic analysis of joint assemblies is a complex problem that contributes significantly to the performance variation of joined products. Most of the literature has focused on the modeling the joints. In this case study, the effects of the joint assembly on the dynamic response are controlled through the proposed tolerance design procedure. This work is based on previous research, which indicated that the torque of bolts and nuts significantly influences the frequencies of some modes. Figure 16.5 shows the influence of the torque variation on the first 12 modes of interests.

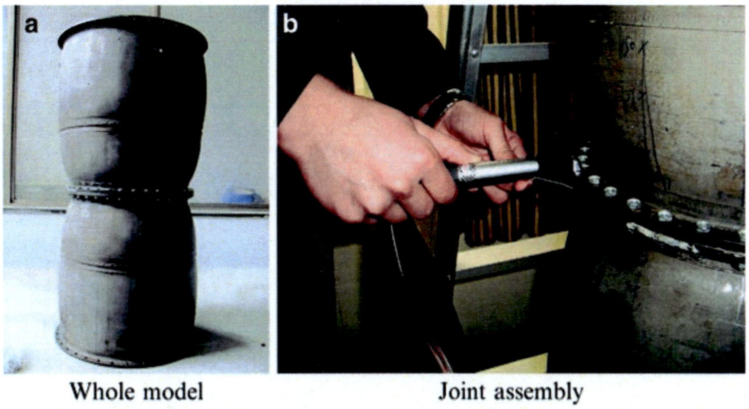

Fig. 16.6 The assembled structure and joint. (**a**) Whole model. (**b**) Joint assembly

Table 16.1 The initial parameter design space

Parameter	Lower bound	Upper bound
Torque/Nm	3	15
Number of bolt-nut pair	18	36
Assembly order	In turn, diagonally, section diagonally (see Fig. 16.7)	

Table 16.2 The targets for modes 2 and 3

–	Nominal design (Hz)	Variation limit (Hz)
Second natural frequency	278	±2
Third natural frequency	391	±2

Table 16.3 The parameter levels

	Level		
Parameter	1	2	3
Torque/Nm	5	9	13
Number of bolt-nut pair	18	27	36
Assembly order	In turn	Diagonally	Section diagonally

16.3.1 Problem Definition

The structure shown in Fig. 16.6 consists of two casings jointed together by bolts and nuts (M6). Three parameters are considered: assembly torque, number of bolt-nut pairs, and assembly order. Assuming that the minimum number of bolt-nut pairs required for strength, sealing and other aspects is 18, and maximum number allowed in the structure is 36. The initial design space is shown in Table 16.1. Since the first natural frequency remains relatively constant regardless of the torque variation (see Fig. 16.5), the second and third natural frequencies are considered as targets here, shown in Table 16.2. For all products, the mean of the second natural frequency should close to 278 Hz while the variation limit should be ±2 Hz. For the third natural frequency, the mean value should be close to 391 Hz, while the variation limit is ±2 Hz.

16.3.2 Determine the Nominal Design

Parameters are determined to have three levels, and thus an L9 array is chosen. The levels of the parameters are shown in Table 16.3. Experiment planning and the results are listed in Table 16.4, and the FRFs of the nine experiments are shown in Fig. 16.8. Clearly the second and third natural frequencies vary dramatically in design space.

Sensitivity analysis for the first three natural frequencies is shown in Fig. 16.9. Each point on x-axis reflects a level of a parameter. For example, "T_1" means the first level of Torque, "N_1" means the first level of the Number of bolt-nut pair, and "A_1" means the first level of Assembly order. From Table 16.4 and Fig. 16.6 we can see that in design space, the first natural frequency hardly varies, while for both the second and third natural frequencies the most influential parameter is the torque. The other two parameters (the number of bolt-nut pairs and the assembly order) have some limited effects that are further evaluated by ANOVA (see Table 16.5). Table 16.5 shows that for both the second and third natural frequencies, the torque has significant effects compared with error, with F values larger than 6. The number of bolt-nut pairs has a strong effect for the second natural frequency but a relatively moderate effect for the third natural frequency. However, for both

Table 16.4 Experiment planning and results

No	Torque/Nm	Number of bolt-nut pair	Assembly order	Fl/Hz	F2/Hz	F3/Hz
1	5	18	In turn	238.12	272.5	392.19
2	5	27	Diagonally	237.81	273.12	388.75
3	5	36	Section diagonally	238.44	276.88	387.5
4	9	18	Diagonally	238.44	273.44	392.81
5	9	27	Section diagonally	238.44	277.19	391.56
6	9	36	In turn	238.75	280.00	390.00
7	13	18	Section diagonally	238.44	275.62	393.75
8	13	27	In turn	239.06	286.56	394.69
9	13	36	Diagonally	239.06	292.81	394.38

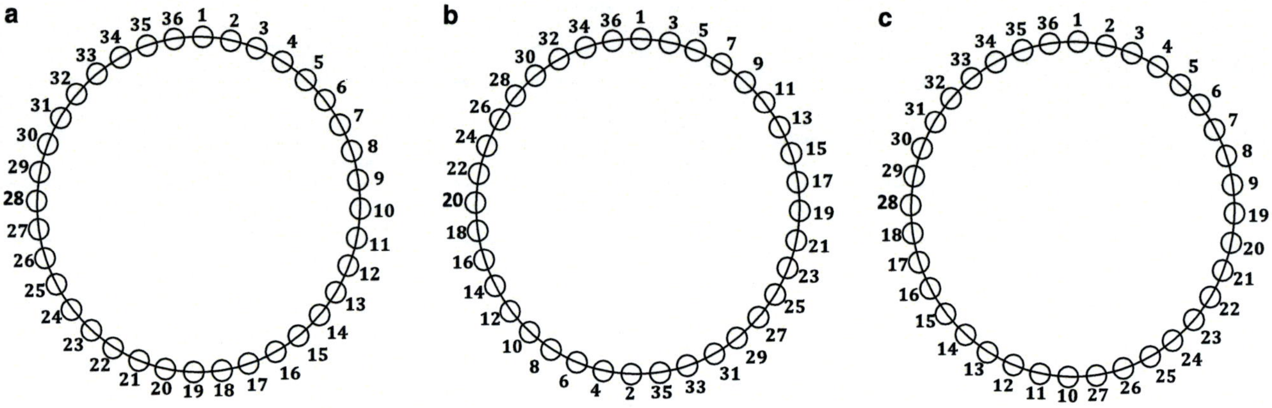

Fig. 16.7 The assembly order for the bolts. (a) In turn. (b) Diagonally. (c) Section diagonally

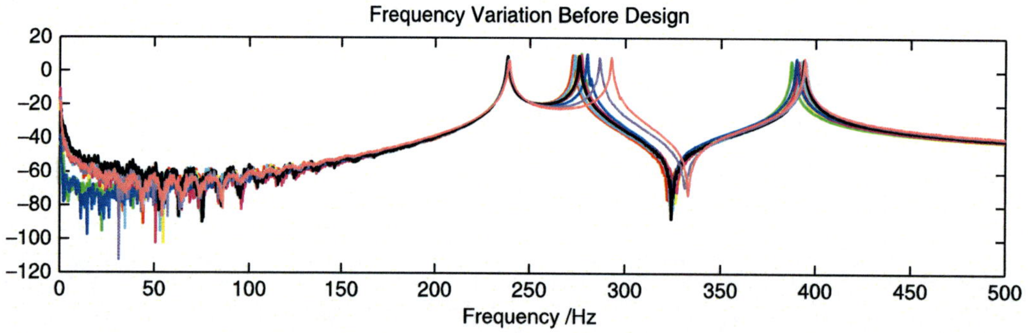

Fig. 16.8 FRFs of the L9 experiments

natural frequencies, the F values for the assembly order are less than 1, outweighed by the experimental error, which cannot serve as a control parameter.

In summary, the torque and the number of bolt-nut pairs are two significant parameters that should be taken into further consideration in design procedure. Regarding the targets listed above, the parameter combination T2N2 is chosen. As for assembly order, the diagonal order is chosen according to the normal practice described in the aero-engine assembly handbook.

16.3.3 Tolerance Design: Multi-objective Optimization

As determined in step (2), the torque for the joint assembly is 9 Nm, the number of bolt-nut pairs is 27, which means that during assembly 27 out of 36 pairs should be guaranteed to work, and diagonally assembly is used. To examine the second and third natural frequencies close to this design vector, another orthogonal array experiment is set up and carried out, and

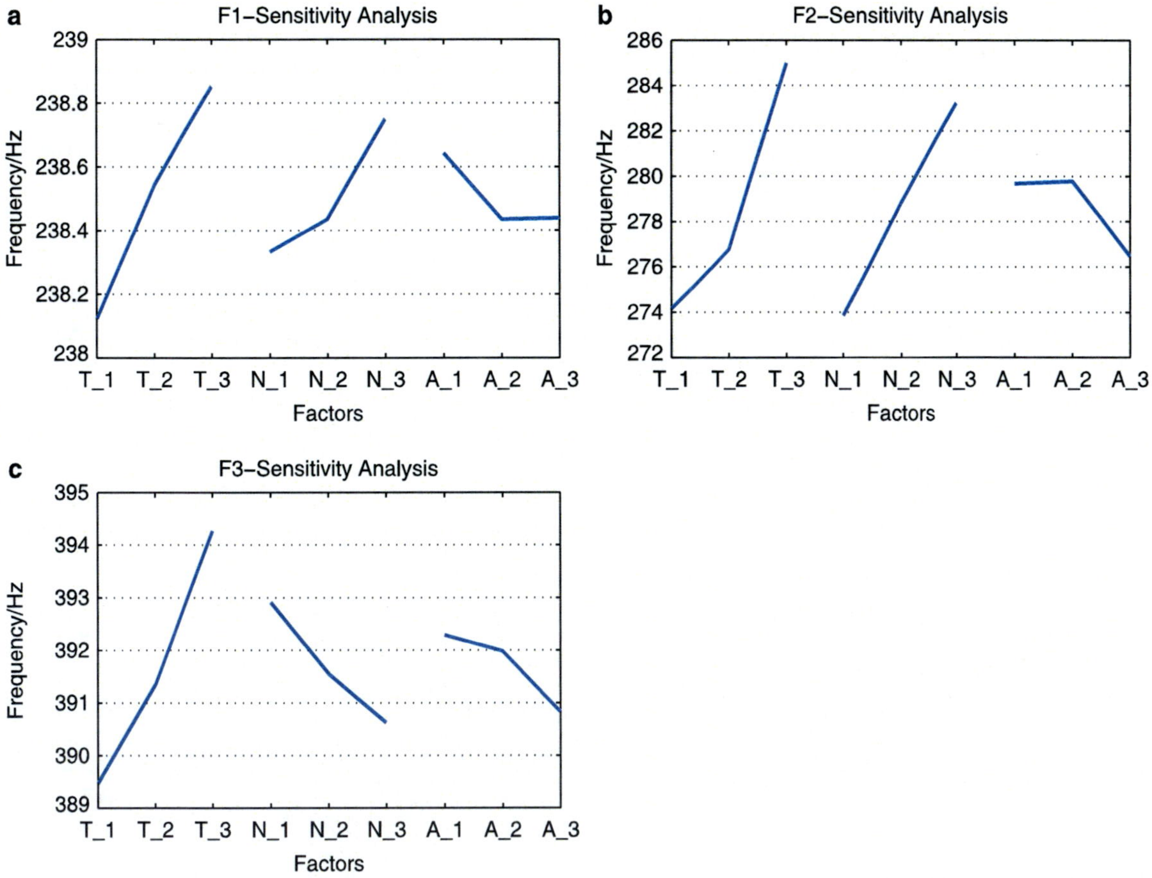

Fig. 16.9 Sensitivity analysis of first three natural frequencies. (**a**) Sensitivity analysis for natural frequency 1. (**b**) Sensitivity analysis for natural frequency 2. (**c**) Sensitivity analysis for natural frequency 3

Table 16.5 ANOVA for the assembled casings

	F2				F3			
Factor	SS	d.o.f	MS	F	SS	d.o.f	MS	F
Torque	191.71	2	95.86	6.27	35.01	2	17.51	7.41
Number of bolt-nut pair	132.08	2	66.04	4.32	7.95	2	3.98	1.68
Assembly order	21.51	2	10.76	0.70	3.54	2	1.77	0.75
–								
Error	30.59	2	15.29	–	4.72	2	2.36	–

regression analysis is performed afterwards. Again, an L9 orthogonal array is adopted, as shown in Table 16.6 along with experiment results. To simulate the second and third natural frequencies in this parameter region, regression analysis is performed for the two frequencies. The generated response surfaces are shown in Fig. 16.10 and the two functions are

$$\textit{Second natural frequency}: 276.06 + 0.18^{*}T^{2} - 3.91^{*}T - 0.01^{*}N^{2} + 0.46^{*}N + 0.07^{*}T^{*}N \quad (16.8)$$

$$\textit{Third natural frequency}: 398.16 + 0.05^{*}T^{2} - 1.05^{*}T - 0.003^{*}N^{2} - 0.22^{*}N + 0.03^{*}T^{*}N \quad (16.9)$$

Table 16.6 shows that an initial torque tolerance of 2 Nm about a nominal torque of 9 Nm, and 9 bolt-nut pairs about 27 bolt-nut pairs, the largest variation in the second natural frequency can be as high as 15 Hz and the variation in the third natural frequency is 5 Hz, far beyond the targets for the variation limit. A multi-objective optimization problem is then formed to determine the final parameters. The design parameters and objectives of the optimization are listed in Table 16.7. Assuming that the uncertain parameters both agree with 3σ theory, the input parameters are converted into probabilistic form, as shown in Table 16.8. The resulting Pareto set is given in Table 16.9. Designers can choose the final solution from the set according to the practical needs for the four objectives. For example, if Solution 6 is chosen, the assembly torque should meet a

Table 16.6 Orthogonal array experiment for tolerance design

–	Torque/Nm	Number of bolt-nut pair	Fl/Hz	F2/Hz	F3/Hz
1	7	18	238.12	270.31	390.94
2	7	27	238.44	274.69	390.00
3	7	36	238.44	279.06	389.06
4	9	18	238.44	273.44	392.81
5	9	27	238.44	276.88	391.25
6	9	36	238.75	280.00	390.00
7	11	18	238.44	276.25	394.06
8	11	27	238.75	282.19	393.44
9	11	36	239.06	285.31	392.19

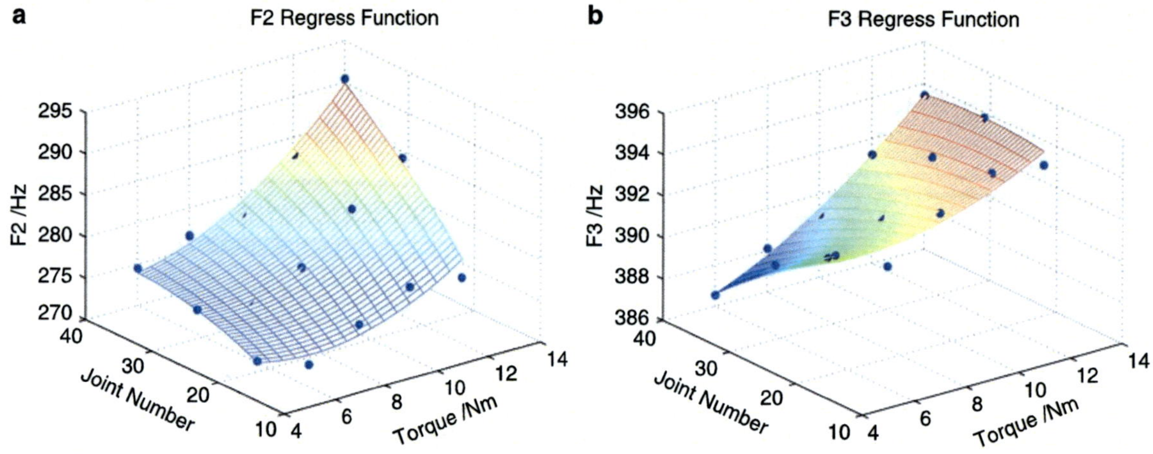

Fig. 16.10 Regression analysis (**a**) Regression analysis for natural frequency 2 (**b**) Regression analysis for natural frequency 3

Table 16.7 Parameters for tolerance design

Inputs				
–	Lower bound	Upper bound	Objectives	
Torque/Nm	7	11	F2/Hz	278
Tolerance of torque	0	2	ΔF2/Hz	< 2
Number of bolt-nut pair	24	30	F3/Hz	391
Tolerance of number of bolt-nut pair	0	6	ΔF3/Hz	< 2

Table 16.8 Converted input parameters

–	Lower bound	Upper bound
E(T)	7	11
σ	0	0.67
E(N)	24	30
σ	0	0.67

probabilistic distribution with a mean of 7 Nm and a standard deviation of 0.6 Nm, and number of bolt-nut pair should have a mean of 29 and a standard deviation of 1. Such input parameters, lead to the second natural frequency with a mean of 276.48 Hz and a standard deviation of 0.57 Hz, and the third natural frequency with a mean of 390.21 Hz and a standard deviation of 0.35 Hz.

Table 16.9 Results of the optimization

–	E(T)	σ(T)	E(N)	σ(N)	F2/Hz	F3/Hz	σ(F2)/Hz	σ(F3)/Hz
1	8	28	0.4	1	276.93	390.86	0.58	0.26
2	7	24	0.6	1	274.54	391.10	0.52	0.25
3	7	25	0.5	1	274.95	390.94	0.50	0.25
4	7	26	0.5	1	275.35	390.75	0.48	0.26
5	7	25	0.4	1	274.98	390.92	0.49	0.22
6	7	29	0.6	1	276.48	390.21	0.57	0.35
7	8	26	0.4	1	276.14	391.17	0.62	0.24
8	9	27	0.2	1	277.62	391.55	0.57	0.17
9	7	26	0.6	1	275.40	390.76	0.51	0.28
10	7	28	0.6	1	276.10	390.41	0.57	0.32

Table 16.10 Verification experiments results

–	Mean	Deviation of target (%)	Standard deviation	Deviation of target (%)	Number of experiments beyond ±2 Hz
F2	276.76	0.36	1.05	37	2
F3	390.82	0.04	0.68	1.5	0

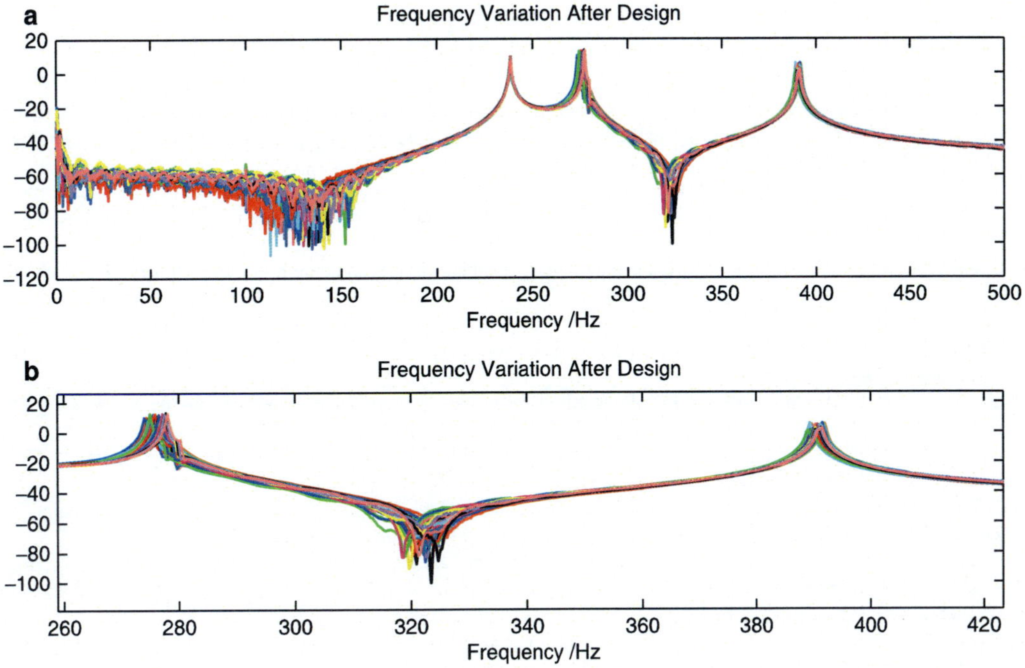

Fig. 16.11 FRFs for the verification experiments. (**a**) Whole frequency range. (**b**) Zoomed frequency range about the second and third natural frequencies

16.3.4 Optimization Verification

To demonstrate the feasibility of the tolerance design, 50 verification experiments are performed for solution 6 in Table 16.9. The experimental results are listed in Table 16.10, which shows that after tolerance design, the means of the second and third natural frequencies are very close to the targets. The FRFs are shown in Fig. 16.11. Compared to Fig. 16.8, the variation in the second and third natural frequencies has been considerably reduced. The standard deviation of the second natural frequency is relatively large, but out of 50 experiments only 2 are outside the target limits.

16.4 Conclusion

This paper has discussed a novel strategy, where tolerance design is used to limit the variation in the dynamic responses. The easy application in practical engineering problems is an important feature of the approach. Hence, the orthogonal array and analysis of variance, through the classic Taguchi's method, are applied. This approach has easy implementation and high efficiency. High accuracy in the optimization is also required, and so a multi-objective genetic algorithm is used in the second stage. Thus the method has two stages: the first uses Taguchi's method to find the best parameter levels and then a multi-objective optimization is used to determine accurate nominal values and tolerances for each design parameter. This involves three steps: the input and output parameters for the problem are defined; then the effects of the control factors on the dynamic responses are determined and their possible nominal values decided; finally factors whose tolerance should be adjusted identified and the problem is transformed into a multi-objective optimization problem. The paper has demonstrated the feasibility of this approach through a benchmark case study.

Acknowledgements The financial support of the National Natural Science Foundation of China (Project No. 51175244), Research Fund for the Doctoral Program of Higher Education of China (Project No. 20093218110008) and the Priority Academic Program Development of Jiangsu Higher Education Institutions (PAPD) are gratefully acknowledged. J. Yang also acknowledges the support of Fundamental Research Funds for the Central Universities of China and Funds of the Graduate Innovation Center in NUAA (Project No. kfjj120104).

References

1. Rout BK, Mittal RK (2006) Tolerance design of robot parameters using Taguchi method. Mech Syst Signal Process 20:1832–1852
2. Fowlkes WY, Creveling CM (1995) Engineering methods for robust product design: using Taguchi methods in technology and product development. Addision-Wesley, Reading
3. Zang C, Friswell MI, Mottershead JE (2005) A review of robust optimal design and its application in dynamics. Comput Struct 83:315–326
4. Marano GC, Greco R (2010) A comparison between different robust optimum design approaches: application to tuned mass dampers. Probab Eng Mech 25:108–118
5. Marano GC, Sgobba S (2008) Robust optimum design of tuned mass dampers devices in random vibrations mitigation. J Sound Vib 313:472–492
6. Marano GC, Quaranta G (2010) Fuzzy-entropy based robust optimization criteria for tuned mass dampers. Earthq Eng Eng Vib 9:285–294
7. Daneshvar N, Khataee AR (2007) Biodegradation of dye solution containing Malachite Green: optimization of effective parameters using Taguchi method. J Hazard Mater 143:214–219
8. Ghani JA, Choudhury IA (2004) Application of Taguchi method in the optimization of end milling parameters. J Mater Process Technol 145:84–92
9. Fusayasu H, Yokota Y (1998) Optimization of a magnetic actuator with Taguchi method and multivariate analysis method. IEEE Trans Magn 34(4):2138–2140
10. Chang HH (2008) A data mining approach to dynamic multiple responses in Taguchi experimental design. Expert Syst Appl 35:1095–1103
11. Mace BR, Vandepitte D, Lardeur P (2011) Editorial: uncertainty in structural dynamics. Finite Elem Anal Des 47:1–3
12. Creveling CM (1996) Tolerance design – a handbook for developing optimal specifications. Addison Wesley Longman, Reading
13. Das I, Dennis JE (1997) A closer look at drawbacks of minimizing weighted sums of objectives for Pareto set generation in multicriteria optimization problems. Struct Optim 14:63–69
14. Goldberg R (2005) Evolutionary multiobjective optimization. Springer, London
15. Zitzler E, Thiele L (1999) Multiobjective evolutionary algorithms: a comparative case study and the strength Pareto approach. IEEE Trans Evol Comput 3(4):257–271
16. Srinivas N, Deb K (1994) Multi-objective optimization using non-dominated sorting in genetic algorithms. J Evol Comput 2(3):221–248

Chapter 17
Uncertainty Propagation in Floating Raft System by FRF-Based Substructuring Method for Elastic Coupling

Huang Xiuchang, Hua Hongxing, Chen Feng, and Xu Shiyin

Abstract The FRF-based substructuring method considering elastic coupling is investigated for predicting vibration transmission and uncertainty propagation in floating raft system. The effects of various commonly encountered errors of substructure FRFs, such as pole shifting, reciprocity violation, negative imaginary part of driving-point FRFs, on the modeling results are investigated. The simulation results demonstrate that the noise on FRFs of raft can influence the assembly FRFs of base and raft; while, some kinds of noises cannot be propagated to the assembly FRFs of machines for the rigid machines are coupled through elastic isolators, bringing about impedance mismatch. The SVD cannot eliminate the studied noises. The uncertainty in substructure FRFs can be amplified due to the coupling process. Propagation of uncertainties of FRFs is quantified based on an analytical approximate moment method using the derivatives of substructure FRFs and variance defined on these parameters. The moment method is of high efficiency and accuracy compared with Monte Carlo probabilistic method.

Keywords Uncertainty propagation • Floating raft system • FRF-based substructuring method • Elastic coupling • Moment method

17.1 Introduction

Floating raft system is a complex dynamic system composed of multiple flexible substructures connected by elastic media, which is widely used in vibration isolation of the important equipment on ships or other industries. For dynamic modeling of such a combined structure, Dynamic Substructuring (DS) methods were usually employed in order to reduce the complexity of dynamical models and to reduce the size of computational models. One of the most widely used DS methods is Frequency Based Substructuring (FBS). One desirable feature of FBS method is that either analytically derived FRFs or experimentally measured FRFs may be used as the substructure model.

Experimental substructuring based on FBS approaches have become an important research issue due to their numerous advantages. While, for the FRFs of substructures are always polluted with random measurement noise and several matrix inversion operations are inherently required by the FBS approach, small errors can be significantly amplified, resulting in large errors in the FRFs of the coupled system. Thus, researches are carried out to deal with the effect of the random inaccuracieson the FRFs of the coupled systemand the uncertainty propagation of the inaccuracies; also, much effort has been devoted to develop well-conditioned FBS methods, such as the coupling method proposed by Jetmundsen et al. [1], which reduced the number of inversion computation to only once and greatly enhanced the robustness and tolerance to data pollution; the Lagrange Multiplier Frequency Based Substructuring (LM FBS) method [2], which was proved to be quite robust to errors in the measurements for the kernel matrix which is to be inversed is of good condition number; and FRF-based substructuring method that takes joint effects into consideration by Liu [3], which is shown to have tolerance to measurement noise for separating the joint apart and adopting the SVD (Singular Value Decomposition) algorithm.

H. Xiuchang (✉) • H. Hongxing • C. Feng • X. Shiyin
State Key Laboratory of Mechanical System and Vibration, Shanghai Jiao Tong University, No.800 Dongchuan Road, Min Hang District, Shanghai, People's Republic of China (200240)
e-mail: xchhuang@sjtu.edu.cn; hhx@sjtu.edu.cn; chenfeng_me@sjtu.edu.cn; xushiyin0628@sjtu.edu.cn

T. Simmermacher et al. (eds.), *Topics in Model Validation and Uncertainty Quantification, Volume 5*,
Conference Proceedings of the Society for Experimental Mechanics Series 41,
DOI 10.1007/978-1-4614-6564-5_17, © The Society for Experimental Mechanics, Inc. 2013

In the case of DS using experimental data, unaviodable measurement errors can directly affect the accuracy of the DS analysis and the response of the coupled system. From previous experiences related to FBS, several common problems seem to exist. Such as errors arising from additional stiffness caused by the suspension, added mass effect induced by the attached measurement equipment and from measurement errors due to sensor and actuator location and orientation. Such errors will result in pole shifting, reciprocity violation and offset antiresonances as well as noise polluted data in FRFs of substructures. Investigation of effects of these typical problems can shed some light on the treatment of measurement.

On the other hand, it is of high interest to know the uncertainty on the FRFs of the coupled system given the uncertainties of components. The hitherto developed approaches to investigate the uncertainty propagation from a number of inputs are based on the probabilistic methods: the statistical probabilistic methods, such as Monte Carlo simulation method; and the analytical probabilistic methods, such as the moment method and stochastic differential equation method. Iankov et al. [4] presented a technique based on variance propagation through FRF-based coupling techniques to predict the uncertainty propagation and applied it to a two-component carbody-subframe assembly model. Klerk and Voormeeren et al. [5, 6] introduced a uncertainty propagation method based on the statistical moment method and uses the measured time data of the subsystems as input to qualify the uncertainty of the coupled system's FRFs.

The paper is to investigate some of these issues for floating raft system with the FRF-based substructuring method by Liu [3]. Although extensive research has been carried out for substructures connected with rigid joints, problems concerning complex flexible joints (rubber isolators), especially for floating raft system with two layers isolation and employing the independent impedance characterization of complex joints, are rarely investigated. Whether the uncertainty propagation is different for such a case is worth studying, also, the effects of mechanism of impedance mismatch introduced by the isolators on the uncertainty propagation is to be unveiled.

17.2 Derivation of FRF-Based Substructuring Method for Floating Raft System

The floating raft system is shown in Fig. 17.1 and the modeling is based on the FRF-based substructuring method which uses FRFs or impedance representations of these separate substructures [3]. The FRF representation of machines, raft and base can be written as follows

$$\left\{\begin{matrix}\mathbf{x}_i^U\\ \mathbf{x}_c^U\end{matrix}\right\} = \begin{bmatrix}\mathbf{H}_{ii}^U & \mathbf{H}_{ic}^U\\ \mathbf{H}_{ci}^U & \mathbf{H}_{cc}^U\end{bmatrix}\left\{\begin{matrix}\mathbf{f}_i^U\\ \mathbf{f}_c^U\end{matrix}\right\} \ (U = M, R \text{ or } B) \tag{17.1}$$

where the subscript i and c represents the internal DOFs (degrees of freedom) and the coupled DOFs, respectively.

The impedance matrix of lower and upper isolators can be analytically described as

$$\left\{\begin{matrix}\mathbf{f}_{p1}\\ \mathbf{f}_{p2}\end{matrix}\right\} = \begin{bmatrix}\mathbf{Z}_{p11} & \mathbf{Z}_{p12}\\ \mathbf{Z}_{p21} & \mathbf{Z}_{p22}\end{bmatrix}\left\{\begin{matrix}\mathbf{x}_{p1}\\ \mathbf{x}_{p2}\end{matrix}\right\}_{(p=l,u)} \tag{17.2}$$

The modeling process includes two synthesis processes as shown in Fig. 17.1 [7]. For the first synthesis process of raft and base connected through lower isolators, the result is

$$\left\{\begin{matrix}\mathbf{X}_I^R\\ \mathbf{X}_I^B\\ \mathbf{X}_C^R\\ \mathbf{X}_C^B\end{matrix}\right\} = \begin{bmatrix}\mathbf{H}_{II}^R & \mathbf{H}_{II}^{RB} & \mathbf{H}_{IC}^{RR} & \mathbf{H}_{IC}^{RB}\\ & \mathbf{H}_{II}^B & \mathbf{H}_{IC}^{BR} & \mathbf{H}_{IC}^{BB}\\ & & \mathbf{H}_{CC}^R & \mathbf{H}_{CC}^{RB}\\ sym & & & \mathbf{H}_{CC}^B\end{bmatrix}\left\{\begin{matrix}\mathbf{F}_I^R\\ \mathbf{F}_I^B\\ \mathbf{F}_C^R\\ \mathbf{F}_C^B\end{matrix}\right\} \tag{17.3}$$

$\mathbf{H}_{II}^R = \mathbf{H}_{ii}^R + \mathbf{H}_{ic}^R\mathbf{D}_1^+\left(\mathbf{H}_{cc}^B\mathbf{Z}_{l21} - \mathbf{H}_{cc}^B\mathbf{Z}_{l22}\mathbf{Z}_{l12}^+\mathbf{Z}_{l11} - \mathbf{Z}_{l12}^+\mathbf{Z}_{l11}\right)\mathbf{H}_{ci}^R$, $\mathbf{H}_{II}^{RB} = -\mathbf{H}_{ic}^R\mathbf{D}_1^+\mathbf{H}_{ci}^B$,

$\mathbf{H}_{IC}^{RB} = \mathbf{H}_{ic}^R\mathbf{D}_1^+\left(\mathbf{H}_{cc}^B\mathbf{Z}_{l22} + \mathbf{I}\right)\mathbf{Z}_{l12}^+$,

$\mathbf{H}_{IC}^{RB} = \mathbf{H}_{ic}^B\mathbf{D}_1^{+T}\left(\mathbf{H}_{cc}^R\mathbf{Z}_{l11} + \mathbf{I}\right)\mathbf{Z}_{l21}^+, \mathbf{H}_{CC}^R = \mathbf{H}_{cc}^R\mathbf{D}_1^+\left(\mathbf{H}_{cc}^B\mathbf{Z}_{l22} + \mathbf{I}\right)\mathbf{Z}_{l12}^+$, $\mathbf{H}_{CC}^{RB} = -\mathbf{H}_{cc}^R\mathbf{D}_1^+\mathbf{H}_{cc}^B$,

$\mathbf{H}_{CC}^B = \mathbf{H}_{cc}^B\mathbf{D}_1^{+T}\left(\mathbf{H}_{cc}^R\mathbf{Z}_{l11} + \mathbf{I}\right)\mathbf{Z}_{l21}^+$,

$\mathbf{H}_{IC}^{RB} = -\mathbf{H}_{ic}^R\mathbf{D}^+\mathbf{H}_{cc}^B$, $\mathbf{H}_{II}^B = \mathbf{H}_{ii}^B + \mathbf{H}_{ic}^B\mathbf{D}_1^{+T}\left(\mathbf{H}_{cc}^R\mathbf{Z}_{l12} - \mathbf{H}_{cc}^R\mathbf{Z}_{l11}\mathbf{Z}_{l21}^+\mathbf{Z}_{l22} - \mathbf{Z}_{l21}^+\mathbf{Z}_{l22}\right)\mathbf{H}_{ci}^B$, $\mathbf{H}_{IC}^{BR} = -\mathbf{H}_{ic}^B\mathbf{D}_1^{+T}\mathbf{H}_{cc}^R$,

$\mathbf{D}_1 = (\mathbf{H}_{cc}^B(\mathbf{Z}_{l22}\mathbf{Z}_{l12}^+\mathbf{Z}_{l11} - \mathbf{Z}_{l21})\mathbf{H}_{cc}^R + (\mathbf{H}_{cc}^B\mathbf{Z}_{l22}\mathbf{Z}_{l12}^+ + \mathbf{Z}_{l12}^+\mathbf{Z}_{l11}\mathbf{H}_{cc}^R) + \mathbf{Z}_{l12}^+$.

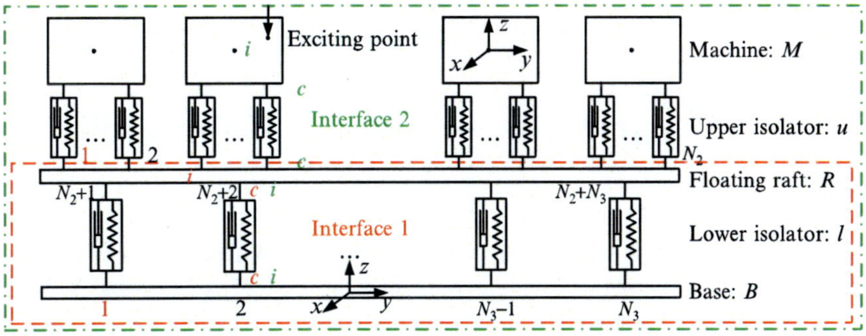

Fig. 17.1 The sketch of floating raft system and the synthesis process

For the second synthesis process of raft-base and machines connected through upper isolators, the FRF representation of the complete floating raft system and some of the submatrices are listed as follows

$$\begin{Bmatrix} \mathbf{X}_I^M \\ \mathbf{X}_C^R \\ \mathbf{X}_C^B \\ \mathbf{X}_C^M \\ \mathbf{X}_I^R \end{Bmatrix} = \begin{bmatrix} \mathbf{H}_{II}^M & \mathbf{H}_{IC}^{MR} & \mathbf{H}_{IC}^{MB} & \mathbf{H}_{IC}^{MM} & \mathbf{H}_{II}^{MR} \\ & \mathbf{H}_{CC}^{R-R} & \mathbf{H}_{CC}^{R-B} & \mathbf{H}_{CC}^{RM} & \mathbf{H}_{CI}^{R-R} \\ & & \mathbf{H}_{CC}^{B-B} & \mathbf{H}_{CC}^{BM} & \mathbf{H}_{CI}^{B-R} \\ & & & \mathbf{H}_{CC}^{MM} & \mathbf{H}_{CI}^{MR} \\ sym & & & & \mathbf{H}_{II}^{R-R} \end{bmatrix} \begin{Bmatrix} \mathbf{F}_I^M \\ \mathbf{F}_C^R \\ \mathbf{F}_C^B \\ \mathbf{F}_C^M \\ \mathbf{F}_I^R \end{Bmatrix} \tag{17.4}$$

$$\mathbf{H}_{II}^M = \mathbf{H}_{ii}^M + \mathbf{H}_{ic}^M \mathbf{D}_2^+ (\mathbf{H}_{II}^R \mathbf{Z}_{u21} - \mathbf{H}_{II}^R \mathbf{Z}_{u22} \mathbf{Z}_{u12}^+ \mathbf{Z}_{u11} - \mathbf{Z}_{u12}^+ \mathbf{Z}_{u11}) \mathbf{H}_{ci}^M, \quad \mathbf{H}_{IC}^{MR} = -\mathbf{H}_{ic}^M \mathbf{D}_2^+ \mathbf{H}_{IC}^{RR},$$

$$\mathbf{H}_{IC}^{MM} = \mathbf{H}_{ic}^M \mathbf{D}_2^+ (\mathbf{H}_{II}^R \mathbf{Z}_{u22} + \mathbf{I}) \mathbf{Z}_{u12}^+,$$

$$\mathbf{H}_{IC}^{MB} = -\mathbf{H}_{ic}^M \mathbf{D}_2^+ \mathbf{H}_{IC}^{RB}, \quad \mathbf{H}_{II}^{MR} = -\mathbf{H}_{ic}^M \mathbf{D}_2^+ \mathbf{H}_{II}^R, \quad \mathbf{H}_{CC}^{B-B} = \mathbf{H}_{CC}^B + \mathbf{H}_{CI}^{BR} \mathbf{D}_2^{+T} (\mathbf{H}_{cc}^M \mathbf{Z}_{u12} - \mathbf{H}_{cc}^M \mathbf{Z}_{u11} \mathbf{Z}_{u21}^+ \mathbf{Z}_{u22} - \mathbf{Z}_{u21}^+ \mathbf{Z}_{u22}) \mathbf{H}_{IC}^{RB},$$

$$\mathbf{D}_2 = (\mathbf{H}_{II}^R (\mathbf{Z}_{u22} \mathbf{Z}_{u12}^+ \mathbf{Z}_{u11} - \mathbf{Z}_{u21}) \mathbf{H}_{cc}^M + (\mathbf{H}_{II}^R \mathbf{Z}_{u22} \mathbf{Z}_{u12}^+ + \mathbf{Z}_{u12}^+ \mathbf{Z}_{u11} \mathbf{H}_{cc}^M) + \mathbf{Z}_{u12}^+.$$

17.3 Uncertainty Propagation in Floating Raft System by FRF-Based Substructuring Method

The derivation of the uncertainty propagation method was addressed in detail in [6]. The theory is applied to the FBS results in Sect. 17.2. Taking uncertainty propagation of FRFs on the bases of the coupled system from uncertainties on the FRFs of bases as an example, the uncertainty propagation yields as follows according to Eqs. (17.3) and (17.4)

$$\sigma(\mathbf{H}_{CC}^{B-B}) = \sqrt{\sum_{i=1}^{N_3} \sum_{j=1}^{N_3} \left(\frac{\partial(\mathbf{H}_{CC}^{B-B})}{\partial(\mathbf{H}_{cc}^B)} \sigma(\mathbf{H}_{cc}^B(i,j)) \right)^2},$$

$$\frac{\partial(\mathbf{H}_{CC}^{B-B})}{\partial(\mathbf{H}_{cc}^B)} = \frac{\partial(\mathbf{H}_{CC}^B)}{\partial(\mathbf{H}_{cc}^B)} + \frac{\partial(\mathbf{H}_{CI}^{BR})}{\partial(\mathbf{H}_{cc}^B)} \mathbf{D}_2^{+T} \Xi \mathbf{H}_{IC}^{RB} + \mathbf{H}_{CI}^{BR} \frac{\partial(\mathbf{D}_2^{+T})}{\partial(\mathbf{H}_{cc}^B)} \Xi \mathbf{H}_{IC}^{RB} + \mathbf{H}_{CI}^{BR} \mathbf{D}_2^{+T} \Xi \frac{\partial(\mathbf{H}_{IC}^{RB})}{\partial(\mathbf{H}_{cc}^B)} \tag{17.5}$$

where

$$\Xi = (\mathbf{H}_{cc}^M \mathbf{Z}_{u12} - \mathbf{H}_{cc}^M \mathbf{Z}_{u11} \mathbf{Z}_{u21}^+ \mathbf{Z}_{u22} - \mathbf{Z}_{u21}^+ \mathbf{Z}_{u22}), \quad \frac{\partial(\mathbf{H}_{CI}^{BR})}{\partial(\mathbf{H}_{cc}^B)} = -\left(\mathbf{H}_{ic}^R \mathbf{D}_1^+ \mathbf{P}_{ij}\right)^T, \quad \frac{\partial(\mathbf{H}_{IC}^{RB})}{\partial(\mathbf{H}_{cc}^B)} = -\mathbf{H}_{ic}^R \mathbf{D}_1^+ \mathbf{P}_{ij},$$

$$\frac{\partial(\mathbf{H}_{CC}^B)}{\partial(\mathbf{H}_{cc}^B)} = \mathbf{P}_{ij} \mathbf{D}_1^{+T} (\mathbf{H}_{cc}^R \mathbf{Z}_{l11} + \mathbf{I}) \mathbf{Z}_{l21}^+ + \mathbf{H}_{cc}^B \frac{\partial(\mathbf{D}_1^{+T})}{\partial(\mathbf{H}_{cc}^B)} (\mathbf{H}_{cc}^R \mathbf{Z}_{l11} + \mathbf{I}) \mathbf{Z}_{l21}^+,$$

$$\frac{\partial(\mathbf{D}_2^{+T})}{\partial(\mathbf{H}_{cc}^B)} = \left(\frac{\partial(\mathbf{D}_2^+)}{\partial(\mathbf{H}_{cc}^B)} \right)^T, \frac{\partial(\mathbf{D}_2^+)}{\partial(\mathbf{H}_{cc}^B)} = -\mathbf{D}_2^+ \frac{\partial(\mathbf{D}_2)}{\partial(\mathbf{H}_{cc}^B)} \mathbf{D}_2^+,$$

$$\frac{\partial(\mathbf{D}_2)}{\partial(\mathbf{H}_{cc}^B)} = \frac{\partial(\mathbf{H}_{II}^R)}{\partial(\mathbf{H}_{cc}^B)}(\mathbf{Z}_{u22}\mathbf{Z}_{u12}^{+}\mathbf{Z}_{u11} - \mathbf{Z}_{u21})\mathbf{H}_{cc}^M + \frac{\partial(\mathbf{H}_{II}^R)}{\partial(\mathbf{H}_{cc}^B)}\mathbf{Z}_{u22}\mathbf{Z}_{u12}^{+}, \frac{\partial(\mathbf{D}_1)}{\partial(\mathbf{H}_{cc}^B)} = \mathbf{P}_{ij}(\mathbf{Z}_{l22}\mathbf{Z}_{l12}^{+}\mathbf{Z}_{l11} - \mathbf{Z}_{l21})\mathbf{H}_{cc}^R + \mathbf{P}_{ij}\mathbf{Z}_{l22}\mathbf{Z}_{l12}^{+},$$

$$\frac{\partial(\mathbf{H}_{II}^R)}{\partial(\mathbf{H}_{cc}^B)} = \mathbf{H}_{ic}^R\frac{\partial(\mathbf{D}_1^{+})}{\partial(\mathbf{H}_{cc}^B)}(\mathbf{H}_{cc}^B\mathbf{Z}_{l21} - \mathbf{H}_{cc}^B\mathbf{Z}_{l22}\mathbf{Z}_{l12}^{+}\mathbf{Z}_{l11} - \mathbf{Z}_{l12}^{+}\mathbf{Z}_{l11})\mathbf{H}_{ci}^R + \mathbf{H}_{ic}^R\mathbf{D}_1^{+}(\mathbf{P}_{ij}\mathbf{Z}_{l21} - \mathbf{P}_{ij}\mathbf{Z}_{l22}\mathbf{Z}_{l12}^{+}\mathbf{Z}_{l11})\mathbf{H}_{ci}^R,$$

$$\partial\mathbf{G}/\partial G_{ij} = \mathbf{P}_{ij},$$

$\mathbf{P}_{ij}$ is a matrix with the same size of $\mathbf{G}$ but only with one nonzero entry (i, j) as 1. And the square operations must be performed elementwise (i.e. for $N_3 \times N_3$ elements in the FRFs of base and N_3 is the number of lower isolators) for Eq. (17.5).

17.4 Numerical Simulation

The floating raft system is composed of two machines, a free-free boundary plate as raft and a free-free boundary plate as base. Each machine connects with the raft through four isolators and raft connects with base through four isolators. Dimensions of both machines are width ×length × height = 0.8 m × 0.8 m × 0.4 m and dimensions of raft and base are 2.4 m × 2 m × 0.002 m and 2.4 m × 2 m × 0.003 m, respectively. All of them are made of steel with density, damping factor and Young's Modulus as $7{,}850\,\mathrm{kg/m^3}$, 0.002 and $2.1 \times 10^{11}\,\mathrm{N/m^2}$, respectively. Stiffness of upper and lower isolators is $2 \times 10^6\,\mathrm{N/m^2}$ and $3.2 \times 10^6\,\mathrm{N/m^2}$, respectively. FE models of machines, raft and base are solved by using ANSYS to generate the FRFs for all substructures. The modeling method has been verified by comparing the FRFs with those obtained from the coupled system in ANSYS.

17.4.1 Noisy Simulated Data

The effect of the inaccuracies due to measurement noise in substructure FRFs is illustrated on the studied model where five kinds of noise are artificially added to the exact data generated theoretically by using the modal data obtained from ANSYS. Five perturbed FRFs for the substructures are generated and the FBS outcome is compared to those with the exact FRFs. The exact FRFs are obtained according to the mode superposition method

$$H_{pq}(\omega) = -\omega^2 \sum_{r=1}^{R} \frac{\phi_{pr}\phi_{qr}}{\omega_r^2 - \omega^2 + j2\omega\xi_r\omega_r}, j = \sqrt{-1} \tag{17.6}$$

where R is the number of the retained eigenmodes in the substructure, ω is the excitation frequency, ω_r is the associated eigenfrequency for the eigenmode r of substructure.

(1) The effect of pole shifting in the substructure FRFs

To mimic measurement errors leading to pole shifting errors, the perturbed FRFs are constructed through modifying $\omega_r$in Eq. (17.6) by the perturbed eigenfrequency $\omega_r' = \omega_r(1+0.01)$. Figure 17.2 shows the comparison of assembly FRFs synthesized from noise-contaminated FRFs with added noise for pole shifting in raft FRF. The inversion of $\mathbf{D}_1$ and $\mathbf{D}_2$ is performed by SVD with an attempt to eliminate noisy data for raft and the threshold value is 0.002 [3]. The results are also plotted in Fig. 17.2. It can be seen that the shifting of poles in raft itself can be reflected in the FRFs of coupled system, and SVD cannot eliminate such kinds of errors. Also, the assembly FRFs from the exciting point to the points on both the excited machine and the unexcited machine are compared. The results show that the former is hardly influenced, while the latter is affected as those for points on raft and base.

(2) The effect of the error on substructure modeshapes

We assume that the FRF of raft is related to a slightly different perturbed modal basis $\phi_{pr}' = (1+0.01n_{pr})\phi_{pr}$, with 0.01 as the perturbation amplitude, and n_{pr} a randomly distributed set between -1 and 1. And the FRF of raft is generated anew according to Eq. (17.6). FBS was performed and the assembled FRFs are compared to the reference FRFs in Fig. 17.3. It can be seen that, the FRFs on the raft and base demonstrate some level of noise, while, it has little influence on the FRFs of both machines. Thus the uncertainty in the modeshapes of raft can be propagated to the FRFs on base. Also, it is observed that the FRFs at anti-resonances is severely changed due to this kind of noise, while the FRFs at resonances changes little,

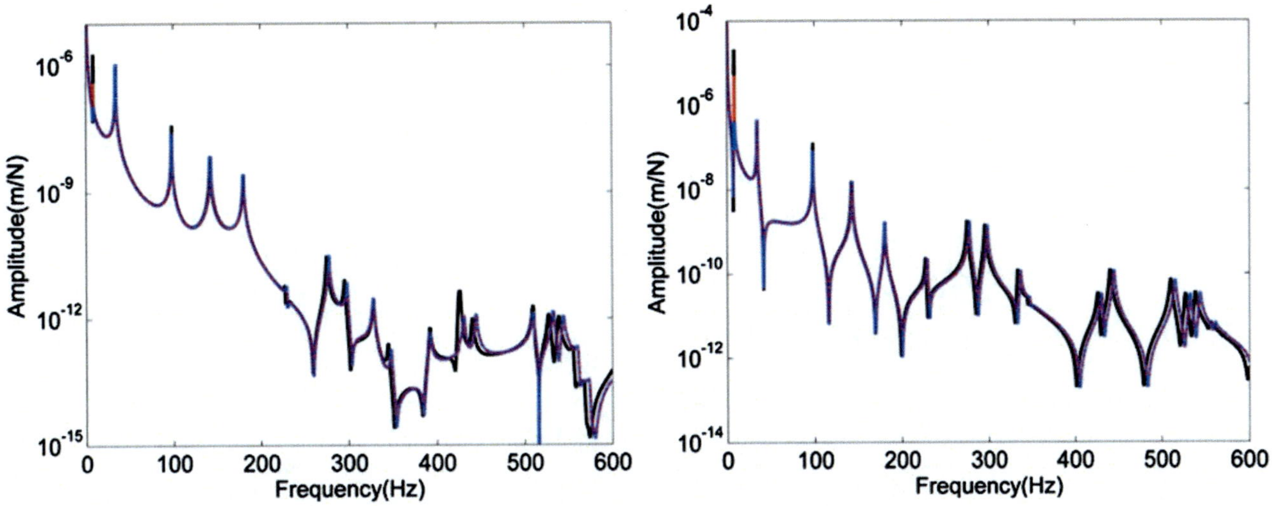

Fig. 17.2 FRFs from the exciting point to the first point on base (*left*) and on raft (*right*) for pole shifting of raft FRFs (—: is the reference result; —: the result with noise; -.-.-.: the result with noise by SVD, the same for all figures)

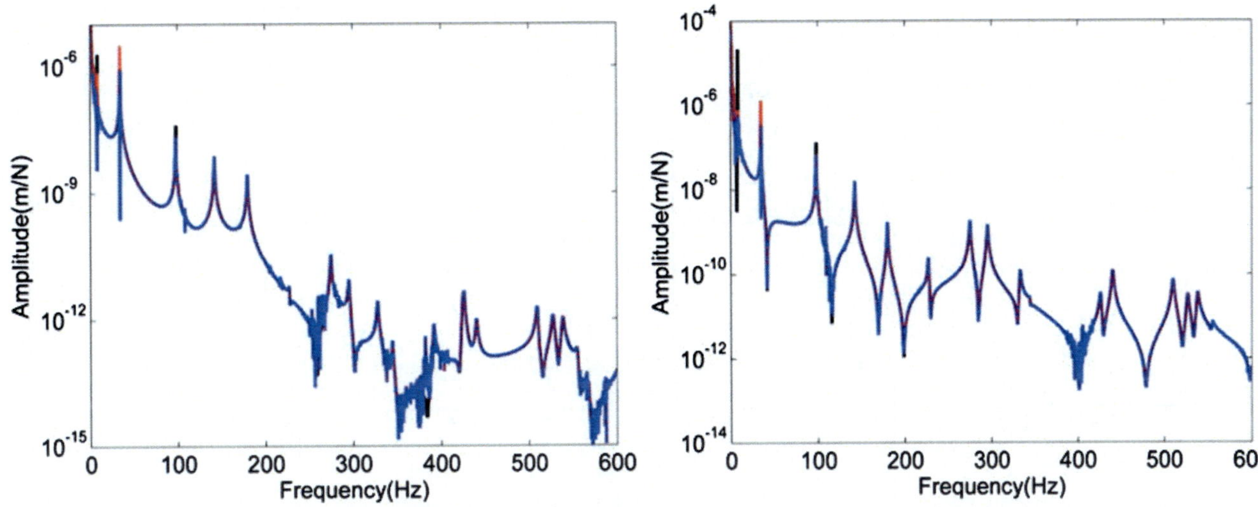

Fig. 17.3 FRFs from the exciting point to the first point on base (*left*) and on raft (*right*) for errors on mode shapes of raft

which means that the FRFs at antiresonances are more sensitive and severely affected by this kind of noise. Noise is most detrimental to FRFs at antiresonances of the assembled system for the reason that FRFs at anti-resonances are relatively small. The noise simulated here is approximately uniform across all frequencies, therefore the noise at anti-resonances is relatively larger. Thus, there is a need for the reconstruction of the FRFs from extracted modal parameters to smooth the noisy data. The measurements with offset connection DOFs are reasonably a serious problem.

(3) Measurement noise

To simulate measurement noise, a third data set for the raft FRFs is created by adding to the exact FRFs both additive and multiplicative noise as follows

$$H_{pq}(\omega) = ((1 + 0.02n_a)abs(H'_{pq}(\omega)) + 0.001n_t norm(\mathbf{H}(\omega)))\exp(j \cdot ang(H_{pq}(\omega)) + 0.02n_p\pi) \quad (17.7)$$

where, *abs*() and *ang*() represent the absolute value and the angle of a plurality; *norm* denotes 2-norm of a matrix; 0.02, 0.001, 0.02 π are respectively the proportional error amplitude, the additive error amplitude and the phase error; n_a, n_t, $n_p$are random numbers between -1 and 1. The synthesized FRFs by a noise-free base FRF and a raft FRF applied with noise level as above mentioned are overlaid in Fig. 17.4. It can be seen that, the effect is similar to that of the noise on modeshapes of raft,

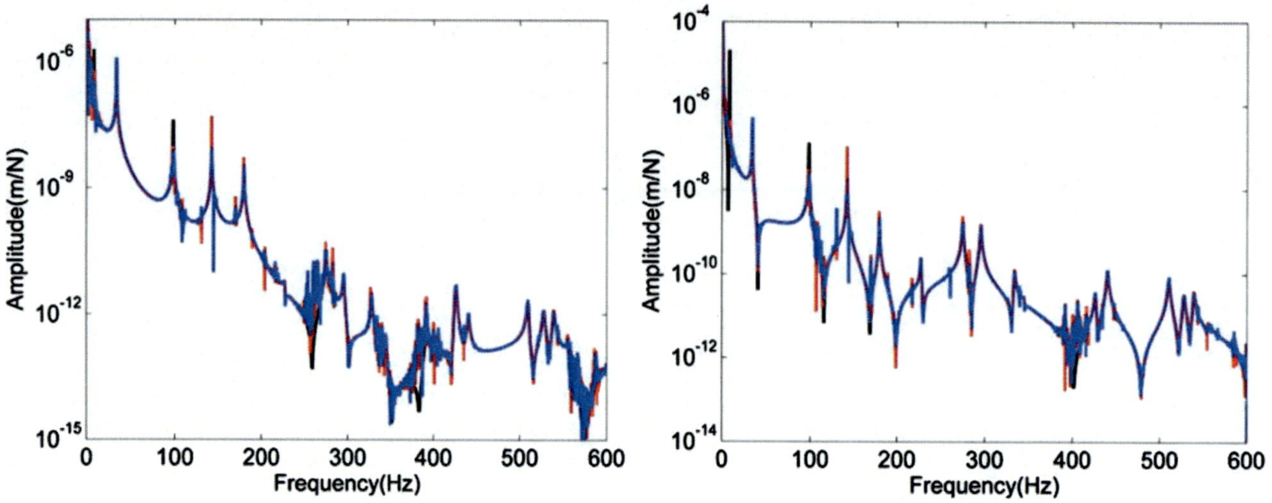

Fig. 17.4 FRFs from the exciting point to the first point on base (*left*) and on raft (*right*) for measurement noise on FRFs of raft

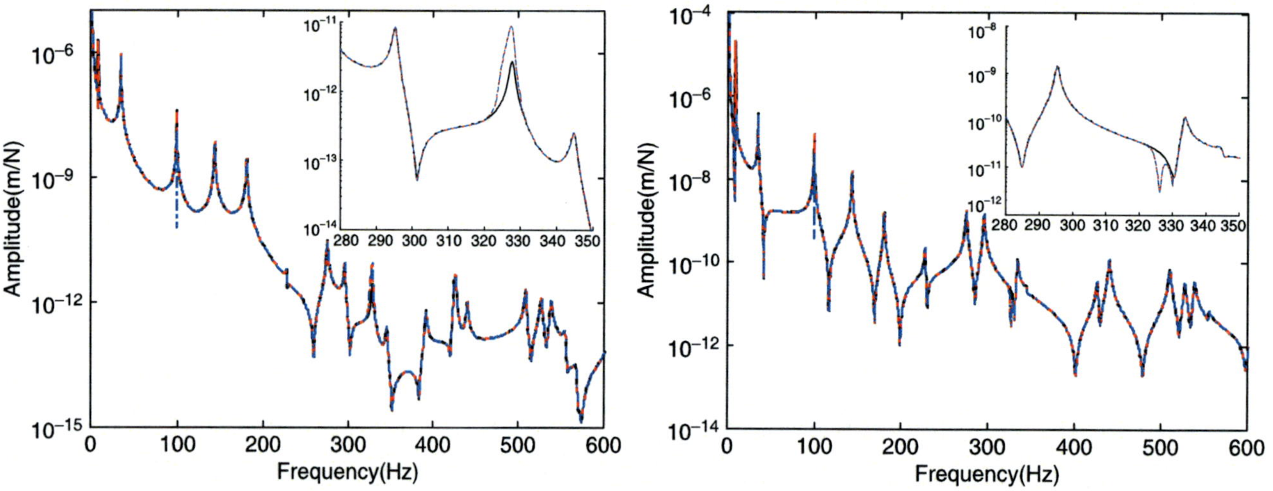

Fig. 17.5 FRFs from the exciting point to the first point on base (*left*) and on raft (*right*) for negative imaginary part of driving point FRFs of raft

while, the assembly FRFs are more severely influenced and the affected frequency range is enlarged to a lower frequency, which makes the first resonance of the floating raft system (8.25 Hz) disappeared. The FRFs on the unexcited machines are also affected to a much large extent. SVD is performed by adopting different threshold values and good outcomes for a chosen value are plotted in Fig. 17.4. It is shown that SVD plays a limited role in this kind of noise.

(4) The effect of negative imaginary part of driving point FRF

As shown by FRFs obtained by modal superposition method according to Eq. (17.6) with $p = q$, the imaginary values of a driving point FRF should always be positive for all frequencies as long as the modal dampings ζ_k are positive. In physical, if the imaginary part of the driving point FRFs is not always positive at some frequency lines, it would imply negative damping. This discovery motivated the investigation of changing the driving point FRFs of raft to its complex conjugate in a frequency range 300~330 Hz, making the imaginary part of the perturbed FRFs negative. The assembly FRFs are computed and the results are plotted in Fig. 17.5 (the insert shows a magnification between 280 ~ 350 Hz). It can be seen that some problems become apparent with larger amplitudes at resonances or spurious antiresonances and flat peaks in the perturbed frequency region. The result differs with that reported for two rigidly coupled structures in Ref. [8] with very sharp glitches or noise spikes. The FRFs on the unexcited machine also displays spurious antiresonances. Another interesting phenomenon is that the spurious antiresonances and flat peaks are varied for different points in raft and base.

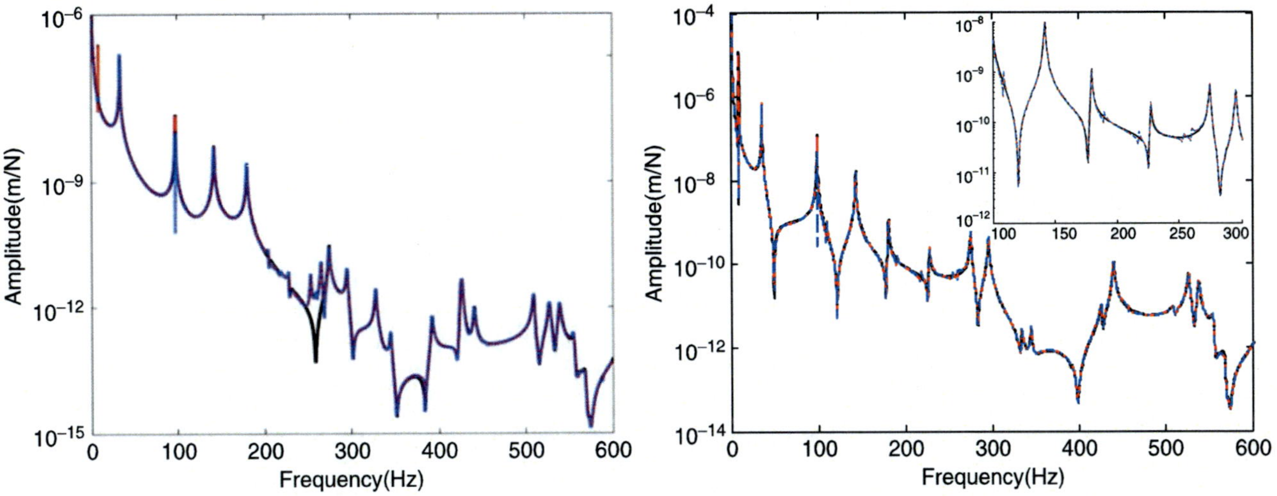

Fig. 17.6 FRFs from the exciting point to the first point on base (*left*) and $(N_3 + 1)$st point on raft (*right*) for antiresonances reciprocity violations of raft FRF

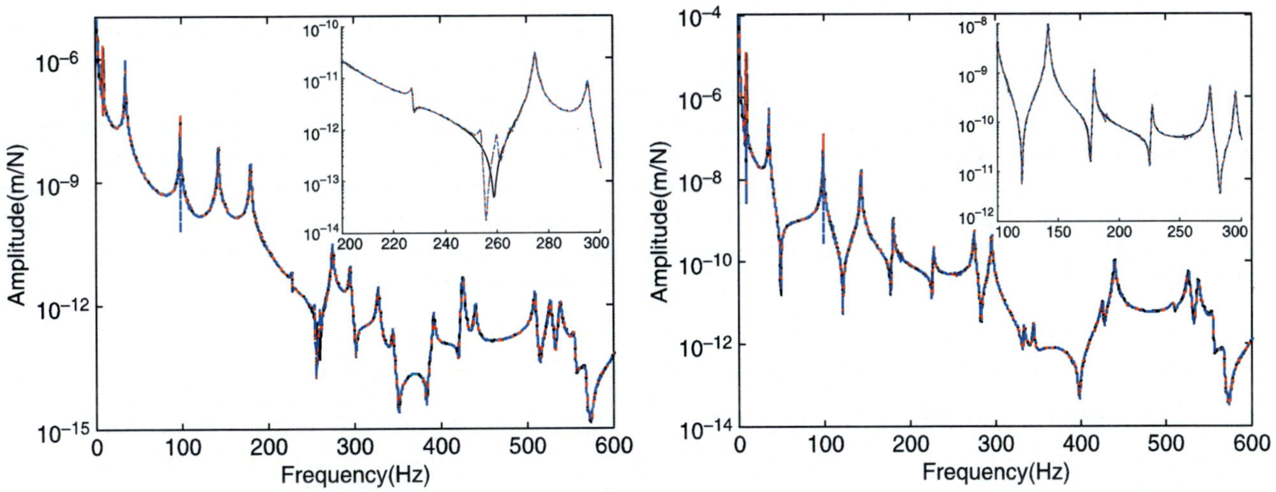

Fig. 17.7 FRFs from the exciting point to the first point on base (*left*) and $(N_3 + 1)$st point on raft (*right*) for resonances reciprocity violations of raft FRF

(5) The effect of reciprocity violations

Reciprocity is important to FBS solutions. Reciprocity violations are explored in two manners: first, by slightly changing the measurement location, resulting in change of antiresonances; second, by slightly violating the linear time invariance (LTI) of the model, resulting in small change of resonances. This could be caused by the suspension system settling or by the roving of a shaker to a different location during testing. For the former, non-reciprocal FRF is created by using a nearby point of point 1 on raft and the outcome FRF of the first row is not consistent with the first column, as the modal superposition method implies, the poles of the outcome FRFs are the same while the antiresonances are slightly different; for the latter, non-reciprocal FRF is created by giving a difference of 0.5 Hz to all resonances when obtaining the FRFs of the first row and the first column, resulting FRFs with different poles. FBS was performed and the results are presented in Figs. 17.6 and 17.7, respectively. For both cases, the assembly FRFs on raft and base are affected most, while those on machines are affected little. It can be seen from the enlarged view in Fig. 17.6 that the predicted FRF is not so smooth and presents some false peaks/antiresonances in some frequency bands. Examining Fig. 17.7, false peaks and antiresonances may also appear. The explorations show that reciprocity violation caused by the experiment set up may introduce some errors.

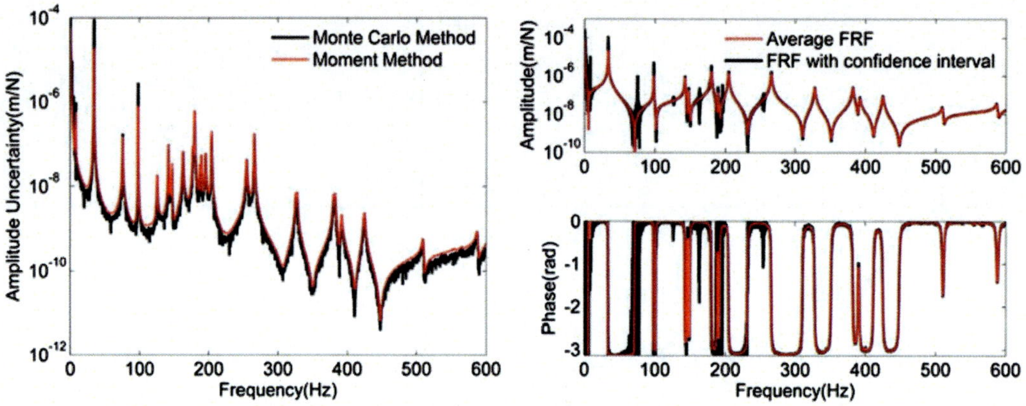

Fig. 17.8 Confidence interval of the amplitude for the assembly FRF of the first point on base (*left*) and the same FRF with confidential interval (*right*)

17.4.2 Verification of the Uncertainty Propagation Method

The uncertainties on the base FRFs are chosen such that the FRF amplitudes have a proportional confidence interval (at the 95 % level) of ±2 % over the frequency range, while the confidence interval on the phase of the FRFs is taken at ±1°. A Gaussian distribution is assumed for the input variables. In the Monte Carlo method the system are coupled 1,000 times, each time with a different, uncorrelated random error on the amplitudes and phases of the FRFs of the base. The results of both methods are shown in Fig. 17.8. It can be seen that, an excellent resemblance between the Monte Carlo method and the moment method is found. The moment method is therefore validated and it is far more efficient than the Monte Carlo method. The CPU time of the moment method is less than 1 s, compared with the CPU time of over 50 min for the Monte Carlo method. From the FRFs with confidential interval shown in Fig. 17.8, it can be seen that small errors in substructures' magnitude result in a large deviation in assembly FRF, which means that the uncertainty is amplified.

17.5 Conclusion

Uncertainty propagation in floating raft system by the FRF-based substructuring method is investigated. The effects of five kinds noises on the modeling results are explored. Simulation results demonstrate that the influence of noises can bring about significant noise or false peaks/antiresonances in the assembly FRFs; the noise on FRFs of raft can influence the assembly FRFs of base and raft; while, some kinds of noises cannot be propagated to the assembly FRFs of machines, this is attributed to the fact that the substructures are connected through elastic isolators and the machines are very rigid, which brings about impedance mismatch between machines and the other substructures. The SVD cannot eliminate the studied noises. Uncertainty propagation of substructure FRFs is quantified based on the moment method, which is of high efficiency compared with Monte Carlo method. The uncertainty can be amplified due to the coupling process.

References

1. Jetmundsen B, Bielawa R, Flannelly W (1988) Generalized frequency domain substructure synthesis. J Am Helicopter Soc 33(1):55–64
2. De Klerk D, Rixen DJ, De Jong J (2006) The frequency based substructuring (FBS) method reformulated according to the dual domain decomposition method. In: Proceedings of the fifteenth international modal analysis conference. Society for Experimental Mechanics, St. Louis
3. Liu W (2000) Structural dynamic analysis and testing of coupled structures. University of London, London
4. Iankov R, Moens D, Sas P, Hermans L (2002) Propagation of variances of FRFs through FRF-based coupling calculation. In: Proceedings of ISMA27, vol IV, Leuven, pp 1845–1852
5. De Klerk D, Voormeeren S (2008) Uncertainty propagation in experimental dynamic substructuring. In: Proceedings of the twenty sixth international modal analysis conference. Society for Experimental Mechanics, Orlando

6. Voormeeren S, De Klerk D, Rixen D (2010) Uncertainty quantification in experimental frequency based substructuring. Mech Syst Signal Process 24(1):106–118
7. Huang X, Zhang Z, Zhang Z, Hua H (2011) Vibro-acoustic optimization of an immersed cylindrical structure with a floating-raft system inside by using frequency-response-function-based substructuring sensitivity analysis. Proc Inst Mech Eng M 225(2):111–132
8. Carne TG, Dohrmann CR (2006) Improving experimental frequency response function matrices for admittance modeling. In: Proceedings of the nineteenth international modal analysis conference. Society for Experimental Mechanics, St. Louis

Chapter 18
Crossing and Veering Phenomena in Crank Mechanism Dynamics

Elvio Bonisoli, Gabriele Marcuccio, and Carlo Rosso

Abstract Modal analysis is widely used both on single components and mechanical complex assemblies and it is recognized to be a fundamental step on the functional design process. From experimental point of view, a change in a system parameter due to the need of describing a different assembly configuration, can require iterative measurements, and can be quite time consuming. On the other hand, by evaluating the dynamic behaviour of the single component instead of the whole system, it is not straightforward to forecast the general dynamics of the entire assembly: inertia and stiffness couplings give rise to curious dynamic phenomena, namely crossing and veering of eigenvalue loci. Many theoretical studies on eigenvalue curve crossing and curve veering, i.e. the coincidence of two eigenfrequencies or the abrupt divergence of natural frequencies trends, have been carried out in recent years, but only few references on detailed test sessions and practical applications are available. The present paper wants to give a better overview on the change of the dynamic properties of a system by comparing global mode shapes to single component mode shapes. The examined structure is a crank mechanism, made of a crankshaft joined to four connecting rods and four pistons. The chosen control parameter that is responsible of a change in the dynamic properties of the system is the crank angle. Numerical models have been used to compute eigenvalues and eigenvectors of the analysed structure, considering both FEM models and multibody approach. Finally, an original graphical interpretation of the transition from component to system dynamics is presented by means of the MAC index.

Keywords Crossing and veering phenomena • Crank mechanism • Linearized dynamics • Coupled systems • LUPOS

18.1 Introduction

Crossing or veering phenomena [1] are becoming a growing cause of concern in many weakly interconnected mechanical systems and practical applications. The parametric variation under which the modes trajectories tend to converge resulting in mode switching in the transition zone can be chosen between system sizes, boundary conditions, materials or applied loads. In all these cases, the essential requirement is a coupling factor either in mass matrix or stiffness matrix. Varying a system parameter in a wide range might be seen unrealistic, but sometimes components show unexpected behaviours due to geometric tolerances, variable operating conditions or kinematic reasons.

The effects of the switching point can be seen on practical applications concerning the design of long span bridges, aircrafts wings, or bladed discs [2–5]. The frequency and damping of aforementioned structures are commonly portrayed in terms of a family loci as a function of mean wind velocity. Aeroelastic characteristics of a wing with control surface are influenced by a coupling factor represented by a hinge stiffness, and jumps in the flutter speed are strictly linked to the behaviour in the transition zone where frequency curves veer away. In [6] a similar phenomenon for the tail wing is observed between the first and second structural modes, dominated respectively by the first bending and the first torsion. When the

E. Bonisoli (✉) • G. Marcuccio • C. Rosso
Politecnico di Torino, Corso Duca degli Abruzzi 24, 10129 Torino, Italy
e-mail: elvio.bonisoli@polito.it; gabriele.marcuccio@polito.it; carlo.rosso@polito.it

T. Simmermacher et al. (eds.), *Topics in Model Validation and Uncertainty Quantification, Volume 5*,
Conference Proceedings of the Society for Experimental Mechanics Series 41,
DOI 10.1007/978-1-4614-6564-5_18, © The Society for Experimental Mechanics, Inc. 2013

tail is cantilevered, the corresponding rigid body torsional frequency $\omega_\phi = \infty$, and the coupling between bending and torsion modes is weak. For lower frequency range stiffness the flutter speed decreases. The rotation frequency at which it reaches a minimum is almost the same as that for the mode shapes veering.

In [7] an experimental investigation on the vibrational characteristics of cantilevered angle-ply laminates made of composite material is presented and the effects on the obtained mode shapes are underlined. The parameters under analysis are two, namely the fibre angle and the aspect ratio, and both are responsible for a change in the degree of coupling and in the slope of the flexural nodal line. The flexural-torsional coupling stiffness changes according to the different stackings considered and its effect could be identified by analyzing the fringe patterns through holographic interferometry.

Such a modal interaction comes also from introducing in the model a possible cause of symmetry breaking. Following this idea a localized lost of symmetry could be used to forecast a local damage in real structures like bridges and suspended cable [8, 9].

Despite du Bois and Adhikari [10] have developed three normalized criteria for veering quantification (i.e. cross-sensitivity quotient, modal dependence factor and veering index), it is still not clear which is the kind of interaction among the modes. Furthermore, the aforesaid analytic indices are only valid when a linear variation either in the stiffness matrix or mass matrix is considered. An attempt has been made in this paper in order to make more evident the dynamic behaviour in the transition zone for more general and quite complex systems, and modal assurance criterion [11] has been used for describing the entire phenomenon. During curve veering the interacting mode shapes are apparently different from those for which relative eigenfrequencies are distinct, giving rise to modes hybridization [12].

In the present paper, based on numerical simulations, the hybridization of modes is depicted for a complex real structure representing the crank system of an internal combustion engine (ICE). Different cases are taken into account, considering both the effects of variable configurations of the system and the gradual presence of new components in the assembly under examination. In automotive field and especially in the design studies of internal combustion engines, neglecting the coupling between modes shapes and their interactions among reciprocating machines can lead to rough approximations and unexpected breakings. The lack of information is basically due to two main reasons. Mathematical models usually can be employed but they are based on too strong assumptions and cannot always meet the needs of nonlinear problems. Experiments, on the other side, can make up for the need of more reliable results but are time consuming and cost-intensive. Moreover, they are not always able to pinpoint the most important reasons of breakdown. Nevertheless, even if dynamic failures are becoming rare thanks to detailed studies that since the beginning of the century provided on dynamic behaviour of crankshaft [13], nowadays stresses and vibrations must be quantified in order to guarantee the structural integrity of the components and higher durability. New and more strict requirements, namely lightweight structures, environment issues, noise emissions reduction [14] demand for sophisticated analyses aimed to optimize component design [15]. Generally speaking component optimization consists of an analysis of the structure, involved excitations and iterative relevant modifications, by which preferably resonant frequencies have to be avoided during operating conditions. Even if the matrix size and the number of equations are greater, more accurate results can be obtained by considering the coupling effect of axial, flexural and torsional vibrations [16, 17]. Large displacements are expected when excited frequencies and modes shapes belonging to different subspaces are close.

Applying local modifications on structural properties of the analyzed component, may lead to unexpected global variations on assembly modal properties. Designers and analysts would prefer to know a priori which are the main effects on the final product.

The here in discussion offers new perspectives into the dynamics of a mechanical structure for damage detection purposes, allows to improve the understanding of self-excited systems, and provides new insights in robust engineering design. The paper is structured as follow: Sect. 18.2 describes the assembly has been employed and how it has been numerically modelled. In Sect. 18.3 a brief overview on crossing and veering is presented. Section 18.4 deals with the gradual transition from component to system modal properties, and Sect. 18.5 focuses on the identification of local and global modes. Finally, in Sect. 18.6 conclusions are depicted and outcomes are summarized.

18.2 Crank Mechanism Modeling

In many fields finite element method (FEM) is nowadays considered mandatory to simulate, predict and characterize complex systems. Achieving high durability, low level vibrations and restrained acoustic emissions are a main concern for many design engineering problems. In this paper detailed numerical investigations on powertrain automotive components are reported. Consider a four cylinders, four strokes cycle engine.

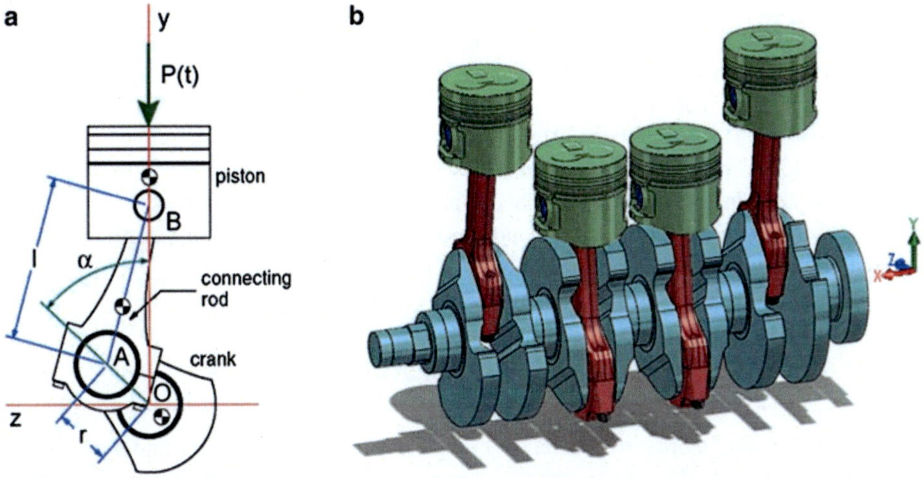

Fig. 18.1 Crankshaft system: (**a**) schematic representation, (**b**) CAD model

Fig. 18.2 Main crank mechanism components

The technical features of the structure (Fig. 18.1) under analysis are listed below:

- Firing order: 1-3-4-2;
- Four-stroke cycle;
- Connecting rod length (l): 145 mm;
- Cylinder bore: 82 mm;
- Piston stroke (r): 44 mm;
- Oscillating masses: 4.0 kg;
- Engine displacement: 1910 cm^3.

All geometrical and inertial data relevant for the analysis of the dynamic behaviour of the crank mechanism and subcomponents have been either measured or computed by means of CAD software (Fig. 18.2). All computations herein shown were performed by using a *LUmped Parametric Open Source* software (LUPOS), developed by the authors in MATLAB®. The detailed model is parameterised, thus enabling virtual prototype testing of various engine designs, as well as allowing the designers to carry out a comprehensive investigation of the engine behaviour during the whole four-stroke cycle.

The complete system can be described with a simplified FE model (~380 dof) made of masses and beams whose section properties are chosen and updated in order to best represent the general dynamic of the same system according to experimental data previously obtained. It is currently common practice to use beams and concentrated masses for representing a crankshaft [13]. Euler-Bernoulli beam elements with axial, flexural and torsional properties, described in textbook such as [18, 19], are used. Rigid roto-translations are applied to all the components. Each bearing is represented by a concentrated spring whose formulation is similar to CELAS elements [20]. Thus both multi-body and finite element benefits are kept in the same code. This approach in engineering design allows to quickly compute and show up the main dynamic properties of the system and their evolution when chosen parameters (relative displacement, size, materials, boundary conditions) are varied.

18.3 Curve Veering and Mode Localization: Overview

It is well known that in order to predict the complete crankshaft dynamics under operating conditions more accurate models are necessary; though, this analysis does not want to take into account crankshaft bent, engine block misboring, structural damping, or bearing hydrodynamics effects, already analyzed in other references [21], but rather aims to show veering phenomena and related effects on the dynamic behaviour along the entire engineering design process either of assemblies or separate components. By performing transient dynamic analyses the real modal evolution of the system is either neglected or kept hidden by the process itself.

In this case, the eigenproblem of the system for different values of a variable parameter, namely the crankshaft angle, is solved. In the FE model structural damping is neglected and the equations for a n-dof system is given by:

$$[\mathbf{M}]\{\ddot{\mathbf{x}}\} + [\mathbf{K}]\{\mathbf{x}\} = \{\mathbf{0}\} \tag{18.1}$$

where the mass matrix **[M]** and the stiffness matrix **[K]** are real, symmetric and positive definite or semi-definite positive respectively. The n eigenvalues ω^2 and eigenvectors $\{\Phi\}$ are calculated by evaluating the non trivial solutions:

$$\det\left([\mathbf{K}] - \omega^2[\mathbf{M}]\right) = 0 \tag{18.2}$$

The model chosen is aimed to put in evidence a real example of veering and crossing phenomena when a physical parameter affecting the system is changed. The variation of this angle takes place every each combustion cycle and it is not due to a variation of an external parameter but is strictly linked to the operating conditions. This means that every four strokes the system will cross the starting configuration twice.

The eigenfrequencies in a range between 0 and 700 Hz for the free-free system are presented in Fig. 18.3. In this case the free-free boundary condition is applied in order to check experimentally the numerical results. Similar results could be evaluated according to actual boundary conditions obtained placing the crank mechanism in the engine block.

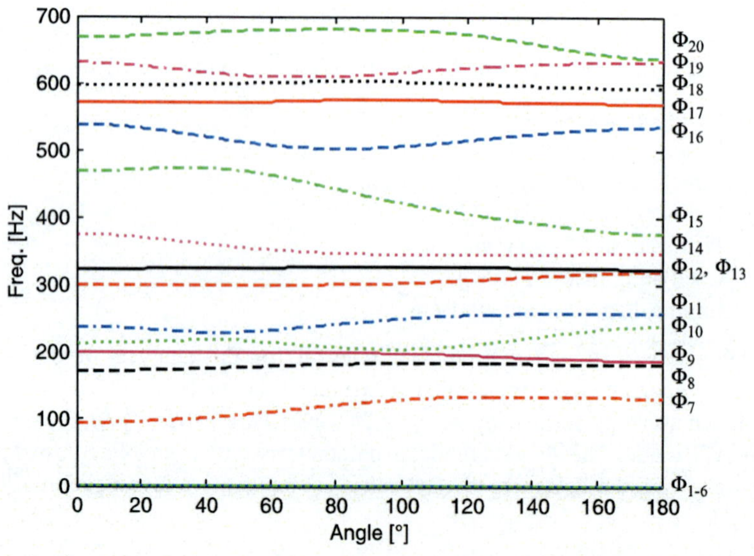

Fig. 18.3 System natural frequencies vs. crankshaft rotation angle

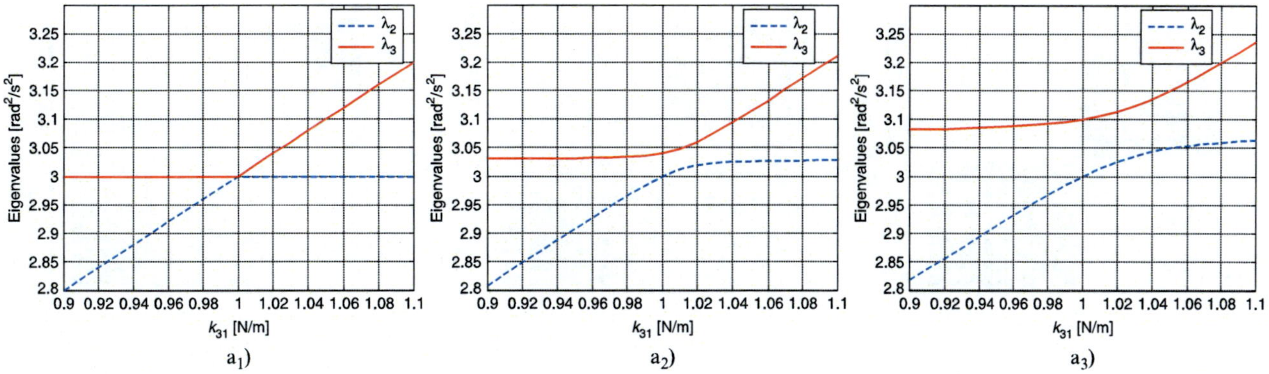

Fig. 18.4 Simplest cyclic spring-mass system ($m_1 = m_2 = m_3 = 1$ kg): natural frequencies versus the variable parameter $k_{31} = 0.9 \div 1.1$ N/m: (a$_1$)$k_{12} = k_{23} = 1$ N/m, (a$_2$) $k_{12} = 1.02$ N/m, $k_{23} = 1$ N/m, (a$_3$) $k_{12} = 1.05$ N/m, $k_{23} = 1$ N/m

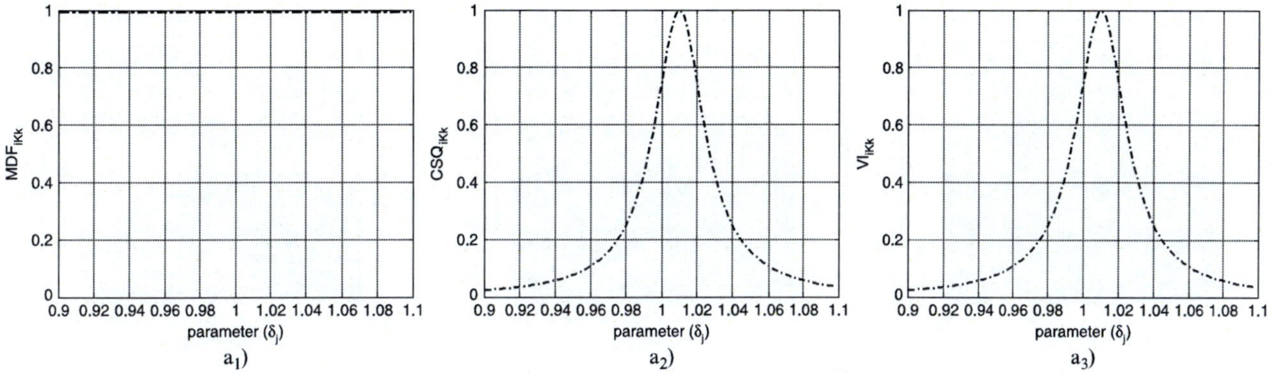

Fig. 18.5 Veering criteria, between Φ_2 and Φ_3, for the simplest cyclic 3 dof spring-mass system, $k_{12} = 1.02$ N/m : (a$_1$)MDF, (a$_2$)CSQ, (a$_3$)VI

The loci appear to veer and cross in several places. Moreover, veering phenomenon seems to involve more than two modes. Usually when this happens modes shape variation can not be easily shown in a graphical way and it is not straightforward to distinguish and quantify the co-partecipation of the modes given at a reference configuration. The mode shape variation throughout veering between two different configurations can be evaluated by means of the modal assurance criterion (MAC) [11], defined as

$$MAC_{i,j} = \frac{\left(\Phi_i^T \Phi_j\right)^2}{\left(\Phi_i^T \Phi_i\right)\left(\Phi_j^T \Phi_j\right)} = \cos^2 \alpha_{i,j} \tag{18.3}$$

in two different ways. The former, as shown in Bonisoli examples [12] compares all the modes for two different parameter values, corresponding to the system just before and after the veering. If crossing happens the off-diagonal terms of MAC[$\Phi_i(\delta_1)$, $\Phi_j(\delta_2)$] shows perfect correlation whereas diagonal terms are proximate to zero. In case of veering a different behaviour for diagonal and off diagonal term is shown, bringing to a *plateau* for those MAC values. The choice of values for quantify the transition is not straightforward. The latter, here in presented, gives the correlation representation between two chosen modes (Φ_i, Φ_j) changing a parameter δ_j over its entire domain. By use of this second approach it is possible to see the transition of the analyzed modes in the two-dimensional subspace spanned by them and quantify the misalignment between mode Φ_j pre-veering mode Φ_k post-veering.

For simple lumped parameters model, as in [22], veering and crossing pattern can be easily described (Fig. 18.4) and veering indices [23] can be applied.

The mass matrix remains constant and serves as an orthonormalising matrix, whereas a linear variation is considered for the stiffness matrix. *Modal dependence factor*, *veering index* and *cross sensistivity quotient* are show in Fig. 18.5.

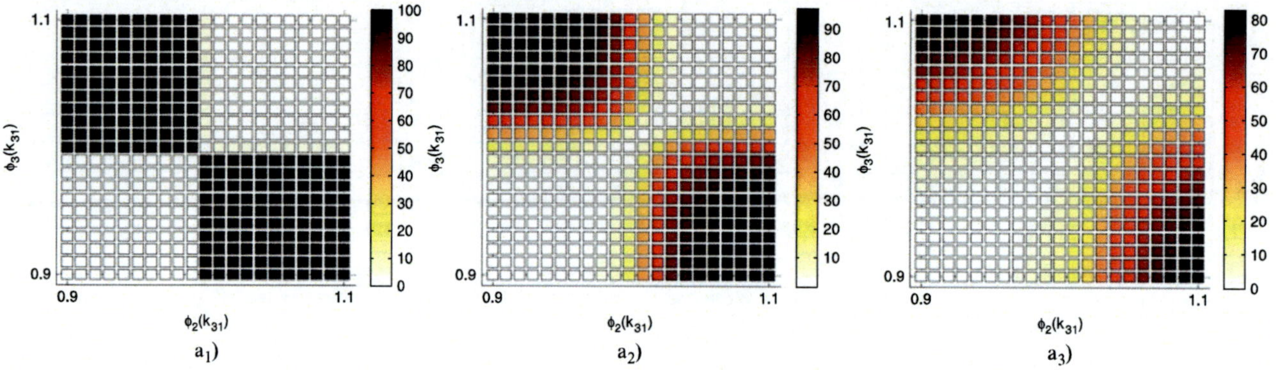

Fig. 18.6 Modal assurance criterion evaluation for the simplest cyclic 3 dof spring-mass system with $k_{31} = 0.9 \div 1.1\,\mathrm{N/m}$: ($a_1$) $k_{12} = k_{23} = 1\,\mathrm{N/m}$, ($a_2$) $k_{12} = 1.02\,\mathrm{N/m}$, $k_{23} = 1\,\mathrm{N/m}$, (a_3) $k_{12} = 1.05\,\mathrm{N/m}$, $k_{23} = 1\,\mathrm{N/m}$

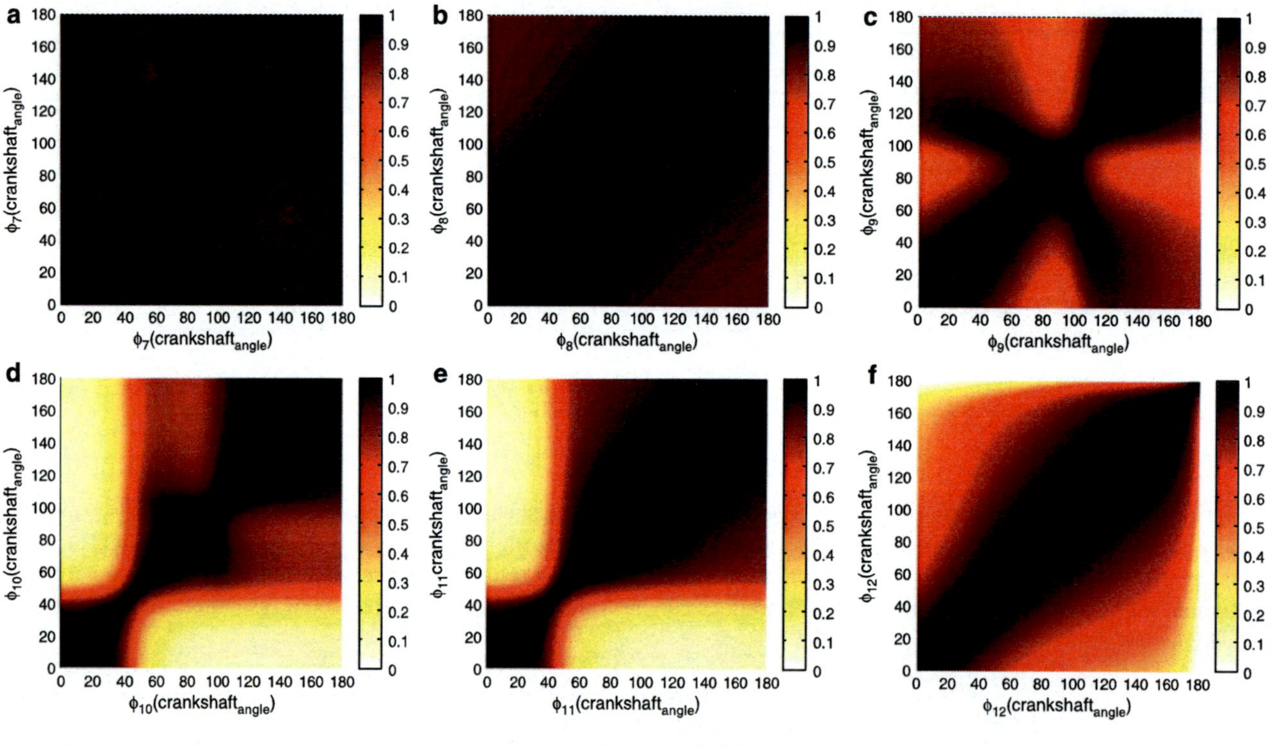

Fig. 18.7 AUTOMAC evaluation: **(a)** 7th mode, **(b)** 8th mode, **(c)** 9th mode, **(d)** 10th mode, **(e)** 11th mode, **(f)** 12th mode

On the main diagonal, the values displayed show the eigenvectors orthogonality, while on the cross diagonal it is possible to distinguish the behaviour of the rotation angle between the two modes considered. Eigenvectors are modal mass unitary normalized. It is interesting to note that the rotation angle spans less than 90° and after veering one mode is always inverted, as recognized by du Bois [23]. By varying two parameters (k_{12}, k_{31}) of this cyclic structure it is possible to increase or decrease the coupling between mode 2 and 3 and consequently change the domain region in which the two modes shows high correlation (MAC > 0.7) (Fig. 18.6).

For a more complex model with free-free boundary conditions, as the one depicted in Fig. 18.1, interaction between modes can be explained in a similar way. An AutoMAC evaluation for the first few modes, since a greater number of modes is involved, allows to look for mode transformation (Fig. 18.7). For each bottleneck on the main diagonal, a veering is acting: the greater the curvature radius, the weaker the veering affecting that mode. Then, an indeep analysis between pairs of interacting modes enlight the exchange of modal properties among them.

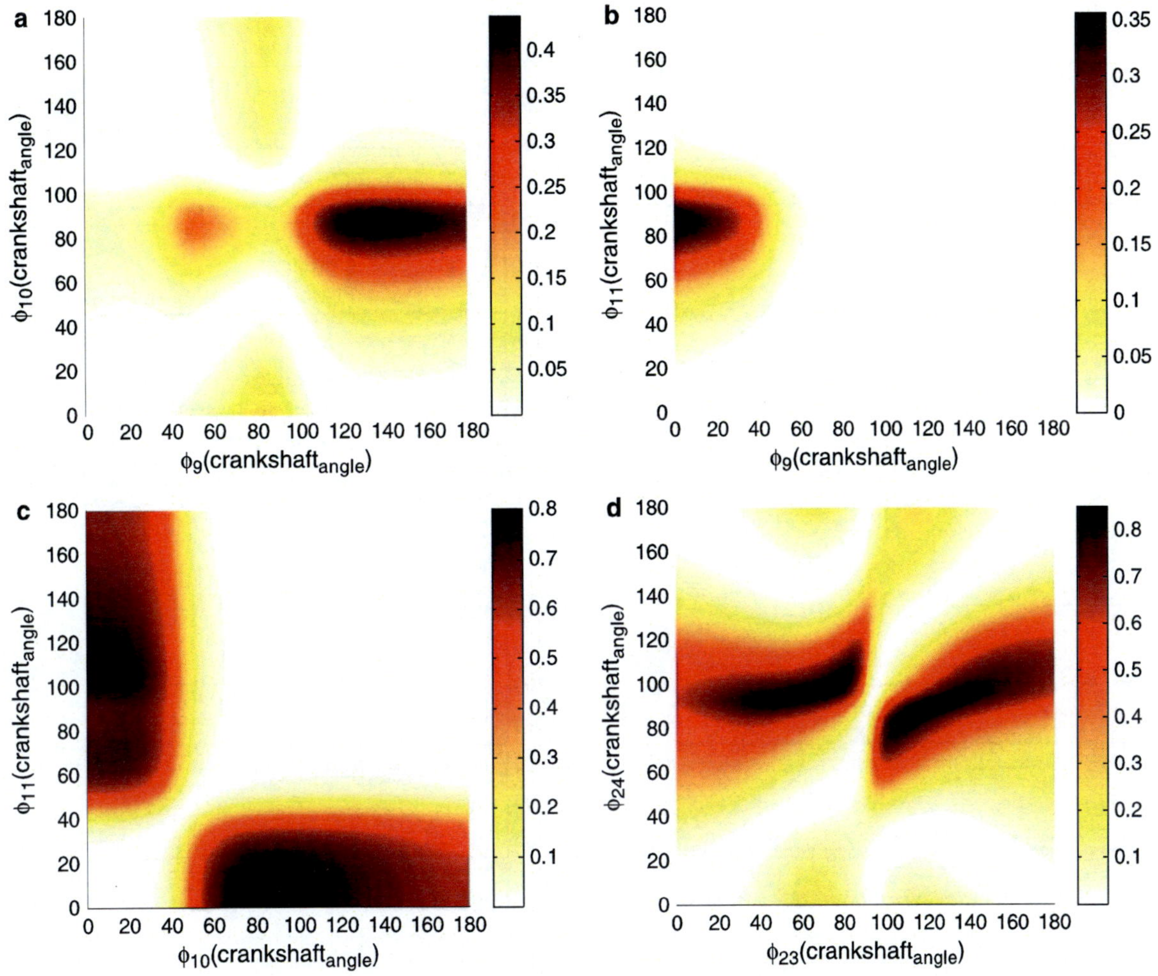

Fig. 18.8 Correlation valuation: (**a**) MAC[$\Phi_9(\alpha)$, $\Phi_{10}(\alpha)$], (**b**) MAC[$\Phi_9(\alpha)$, $\Phi_{11}(\alpha)$], (**c**) MAC[$\Phi_{10}(\alpha)$, $\Phi_{11}(\alpha)$], (**d**) MAC[$\Phi_{23}(\alpha)$, $\Phi_{24}(\alpha)$]

Weak veering takes place for mode 9th, more strong veering affects modes 10th and 11th, 14th and 15th, and at higher frequency modes 23rd and 24th (Fig. 18.8).

In [23] a veering index VI, combination of two normalized criteria, is proposed as a measure of the presence and intensity of curve veering between modes, considering either a variation in the stiffness matrix or mass matrix. It allows to precisely detect where the transformation is taking place, but not to localize the parameter value for which the angle between the modes becomes smaller. Moreover is not valid for systems in which both mass matrix and stiffness matrix are varying. Since the two vectors throughout veering always remain in the same plane, it is possible to quantify the similarities among them by using the Modal Assurance Criterion and comparing the MAC index for each configuration.

18.4 From Components Dynamic Requirements to System Dynamic Properties

Adding components to the system, the number of components modes are simply added to the number of system modes, with complex veering-crossing phenomena, and great interactions are detected.

In a range from 0 to 1,000 Hz, the crankmechanism dynamics might be considered as the modal superposition of the crankshaft modes, namely the first bending mode in plane *xy* at 291 Hz, the first bending mode in plane *yz* at 352 Hz, the first torsional mode around 795 Hz (Fig. 18.9a), and groups of four local modes of piston/connecting-rod dynamics (PCR) at 231 Hz, at 375 Hz, and around 600 Hz (Fig. 18.9b).

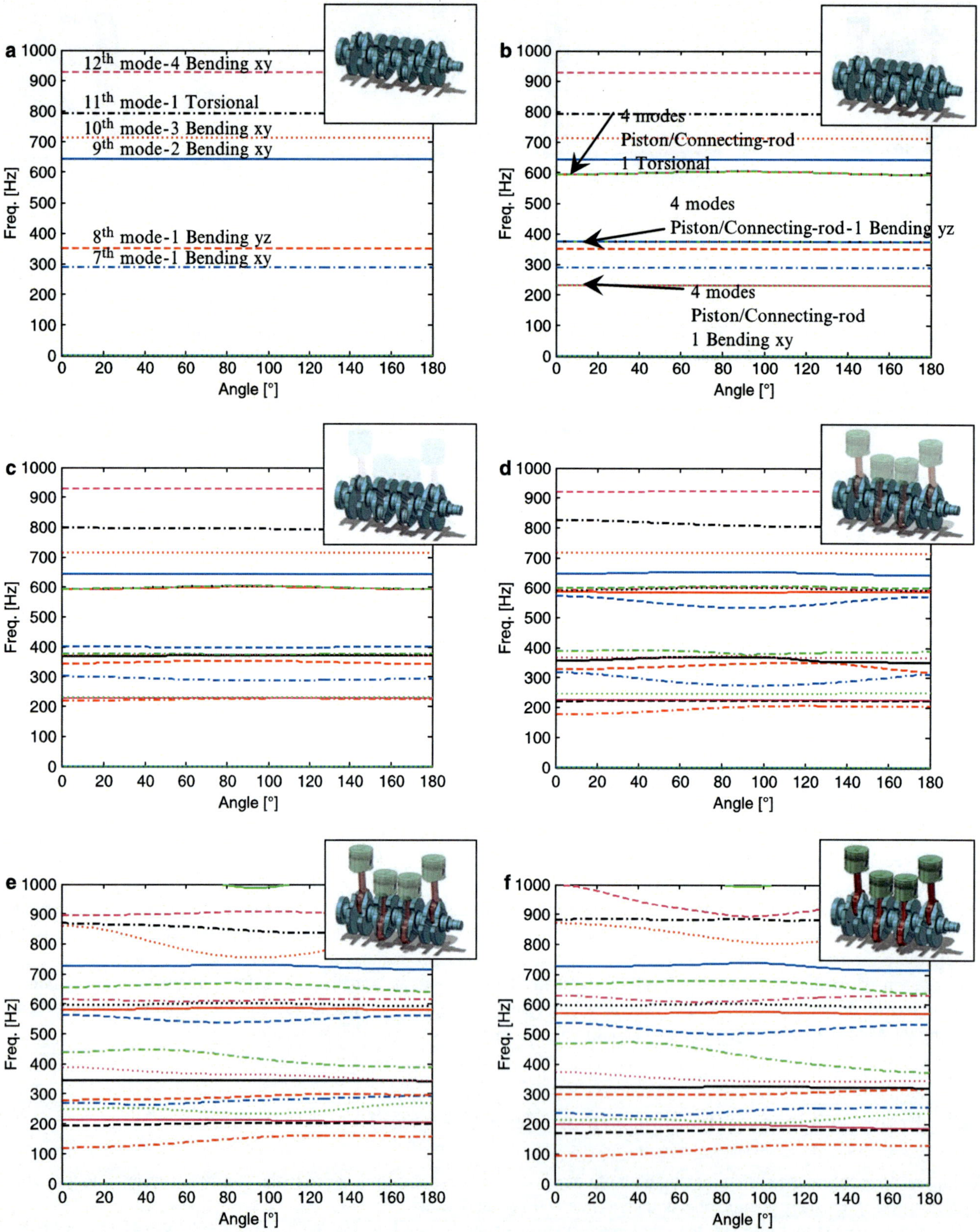

Fig. 18.9 Natural frequencies of the crank mechanism versus crankshaft angle, free-free case. Effects of piston-connecting rod (PCR) density on global properties. (**a**) $\rho_{PCR} = 0$ %, (**b**) $\rho_{PCR} = 0.01$ %, (**c**) $\rho_{PCR} = 1$ %, (**d**) $\rho_{PCR} = 10$ %, (**e**) $\rho_{PCR} = 50$ %, (**f**) $\rho_{PCR} = 100$ %

The transition from component to system and its effect on the eigenvalue problem has been generated by gradually adding the pistons and connecting rod mass matrices to the global system mass matrix, and keeping constant the ratio between density ρ_{PCR} and Young's Modulus E_{PCR}. In Fig. 18.9 veering phenomena become more and more evident by increasing the ρ_{PCR} value.

Generally by only looking at the mode shapes on a FEM post-process and by trying to compare them visually can give rise to mistakes and wrong conclusions. Considering the eigenfrequencies plot for the complete assembly (Fig. 18.9f), the seventh mode is not the first global bending mode of the crankmechanism in plane *xy*, but one of the four piston/connecting-rod bending modes; the modal properties of the first bending mode of the crankshaft alone (291 Hz) and the first bending modes of piston/connecting-rod (231 Hz) in the system configuration are very similar.

18.5 Identification of Local and Global Modes

It becomes evident that modes strongly interact each other and might show different behaviours while a system parameter is varying. Furthermore when converging loci get closer and suddenly veer away, related modes swap their properties. Localization and veering are related to couplings that can be seen as physical (namely a spring in a 4-dof [23] or the PCR component in the analyzed structure), or modal interactions [24]. In both cases these couplings can be translated, from numerical point of view, into off diagonal terms either in the stiffness or mass matrix.

This makes hard to precisely define a component mode, to look for it in a system dynamics perspective through the entire variable parameter range, and to transform the single component design into a more synergetic and globally optimized process. However, modes patterns and frequency limits can be identified.

All the eigenfrequency loci plots (Fig. 18.9) do not allow to emphasize the great eigenvector transformation in act during the four-stroke cycle. Due to the strong modal interaction, the seventh mode in the crank mechanism comes from the hybridization of one of the four bending modes of the piston/connecting-rod in plane *xy* with the first bending mode in plane *xy* of the crankshaft (seventh crankshaft mode); the MAC provides a clear and expected correlation between the 7th crankshaft mode and the 11th system mode (Fig. 18.10a), when crankshaft and piston-connecting rods are weakly coupled, i.e. $\rho_{PCR} = 1\,\%$. As soon as the components become more strongly coupled ($\rho_{PCR} = 10\,\%$), the first bending in *xy* plane shows unambiguous similarities with 11th and 13th system modes (Fig. 18.10b). By increasing the coupling factor ρ_{PCR} up to 100 % the link between system and component modes changes again. In all the cases depicted, the MAC puts in evidence a non unique match between system and component.

Considering the configuration at the top dead centre (TDC = 0°), the 8th mode of the crankshaft – the 1st bending mode in plane *yz* – is similar to the 14th mode of the crank mechanism (Fig. 18.11b); the 9th mode of the crankshaft – the 2nd bending mode in plane *xy* – is correlated with the 20th mode of the crank mechanism (Fig. 18.11c). Analogous reasoning could be extended to higher frequencies (Fig. 18.11d, f–h).

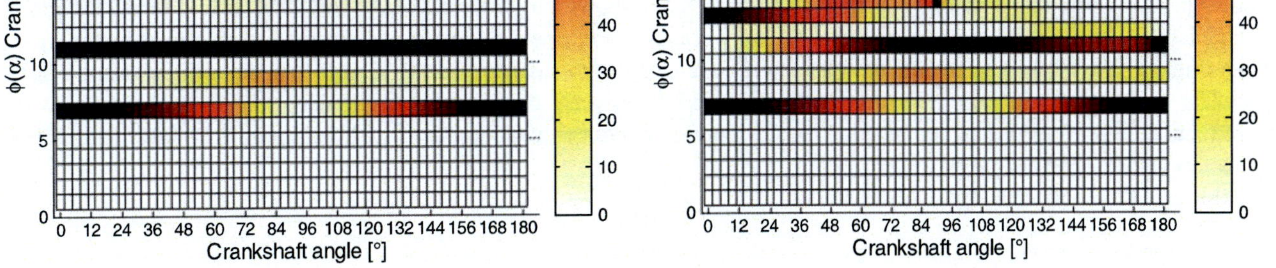

Fig. 18.10 MAC index evaluation between system and component ($\Phi_{7\text{crankshaft}}$, $\Phi_{[1:30]\text{crankmechanism}}$), free-free: (**a**) $\rho_{PCR} = 1\,\%$, (**b**) $\rho_{PCR} = 10\,\%$

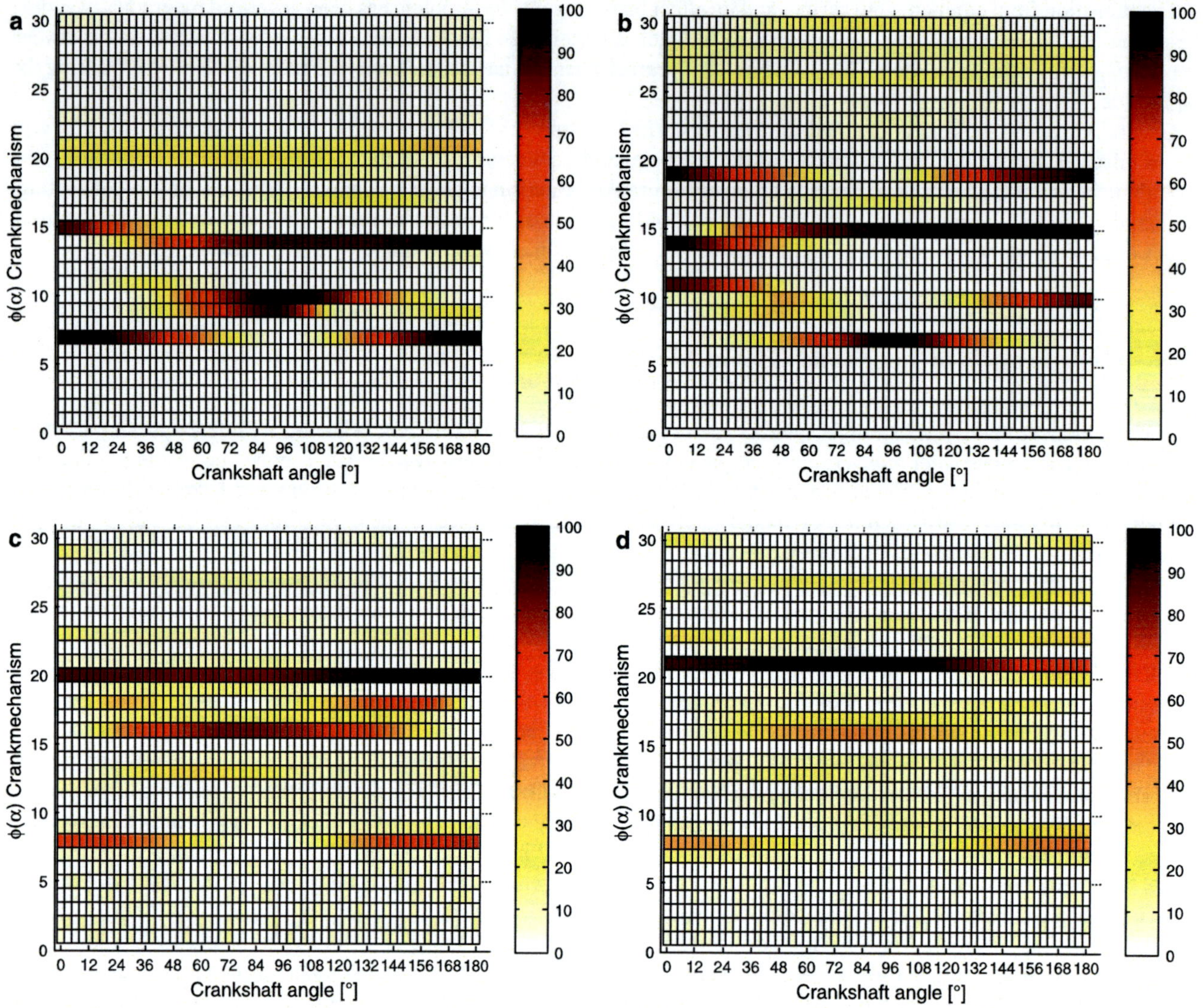

Fig. 18.11 MAC index evaluation between system and component (free-free): **(a)** MAC($\boldsymbol{\Phi}_{7\text{crankshaft}}, \boldsymbol{\Phi}_{[1:30]\text{crankmechanism}}$), **(b)** MAC($\boldsymbol{\Phi}_{8\text{crankshaft}}, \boldsymbol{\Phi}_{[1:30]\text{crankmechanism}}$) **(c)** MAC($\boldsymbol{\Phi}_{9\text{crankshaft}}, \boldsymbol{\Phi}_{[1:30]\text{crankmechanism}}$), **(d)** MAC($\boldsymbol{\Phi}_{10\text{crankshaft}}, \boldsymbol{\Phi}_{[1:30]\text{crankmechanism}}$), **(e)** MAC($\boldsymbol{\Phi}_{11\text{crankshaft}}, \boldsymbol{\Phi}_{[1:30]\text{crankmechanism}}$), **(f)** MAC($\boldsymbol{\Phi}_{12\text{crankshaft}}, \boldsymbol{\Phi}_{[1:30]\text{crankmechanism}}$), **(g)** MAC($\boldsymbol{\Phi}_{13\text{crankshaft}}, \boldsymbol{\Phi}_{[1:30]\text{crankmechanism}}$), **(h)** MAC($\boldsymbol{\Phi}_{14\text{crankshaft}}, \boldsymbol{\Phi}_{[1:30]\text{crankmechanism}}$)

The modal coupling seems to seriously influence the torsional mode, posing difficulties for modal correlation.

The expected correspondence between the 11th component mode and the 23rd system mode vanishes for high values of coupling factor. The MAC values are lower (MAC $\cong$ 0.7), and the correlation involves a wider frequency spectrum (Fig. 18.11e).

This can be explained in more detail by looking at the similarities between system modes and the pure first torsional mode when the coupling factor ρ_{PRC} is changed. Figure 18.12a shows a high correlation (MAC $\cong$ 1) between the 23rd crank mechanism mode and the 7th mode related to the only crankshaft in free-free conditions. By gradually increasing the ρ_{PRC} value (Fig. 18.12b–e), the cosine of the angle between the eigenvectors is lowered (MAC $\cong$ 0.6) and is difficult to find only one clear and distinct similarity. The torsional mode undergoes great changes due to the deep effect of interacting with other local modes of pistons and connecting rods components and it results into two new hybrid modes at lower and higher frequency (mode 22nd and mode 25th).

Table 18.1 summarizes all these observations.

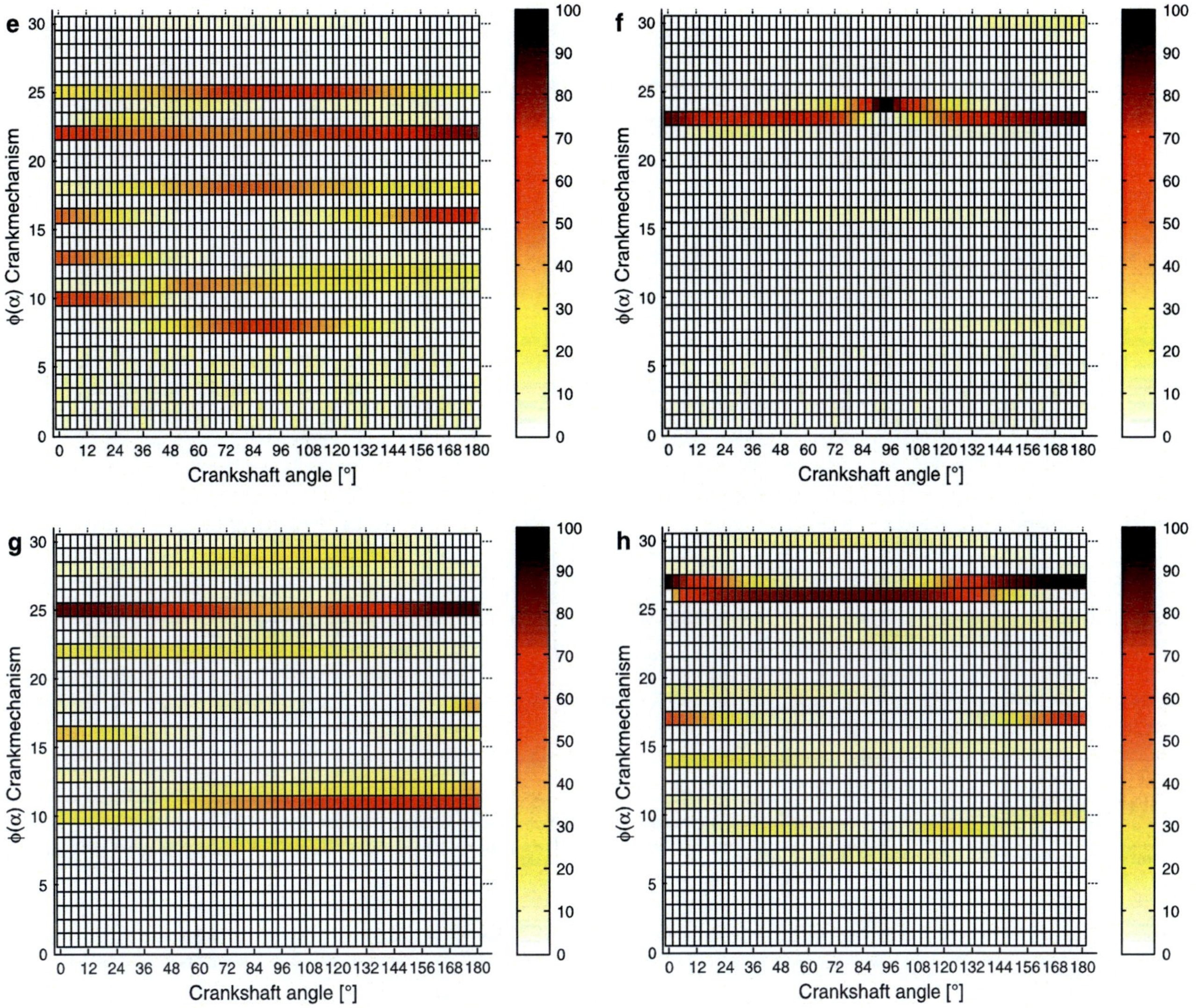

Fig. 18.11 (continued)

18.6 Conclusions

A clear numerical demonstration of crossing and veering phenomena for an actual world application has been given. By means of MAC evaluations it is possible to localize veering parameter values, quantify the coupling between modes, and get a more precise description of modal properties in the transition zone. This approach might be useful in characterizing modified designs and to make more clear how they should be improved under dynamic response restrictions.

During the whole optimization process of design and validation of assemblies, it is important to consider wider frequency range for each mode rather than to avoid single frequency values.

Moreover the example presented highlighted the need to take into account these aspects also in modal correlation and model updating fields without neglecting changes into the assembly configuration.

Further numerical analyses, by means of more sophisticated FE models, followed by experimental validation, could exemplify the benefit of the presented approach into global product design process.

Boundary conditions could play a similar role.

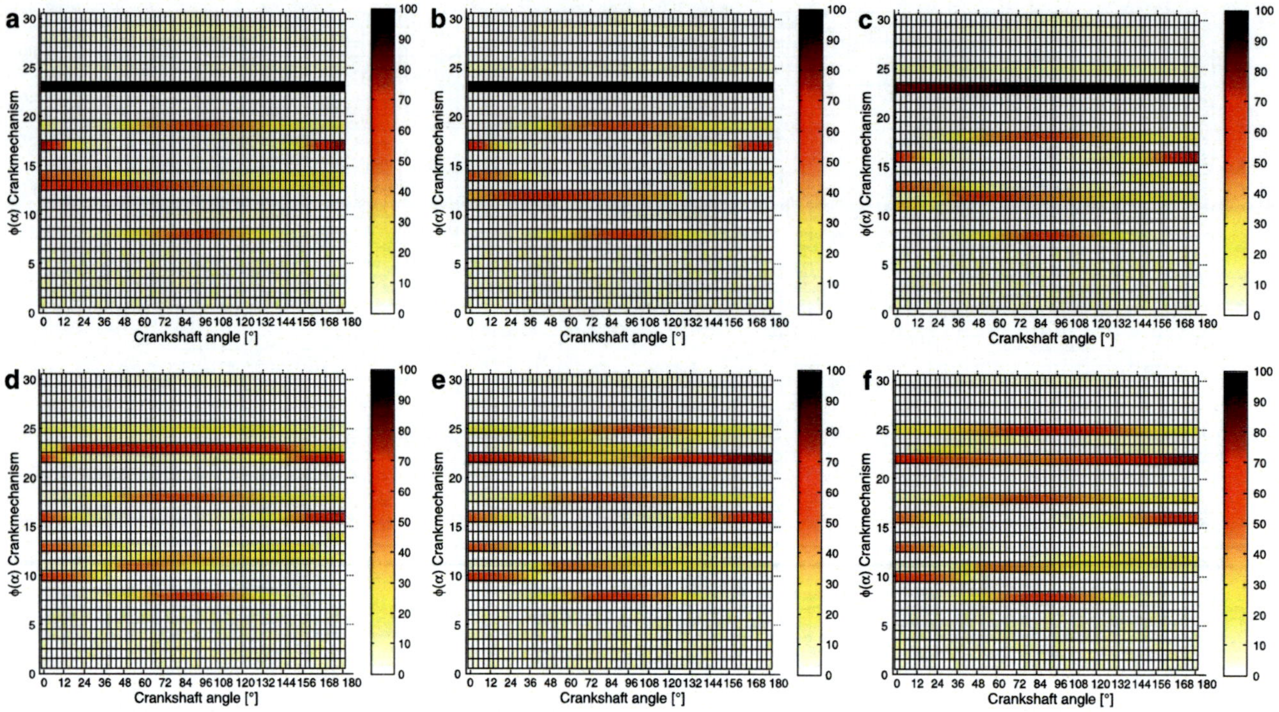

Fig. 18.12 MAC index evaluation between system and component (MAC($\Phi_{7\text{crankshaft}}$, $\Phi_{[1:30]\text{crankmechanism}}$), free-free: **(a)** $\rho_{PCR} = 1\ \%$, **(b)** $\rho_{PCR} = 10\ \%$, **(c)** $\rho_{PCR} = 25\ \%$, **(d)** $\rho_{PCR} = 50\ \%$, **(e)** $\rho_{PCR} = 75\ \%$, **(f)** $\rho_{PCR} = 100\ \%$

Table 18.1 Modes and frequency comparison between system and component

	Component Crankshaft (mode – freq.)	System Crankmechanism (mode – freq. range $[f_l, f_u]$)	$\Delta f_{\max} = \max(f_{u,l} - f_{CS}) = \frac{1}{2\pi}\max\lvert\omega_{CS}(\alpha) - \omega_{CM}(\alpha)\rvert$
Modes	7th(291 Hz)	15th–14th [348 Hz, 475 Hz]	184 Hz
	8th(352 Hz)	14th–15th[348 Hz, 475 Hz]	123 Hz
	9th(645 Hz)	20th[639 Hz, 681 Hz]	36 Hz
	10th(716 Hz)	21st[716 Hz, 740 Hz]	24 Hz
	11th(794 Hz)	~22nd[805 Hz, 873 Hz]	79 Hz
	12th(930 Hz)	23rd–24th[882 Hz, 900 Hz]	48 Hz
	13th(1037 Hz)	25th[1,000 Hz, 1,175 Hz]	138 Hz
	14th(1324 Hz)	27th–26th[1,218 Hz, 1,349 Hz]	106 Hz

References

1. Leissa W (1974) On a curve veering aberration. J Appl Math Physics 25:99–111
2. Chen X, Kareem A, Matsumoto M (2001) multimode coupled flutter and buffeting analysis of long span bridges. J Wind Eng 89:649–664
3. Bae JS, Inman DJ, Lee I (2004) Effects of structural nonlinearity on subsonic aeroelastic characteristics of an aircraft wing with control surface. J Fluids Struct 19:747–763
4. Khodaparast H, Mottershead JE, Badcock K (2010) Propagation of structural uncertainty to lineat aeroelastic stability. J Fluids Struct 88:223–236
5. Chan Y, Inman DJ (2010) Management of the variability of vibration response levels in mistuned bladed discs using robust design concepts. Part 1 parameter design. Mech Syst Signal Process 24:2777–2791
6. Tang D, Dowell E (2010) Aeroelastic response of aircraft with freeplay structural nonlinearity. In: Proceedings of 2nd aircraft structural design conference, London
7. Maeda T, Baburaj V, Ito Y, Koga T (1998) Flexural-torsional coupling effect on vibrational characteristics of angle-ply laminates. J Sound Vib 210(3):351–365
8. Chen X, Kareem A (2003) Curve veering of eigenvalues loci of bridges with aeroelastic effects. J Eng Mech 129(2):146–159
9. Benedettini F, Zulli D, Alaggio R (2009) Frequency-veering and mode hybridization in arch bridges. In: Proceedings of the IMAC-XXVII, Orlando

10. du Bois JL, Adhikari S, Lieven NA (2007) Experimental and numerical investigation of mode veering in a stressed structure. In: Proceedings of 25th IMAC, Detroit
11. Allemang RJ, Brown DL (1982) A correlation coefficient for modal vector analysis. In: Proceedings of first IMAC, Orlando, pp 110–116
12. Bonisoli E, Delprete C, Esposito M, Mottershead JE (2011) Structural dynamics with coincident eigenvalues: modelling and testing. In: Proceedings of 29th IMAC, Detroit, vol 3(6), pp 325–337
13. Wilson K (1963) Torsional vibration problems. Chapman & Hall, London
14. Mourelatos Z. (2001) A crankshaft system model for structural dynamic analysis of internal combustion engines. Comput Struct 79(20–21):2009–2027
15. Mendes AS, Meirelles PS, Zampieri DE (2008) Analysis of torsional vibration in internal combustion engines: modelling and experimental validation. J Multi-body Dyn 222(2):155–178
16. Brusa E, Delprete C, Genta G (1997) Torsional vibration of crankshafts: effects of non-constant moments of inertia. J Sound Vib 205(2):135–150
17. Song XG, Song TX, Xue DX, Li BZ (1991) Progressive torsional-axial continued vibrations in crankshaft systems: a phenomenon of coupled vibration. Trans ASME Rotat Mach Veh Dyn 319–323
18. Paz M, Leigh W (2003) Structural dynamics. Springer, New York
19. Asghar Bhatti M (2005) Fundamental finite element analysis and applications. Wiley, Hoboken
20. Reymond MA (1991) MSC/NASTRAN: user's manual. The McNeal-Schwendler Corp, Los Angeles
21. Mourelatos Z (2000) An efficient crankshaft dynamic analysis using substructuring with Ritz vectors. J Sound Vib 238(3):495–527
22. Balmes E (1993) High modal density, curve veering, localization: a different perspective on the structural response. J Sound Vib 161(2):358–363
23. du Bois J, Lieven EJ, Adhikari S (2009) Localisation and curve veering: a different perspective on modal interactions. In: Proceedings of 27th IMAC, Orlando
24. Pierre C (1988) Mode localization and eigenvalue loci veering phenomena in disordered structures. J Sound Vib 126(3):485–502

Chapter 19
Validating Low-Level Footfall-Induced Vibration Predictions in Steel and Concrete Structures

Michael J. Wesolowsky, Julia M. Graham, J. Shayne Love, Jon K. Galsworthy, and John C. Swallow

Abstract Occupant footfalls are often the most critical source of floor vibration on the elevated floors of buildings. Floor motions can disturb occupants, leading to frequent complaints and loss of functionality. In laboratory and healthcare facilities, this issue can be more critical, as high-resolution imaging equipment with stringent vibration criteria is often employed. Achieving these criteria requires sufficiently stiff and massive floor structures to effectively resist the forces exerted from user traffic. The difficulty for engineers is predicting these low levels of vibration. Two commonly used analysis methods to predict footfall vibration levels in steel buildings are the American Institute of Steel Construction (AISC) Design Guide 11, and The Steel Construction Institute (SCI) P354. The latter is more robust, as it can predict multi-modal time history responses at any point on the floor. Dynamic footfall loading is determined by considering walkers moving along reasonable pathways identified in the architectural floor plans. For concrete structures, The Concrete Centre (CCIP-016) proposes a methodology similar to the SCI. In this study, three steel and one concrete building are instrumented to measure footfall-induced vibrations. The measured values are compared to the predictions of the aforementioned methods, and the superiority of the SCI-P354 and CCIP-016 methods is shown.

Keywords Footfall vibration • Vibration measurements • Model validation • Sensitive floors • Dynamic loading

19.1 Introduction

Advances in materials, construction, and design technology have led to the construction of lighter floor systems, which are more susceptible to footfall-induced vibration. Meanwhile, improvements to imaging technology, including magnetic resonance imaging (MRIs) and electron microscopes, have led to devices which have increasingly stringent vibration requirements to ensure optimal performance. As a result, it is becoming more common for floor vibrations to be the governing factor in the design of the structural system.

Footfall vibrations arise from the dynamic force imparted to the floor during a pedestrian's gait. The resulting vibration is then transmitted to other areas of the floor, within the bay and across bays. Therefore, footfall-induced vibrations originating from walking in a corridor will affect the surrounding area, including bays other than the one containing the walker. The challenge faced by engineers is to predict the magnitude of the vibrations in sensitive areas of the structure and, where necessary, perform design alterations to mitigate vibrations that exceed design criteria.

Several methodologies have been proposed to estimate floor vibrations [1]. Unfortunately, there has been relatively few validation studies published in which vibration levels predicted by a methodology are compared to vibration measurements taken in existing structures. In this study, three commonly used methodologies, the American Institute of Steel Construction (AISC) Design Guide 11 [2], The Steel Construction Institute (SCI) P354 [3], and The Concrete Centre (CCIP-016) [4]

M.J. Wesolowsky (✉) • J.C. Swallow
Swallow Acoustic Consultants Ltd., 23-366 Revus Avenue, L5G 4S5, Mississauga, ON, Canada
e-mail: mwesolowsky@swallowacoustic.ca; jswallow@swallowacoustic.ca

J.M. Graham • J.S. Love • J.K. Galsworthy
Rowan Williams Davies & Irwin Inc., 650 Woodlawn Road West, N1K 1B8, Guelph, ON, Canada
e-mail: Julia.Graham@rwdi.com; Shayne.Love@rwdi.com; Jon.Galsworthy@rwdi.com

T. Simmermacher et al. (eds.), *Topics in Model Validation and Uncertainty Quantification, Volume 5*,
Conference Proceedings of the Society for Experimental Mechanics Series 41,
DOI 10.1007/978-1-4614-6564-5_19,

are used to predict the vibration levels of several steel and concrete buildings. The buildings were then instrumented with accelerometers and/or velocity transducers to record the dynamic response of the floor. Scenarios with walkers of known mass and pace were conducted to represent footfall loading cases.

19.2 Overview of Prediction Methodologies

19.2.1 American Institute of Steel Construction Design Guide 11

The modeling technique outlined by the American Institute of Steel Construction (AISC) Design Guide 11 has been extensively used in North America for the past 15 years. For human comfort, the AISC method assumes the floor behavior is governed by the resonant response of the fundamental floor mode. A frequency-dependent dynamic coefficient, fitted to empirical data, is used to estimate the magnitude of the harmonic loading applied to the floor. A steady-state floor response is calculated assuming the walker and vibration-sensitive receptor are both located at the position of the maximum modal displacement (center of the bay) to produce a worst-case response. The amplitude of the vibration is then determined using the modal damping assumed. A reduction factor is applied to this idealized resonant response to account for the inability of a resonant response to build up in the time it takes a walker to transverse the bay under consideration, and the unlikelihood that the walker and vibration-sensitive receptor are both at the position of the maximum modal displacement. The natural frequency of the floor is estimated from the maximum static deflection of the bay under consideration due to the acting dead and live loads. The total floor deflection is estimated by summing the deflections of the beam and girder elements. Such a simplistic method cannot easily predict the natural frequency when the floor layout is irregular, such as would occur around shafts and/or in non-rectangular buildings.

For sensitive equipment, the AISC uses a slightly different method. Three walking paces are considered: 50, 75, and 100 steps per minute (spm). An empirical force coefficient is estimated as a function of the walking pace, and the weight of the walker. Simple beam theory is employed to calculate the floor deflection, and the natural frequency. The floor response velocity is then calculated as a function of the force coefficient, floor natural frequency, and floor deflection.

19.2.2 Steel Construction Institute and Concrete Centre

The methodologies proposed by The Steel Construction Institute (SCI P354) and The Concrete Centre (CCIP-016) for calculating footfall-induced floor vibrations are similarly derived. The methods recommend utilizing a Finite Element (FE) model to predict the mode shapes and natural frequencies of the floor. FE modeling is advantageous since it allows the calculation of many mode shapes, any of which can contribute significantly to the vibration of a specified floor region. Moreover, irregular floor features (including non-rectangular bays, shafts, and different beams or slab thicknesses from bay-to-bay) can easily be accommodated by the FE model. The predicted mode shapes and natural frequencies are then used to estimate the generalized mass of each mode.

The response of each mode is assumed to be either resonant or impulsive, depending on the associated floor frequency. Empirically determined frequency-dependent dynamic load factors are used to excite the floor along probable walking paths as determined from the architectural drawings. The loading conditions are then used to develop a time series response for each mode. The time series responses for each mode are then linearly superimposed to produce a total floor response at any receptor point on the floor. A range of realistic walking frequencies and walking paths are considered to determine the governing floor response. Spectral analysis can be performed on the predicted time-series responses to express the floor behavior in a format appropriate for comparison with the relevant criteria.

19.3 Vibration Criteria

For human comfort, vibration criterion is normally expressed as the root mean square (RMS) response of each one-third octave band from 1 to 80 Hz [5]. For sensitive equipment, the criteria may be expressed in one-third octave bands, or other formats, including power spectral densities, peak-to-peak levels, etc. Over the past 25 years, generic vibration limits have been developed which provide frequency-dependent sensitivities for wide classes of equipment, and are used extensively in design for healthcare and research facilities [6]. These vibration criterion (VC) curves are internationally accepted as a basis for designing and evaluating the performance of vibration sensitive equipment and the structures that support them. The VC curves range between Workshop (least stringent) through VC-G (most stringent), and are the basis for the comparisons in this study (as seen throughout the comparison figures).

These curves were originally based on the ISO 2631-2 (1989) [7] base curve for human response to whole body vibration, which is the threshold of human perception, but have since evolved. The ISO base curve is often referred to as the ISO-Operating Room criteria. The above noted criteria are also specified as RMS velocities in one-third octave bands. The generic vibration curves should not be used to replace manufacturers' specifications for vibration requirements, but are beneficial where manufacturers' specifications are non-existent, incomplete, or where specific equipment has not yet been selected.

The AISC uses frequency-dependent peak accelerations for human comfort vibration criteria [2]. However, for low-level vibration in sensitive spaces, it employs VC curves as the appropriate criteria.

19.4 Predicted and Measured Floor Vibrations

One concrete and three steel structures were considered for this study. For each structure, accelerometers were mounted to the floor, and walking scenarios were conducted in nearby corridors or open spaces. Generally, the sensor was positioned near the center of the bay, as this is typically where the maximum floor response occurs. A metronome was employed to ensure the walker maintained a constant pace to facilitate direct comparison with the modeled results. Wherever possible, walkers were employed who had masses of approximately 80 kg, which is consistent with the AISC, SCI and CCIP methods. If the walker mass differed significantly from 80 kg, linear factors were applied to the results for comparison purposes. SAP2000 structural analysis software was employed to predict the dynamic properties of the steel and concrete floors for the SCI and CCIP methods. FloorVibe [8] was employed to perform the AISC response predictions.

19.4.1 Steel Structure #1

Steel structure #1 is a hospital in the North-East United States that was undergoing renovations to install an MRI on an elevated floor. Floor vibration measurements were taken to ensure the proposed MRI location satisfied the manufacturer-specified vibration criteria. Figure 19.1 shows the architectural layout of the floor with the walking paths considered, as well

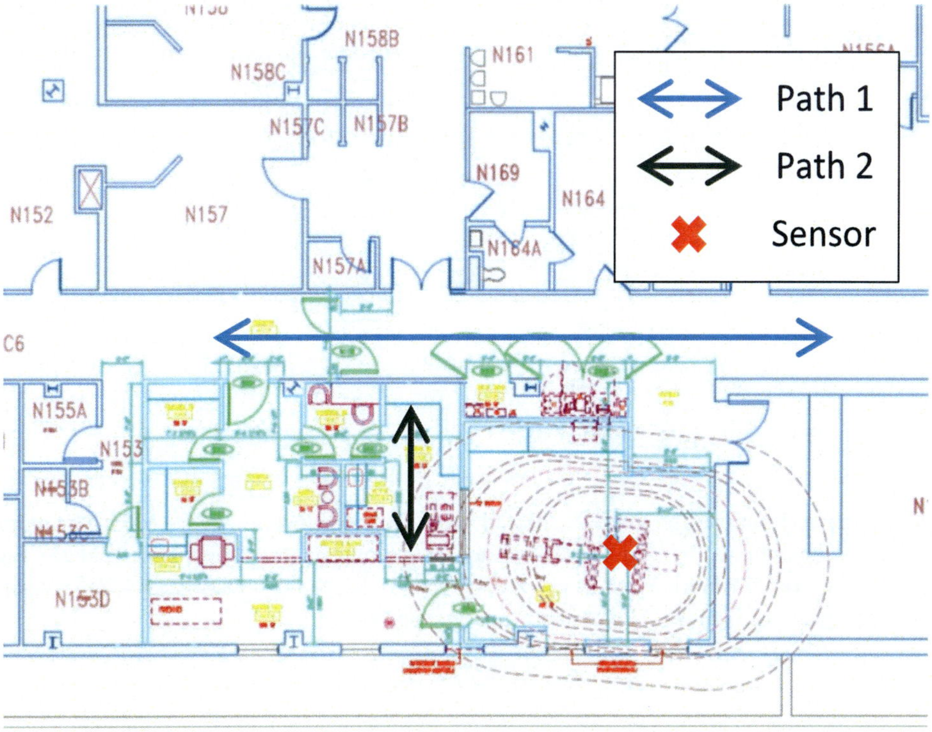

Fig. 19.1 Walking paths for steel structure #1

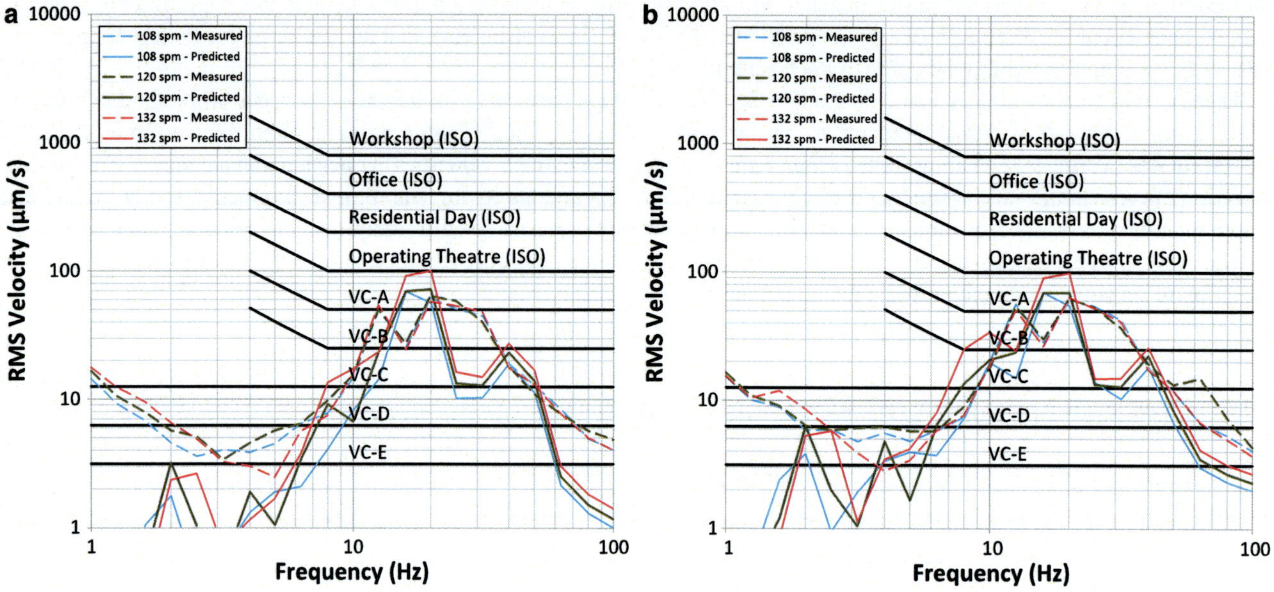

Fig. 19.2 Predicted and measured floor response of steel structure #1 (**a**) walking path 1, (**b**) walking path 2

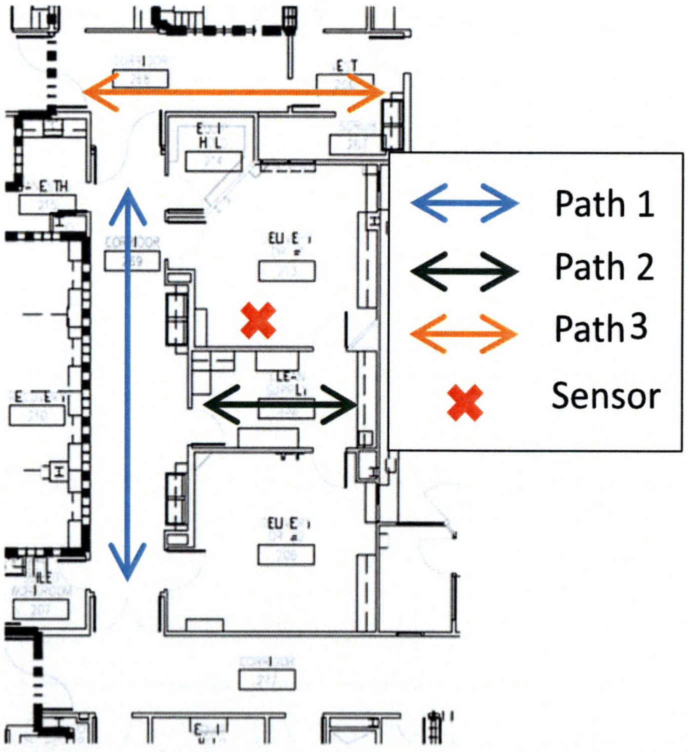

Fig. 19.3 Walking paths for steel structure #2

as the sensor location. The sensor is located at the proposed location of the MRI. The damping ratio was assumed to be 3%, which is consistent with the SCI for fully fitted-out composite steel floors in normal use. Figure 19.2 shows the predicted and measured footfall vibrations for three walking speeds. Since the walking paths and sensor location are not within the same bay, comparisons with the AISC method are not appropriate.

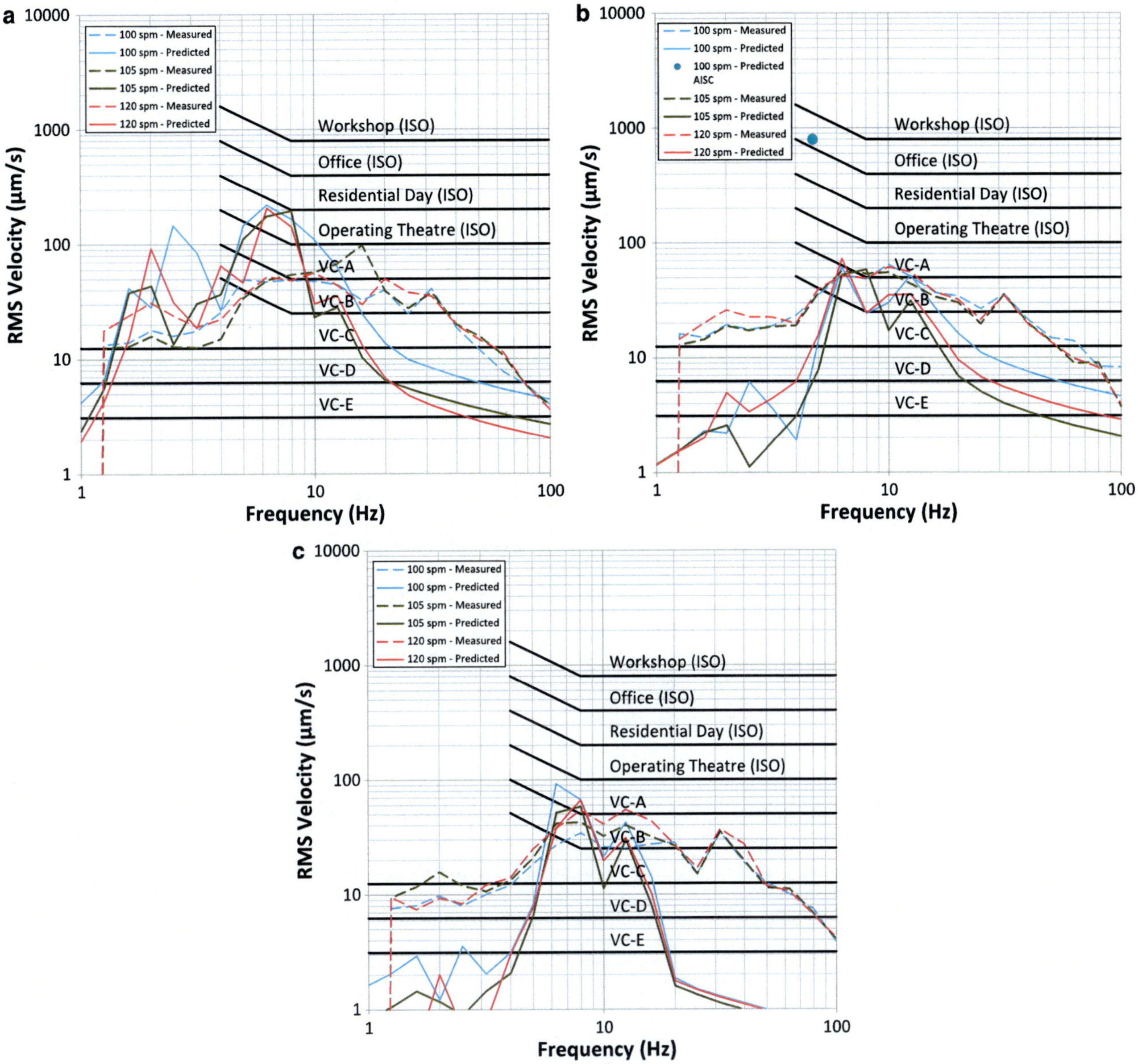

Fig. 19.4 Predicted and measured floor response of steel structure #2 (**a**) walking path 1, (**b**) walking path 2 (**c**) walking path 3

19.4.2 Steel Structure #2

Steel structure #2 is a hospital in the South-East United States that was undergoing major renovations and had been experiencing floor vibration issues. Figure 19.3 shows the architectural layout of the floor, as well as the walking paths and sensor location. The damping ratio was assumed to be 3% for this fitted-out floor. Figure 19.4 shows the predicted and measured vibration levels. Since path 2 was located in the same bay as the sensor, it was possible to conduct a comparison with the AISC methodology, which is done in Fig. 19.4b. Note that the AISC predicted value was plotted at the response frequency predicted by FloorVibe.

19.4.3 Steel Structure #3

Steel structure #3 is long-span public concourse in Eastern Canada. The damping ratio was assumed to be 1.1%, which is consistent with the SCI for buildings with bare floors and few furnishings. The measurements were collected at night to

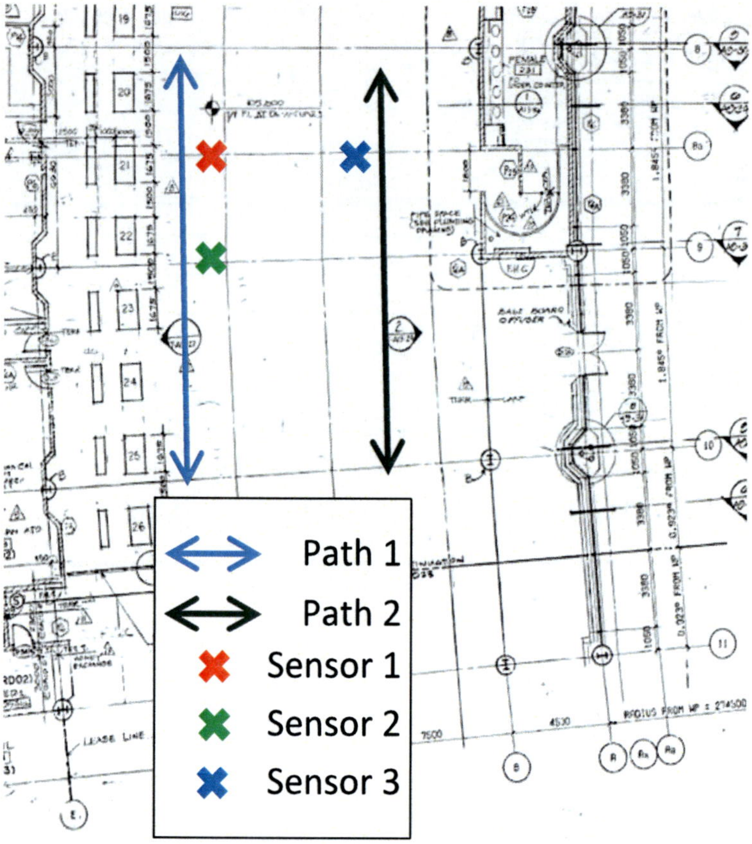

Fig. 19.5 Walking paths for steel structure #3

ensure a controlled testing environment. Two walking paths and three sensor locations are considered as shown in Fig. 19.5. The predicted and measured results are shown in Fig. 19.6. Since paths 1 and 3 are located in the same bay as the sensor, Fig. 19.6a, c show the AISC predicted response in addition to the SCI prediction and the measured results.

19.4.4 Concrete Structure

The concrete structure studied is a six-storey university building in the South-West United States that houses laboratory and classroom space. Measurements taken on the first and third floors are presented herein, and are indicated as "site 1" and "site 2", respectively. The damping is assumed to be 2% of critical, which is consistent with the CCIP for concrete floors that are partially fit-out. Two walking paths were considered for each measurement site, as shown in Fig. 19.7. Figures 19.8 and 19.9 show the predicted and measured floor response at a walking pace of 120 spm. FloorVibe is not valid for concrete structures, so AISC comparisons are not appropriate for this structure.

19.5 Discussion

The three steel structures studied showed that the SCI methodology can generally predict the vibration criteria accurately to within one vibration class of the measured results. The CCIP methodology is also shown to compare well with the measured results. For one-third octave frequencies well above or below the dominant frequency content, some experimental and predicted response plots show discrepancy. At high frequencies, it becomes increasingly difficult to accurately predict mode shapes, and their response to footfall excitation, which may account for this discrepancy. At low frequencies, the floor is responding at frequencies below its predicted fundamental frequency. The narrow frequency bandwidths associated with low-frequency one-third octaves combined with this off-resonant floor response, may partially account for the observed discrepancies. The governing floor response generally occurs when the floor is excited at its natural frequencies, which the

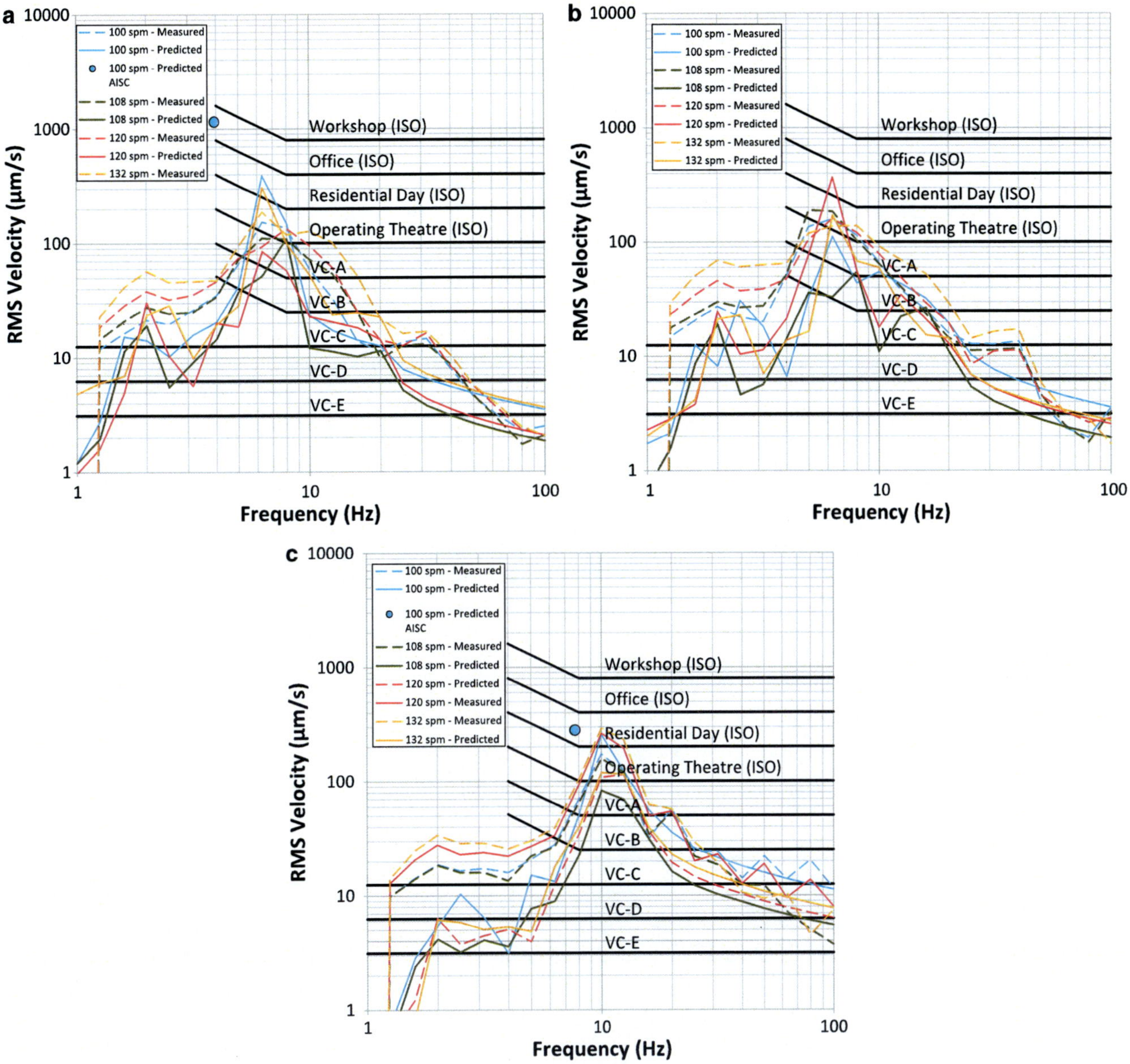

Fig. 19.6 Predicted and measured floor response of steel structure #3 (**a**) walking path 1 and sensor 1, (**b**) walking path 1 and sensor 2 (**c**) walking path 2 and sensor 3

SCI and CCIP predict with sufficient accuracy. The AISC method is shown to significantly overestimate the floor response for two of the three comparisons made. This overestimation may be partially attributed to the afore-mentioned overly-conservative assumptions made by the methodology.

For the SCI and CCIP methodologies, knowledge of the FE predicted mode shapes enables the response at any floor location to be predicted based on a person walking at any other location. Because of this modeling capability, realistic walking paths can be selected based on the architectural layout of the floor. This ability is advantageous, since it often does not consider the unrealistic worst-case response of a walker and vibration-sensitive receptor both being located at the center of the same bay. For example, since nobody aside from the patient is allowed within an MRI room during its operation, considering a walker within the MRI room is an unrealistic and overly-conservative design case. However, when using the AISC, the designer is forced to make this unrealistic assumption and will therefore be likely to over-design the structural system for this bay.

Since it is becoming more common for manufacturers of sensitive equipment to specify their own unique vibration criteria, a prediction methodology must be able to accommodate many forms of output. The SCI and CCIP methodologies generate

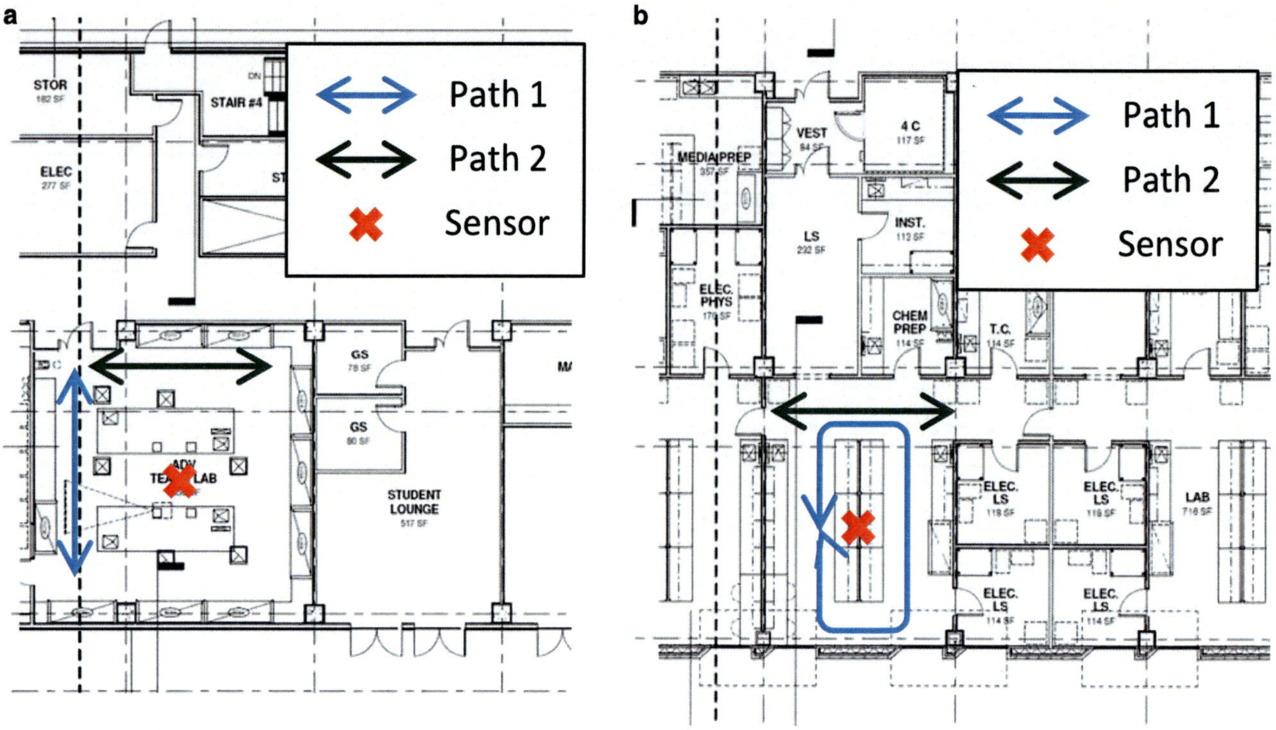

Fig. 19.7 Walking paths for concrete structure, (**a**) site 1, (**b**) site 2

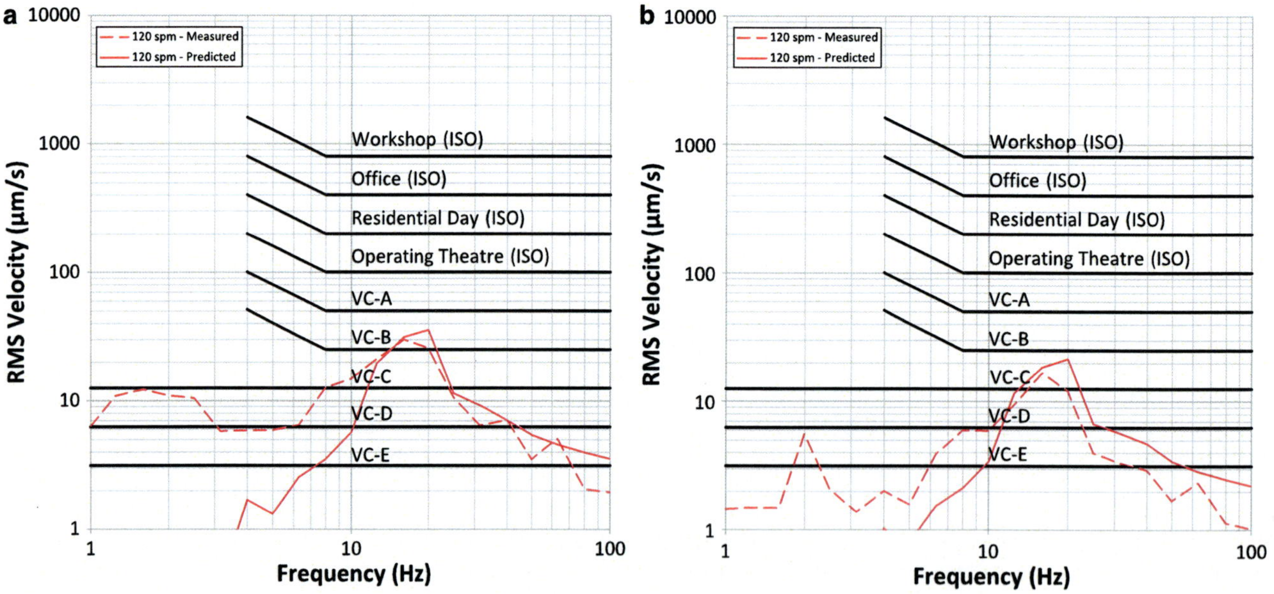

Fig. 19.8 Predicted and measured floor response of concrete structure, site 1 (**a**) walking path 1, (**b**) walking path 2

time series of the floor vibration, which can be post-processed to present the results in any desirable format, including one-third octaves, narrow-band spectra, or peak-to-peak values. Conversely, the AISC methodology determines only the harmonic response amplitude, assuming the floor responds uniquely at its fundamental frequency.

Damping is often challenging to estimate. The SCI and CCIP recommended damping values which depend on the floor construction type, the presence and location of partitions, and the amount of furnishing present. Using these qualitative descriptors, damping values between 1 and 5% of critical are typically recommended. When sensitive spaces are considered, the AISC methodology assumes a floor response which is independent of the damping level; an assumption that the methodology does not justify.

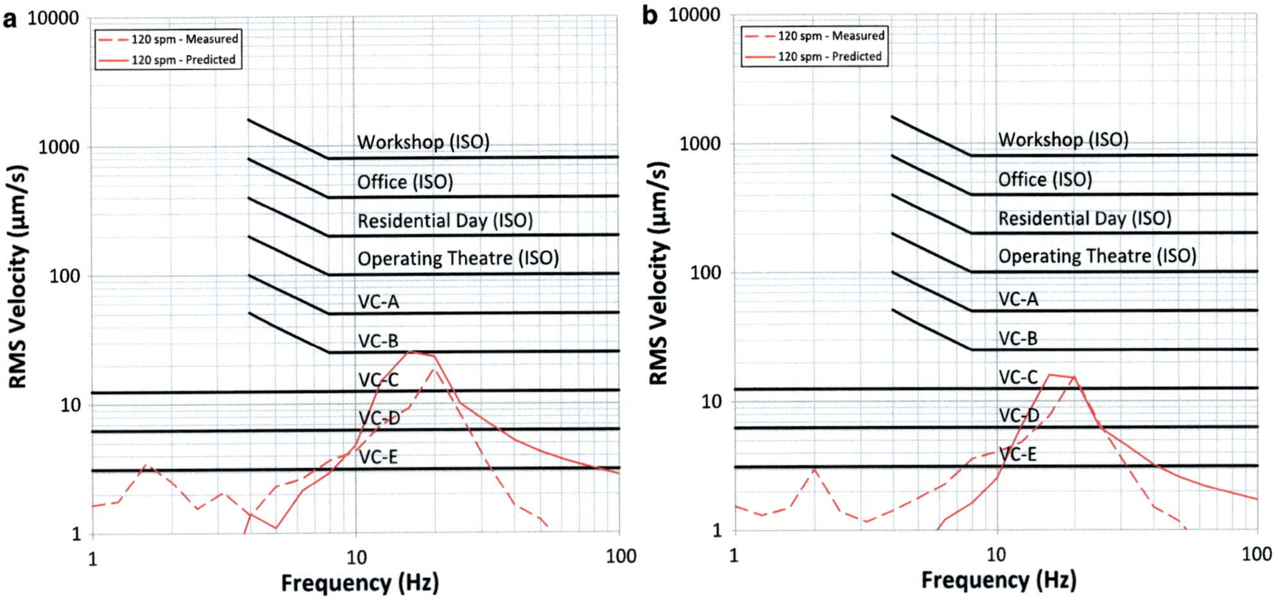

Fig. 19.9 Predicted and measured floor response of concrete structure, site 2 (**a**) walking path 1, (**b**) walking path 2

The SCI and CCIP methodologies are shown to be more robust, and provide results consistent with measurements collected from existing buildings. The measurements collected show that the AISC method can be overly-conservative, and in many cases its limitations restrict its applicability.

19.6 Conclusions

Footfall vibrations on elevated floors can cause significant occupant discomfort and can result in poor performance of sensitive equipment. Therefore, it is essential that suitable methodologies for predicting footfall vibrations are developed and verified. The performance of three popular methodologies is compared to data collected from existing steel and concrete structures. The methodologies considered are: the American Institute of Steel Construction (AISC) Design Guide 11, The Steel Construction Institute (SCI) P354, and The Concrete Centre CCIP-016.

The SCI and CCIP methods are generally able to predict the vibration levels to within one VC class. These methods are robust since they can accommodate irregular floor layouts by employing the Finite Element method to predict the dynamic floor properties. The SCI and CCIP also enable the walker path and sensor location to be specified, which often eliminates the unrealistic and overly-conservative design scenarios. These multimodal methods utilize linear superposition to generate time series of the floor response, which can be post-processed to produce results that are easily compared with generic VC criteria or any manufacturer-specified vibration criteria.

The AISC method uses simple beam theory to predict the fundamental bay frequency, which is assumed to dominate the floor response. When the floor layout is irregular, it becomes cumbersome to estimate the fundamental frequency using the AISC method. The AISC also assumes that the walker and sensor are both located in the center of the bay, regardless of the relative locations of corridors and vibration sensitive spaces. This assumption is generally unrealistic and overly-conservative. The output from the AISC method is expressed as a single RMS velocity value, which is assumed to vibrate at the fundamental frequency of that bay. With this form of output, comparisons with manufacturer-specified vibration criteria may not be possible.

The results of this study indicate that the SCI and CCIP methodologies are robust and provide reasonable estimates of the floor vibration response. Conversely, the AISC method contains several unrealistic assumptions that are only valid for floors with regular layouts.

References

1. Wilford MR, Young P, Field C (2007) Predicting footfall-induced vibration. Struct Build 160(SB2):65–79
2. Murray TM, Allen DE, Ungar EE (1997) Floor vibration due to human activity. AISC design guide, series no 11. American Institute of Steel Construction, Chicago
3. Smith AL, Hicks SJ, Devine PJ (2009) Design of floors for vibration: a new approach – revised edition (SCI P354). Steel Construction Institute, Ascot
4. Willford MR, Young P (2006) A design guide for footfall induced vibration of structures (CCIP-016). The Concrete Centre, Blackwater, Camberley
5. BS 6472:1992 (1992) Guide to evaluation of human exposure to vibration in buildings (1 Hz to 80 Hz). British Standards Institution, London
6. Amick H, Gendreau M, Busch T, Gordon C (2005) Evolving criteria for research facilities: I – vibration. In: Proceeding of SPIE conference 5933: buildings for nanoscale research and beyond, San Diego, 31 July to 1 Aug 2005
7. ISO 2631-2 (1989) Evaluation of human exposure to whole-body vibration – part 2: human exposure to continuous and shock-induced vibrations in buildings (1 to 80 Hz). International Standard, ISO 2631-2, Geneva
8. Murray T (2006) FloorVibe (Version 2.02) [Software], Structural Engineers, Inc., Radford, Virginia, USA

Chapter 20
Finite Element Model Updating of an Assembled Aero-Engine Casing

Chaoping Zang, Shuangchao Ma, and M.I. Friswell

Abstract In this paper, the finite element model updating technique based on first-order optimization is investigated and applied to a jointed aero engine casing. Vibration modal testing is conducted and modal data, i.e. natural frequencies and mode shapes, are obtained to update the FE models of two aero engine casings and their jointed structure. A two-step strategy is proposed in the updating process. In the first step, model updating is carried out on the two single casings separately in order to validate the casing component models. In the second step, the assembled casing is updated with emphasis on updating joint parameters. The joints are modeled using a layer of continuous solid elements that have material properties to be adjusted. The final updated FE model for the aero-engine casing is able to predict natural frequencies and mode shapes close to the measured ones.

Keywords Model updating • Finite element modelling • Modal test • Sensitivity analysis • Jointed casing

20.1 Introduction

With the rapid development of digital computers, the finite element (FE) method has become one of the most important and practical numerical analysis tools in product design and optimization in the mechanical and aerospace engineering industries. Generally speaking, the accuracy of the model predictions based on the FE method largely depends on the quality of the FE model. Once an FE model is generated, it should be subjected to a quality check to assess the accuracy of the predicted results [1]. However, the accurate modeling of a real-world structure is still a challenging task in many cases, especially for complex structures due to difficulties in the modeling of joints and boundary conditions. Thus, the results obtained from an FE model often differ from the test results of an actual structure. This has lead to the development of model updating techniques, which aim to improve the initial prediction capability of the FE model by means of adjusting some model parameters which are considered to be uncertain.

The model updating method has been investigated thoroughly over the past 30 years [2–10] and many techniques have been proposed. In general, model updating methods can be broadly classified into two groups: direct methods and iterative (or parametric) methods. The former group [4–6], which are emerged first, are based on updating the stiffness and mass matrices of elements directly without any iteration, and hence they are very fast in the updating process. However, due to the incompleteness of the measurement data, the updated matrices do not always maintain the structural connectivity, so that the updated model often has nonphysical meaning. In contrast, the iterative methods concentration the updating of design parameters which leads to an indirect update of the stiffness and mass matrices. Therefore, the structural connectivity can be easily maintained and parameter changes in the updated model can be physically interpreted. The iterative methods [8–10] can be implemented easily using existing codes and are now widely used in model updating applications. For example, Ahmadian et al. [8] investigated the model updating of joints in the AWE-MACE structure and obtained linear equivalent

C. Zang (✉) • S. Ma
College of Energy and Power Engineering, Nanjing University of Aeronautics and Astronautics, Nanjing, 210016, China
e-mail: c.zang@nuaa.edu.cn; mashuangchao198854@yahoo.cn

M.I. Friswell
College of Engineering, Swansea University, Swansea, SA2 8PP, UK
e-mail: m.i.friswell@swansea.ac.uk

T. Simmermacher et al. (eds.), *Topics in Model Validation and Uncertainty Quantification, Volume 5*,
Conference Proceedings of the Society for Experimental Mechanics Series 41,
DOI 10.1007/978-1-4614-6564-5_20,

models that are valid within the usual range of vibration tests. Mottershead et al. [9] applied the iterative method to update finite element models of welded joints and boundary conditions of a cantilever plate. The use of nodal offset dimensions results in an updated model of the welded joint with physical interpretation.

Joint modeling or descriptions of joint dynamic properties has a significant role in the modeling of complex structures. Up to now, many methods have been developed to construct predictive models of structures with joints and interfaces [11]. Generally, two groups of joint models are commonly employed in the literature: zero-thickness and thin layer elements [12–14]. Joints modeled with zero-thickness elements, which sometimes refers to a lumped model, have properties that are concentrated at a single point with no dimensions. The prominent examples of zero-thickness elements are lumped springs, and frictional sliders which are commonly used to represent the force–displacement relationship in the joint interface [12]. These models are valid as long as the size of the joint patch is small in comparison to the smallest length of the connection regions. The thin layer element in the second group of joint modeling is considered as a continuous element with very small but finite thickness. They can use hexahedral or tetrahedral elements that are available in commercial finite element software and the elements are assigned special constitutive relations. Recent research [8, 13–16] shows that the thin layer concept with a linear elastic constitutive relation is very well suited to model the linear behavior of joints.

In this paper, two steps are applied to relieve the difficulties in the modeling of complex structures. Firstly, the finite element model updating technique based on first-order optimization is investigated to correct the FE model of each component using the test data. Secondly, the joints are modeled with thin layer elements and the parameters of the joints are updated with the test data from the assembled structure. The approach is validated using an assembly of two aero engine casings as an example.

20.2 Methodology of the Two Step Updating Procedure

Generally speaking, the dynamic predictions from an FE model of a complex structure may differ from the practical measurement of a real structure due to modeling errors, such as uncertainty in the material and geometrical parameters (e.g. modulus, thickness, density and residual strain) and also in the joints and component assembly. A large number of design parameters in the FE model may be required to be updated to correct these errors using the test data. The numbers of transducers on the structure mean that the measured degrees of freedom are limited and therefore the measurements are incomplete compared to the FE model. This creates difficulties to keep the physical meaning in the updating process of such complex structures with many uncertain design parameters. Therefore, a two-step updating procedure is suggested as a means to reduce the number of updating parameters at each step and keep the updating as meaningful as possible in practice. Figure 20.1 shows the framework of the two-step updating procedure to update a joint assembly of two components. The first step is to update each component separately with free-free boundary conditions in order to correct the component modeling errors. The second step is to update the joints of the assembly of two updated components to have the right connection parameters between the two components.

20.2.1 Planning of the Modal Test

In order to obtain effective test information for successful FE model updating, modal tests were conducted on both individual casings and on their assembly under the free-free boundary conditions. Test planning was performed to select the suspension placement of the structures and the transducer locations for informative data collection in order to guarantee efficient modal tests.

Theoretically, under free-free boundary conditions, the test structure is not connected to any other object. However, this is difficult to achieve in practice. In order to reduce the influence of the supports the attachment locations should be situated on, or very close to, the nodal lines of all of the modes excited during testing. The optimum suspension locations for modal testing are generally found to be in the regions of a structure where the average displacement in the modes of interest is the lowest, which can be defined as [17] minimizing

$$\text{Optimumsuspension}: \qquad \min\left(\sum_{r=1}^{m}\frac{\varphi_{jr}^2}{\omega_r^2}\right) \tag{20.1}$$

where m is the number of modes of interest, and ω_r and ϕ_r are the r-th natural frequency and mode shape, respectively.

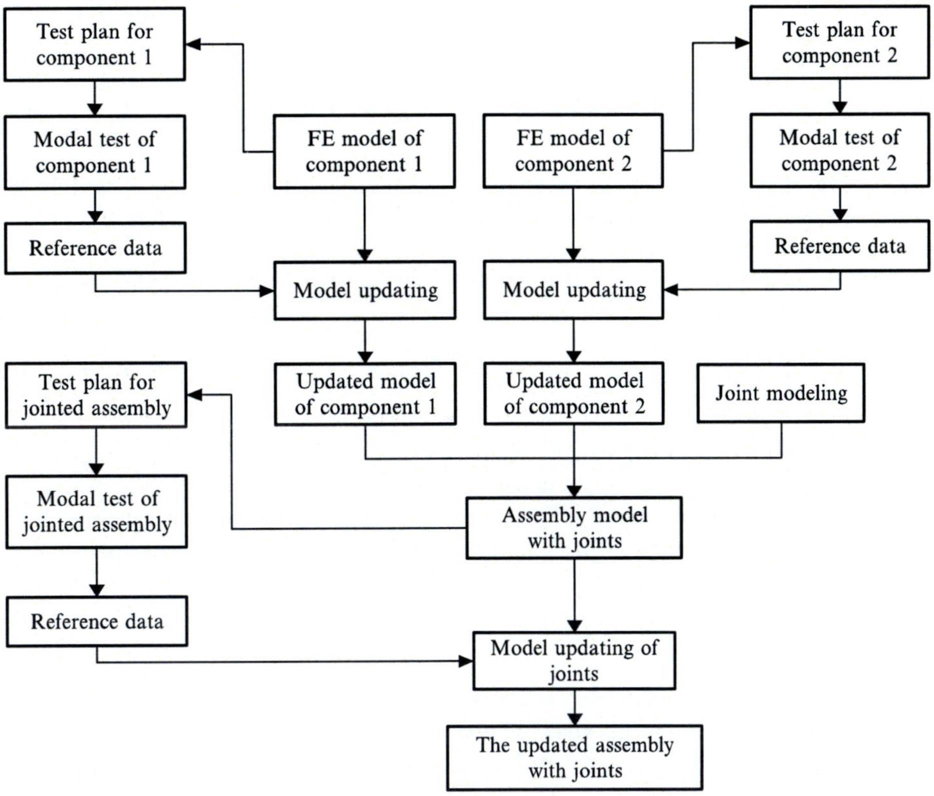

Fig. 20.1 Framework of the two step updating procedure

The Effective Independence (EI) transducer location optimization method, proposed by Kammer in 1991, is perhaps the most popular tool to select the measurement DOFs. Using the EI method, the aim is to find the transducer locations to ensure that the measured modes as linearly independence as possible. The approach does not consider any other requirement for selection of measurement DOFs during the optimization procedure. Therefore, many DOFs selected using the EI method will have a low response which will lead to a low signal-to-noise ratio. In order to overcome this problem, Imamovic [17] proposed a modified method called the Average Driving DOF-Effective Independence method which is applied to select the measurement DOFs having the higher response and meets the EI method requirements. The improved method minimizes a modified Fisher information matrix given by

$$\left[E^{addofa}\right]_{n\times n} = [E]_{n\times n} diag\left(\sum_{r=1}^{m} \phi_{jr}^{2}\right)) \tag{20.2}$$

where m is the number of modes of interest and ϕ_r is the r-th mode shape.

The correlation between the measured and predicted mode shapes is quantified by the SCO (System Equivalent Reduction Expansion Process-Cross-Orthogonality) parameter, which will generally have a higher value for two similar mode shapes than the MAC correlation coefficient and, conversely, a lower value for two dissimilar mode shapes. The SCO is defined by:

$$S_{co}(i,j) = \frac{\left|\{\psi_{X,i}\}^{T}\left([\phi_A]^{+T}[\phi_A]^{+}\right)\{\psi_{A,j}\}\right|^{2}}{\{\psi_{X,i}\}^{T}\left([\phi_A]^{+T}[\phi_A]^{+}\right)\{\psi_{A,i}\}\cdot\{\psi_{X,j}\}^{T}\left([\phi_A]^{+T}[\phi_A]^{+}\right)\{\psi_{A,j}\}} \tag{20.3}$$

where $\psi_{X,i}$ the i-th is test eigenvector and $\psi_{A,j}$ is the j-th analytical eigenvector. ϕ_A is the mass-normalized test or analytical mode shape matrix.

20.2.2 Model Updating Process

The main purpose of FE model updating is to adjust parameters in the FE model to minimize the differences between the analytical and experimental models in order that the predictions of the dynamic characteristics from the FE model match the measured data in the frequency range of interest. The model updating problem is essentially an optimization, where the prediction error is given as

$$g = \min_{x} \|W_i R(x)\|_2^2, \quad \text{where} \quad R(x) = \{f_e\} - \{f_a(x)\}, \quad x_l \leq x \leq x_u, \tag{20.4}$$

g is the objective function, W_i is the weighting matrix, and R is the residual vector. The vector x represents the design parameters that have specified upper and lower bounds. f_e and f_a denote vectors of the test and predicted dynamic properties, respectively.

Using the first-order method [18], an unconstrained objection function outlined can be formulated from Eq. (20.4) as

$$Q(x,q) = g/g_0 + \sum_{i=1}^{n} P_x(x_i) + q\left[\sum_{i=1}^{m_1} P_f(f_i) + \sum_{i=1}^{m_2} P_h(h_i) + \sum_{i=1}^{m_3} P_w(w_i)\right] \tag{20.5}$$

Where Q denotes a dimensionless and unconstrained objective function, g_0 represents the reference prediction error selected from the current group of design parameter sets. P_x are penalties applied to the design variables, i.e. the Young's modulus and density of the structure and P_f, P_h, P_w are penalties based on the state variables, for instance, natural frequencies and mode shapes. The constraint satisfaction is controlled by a response surface parameter, q.

The penalty functions are defined as

$$P_g(g_i) = \left(\frac{g_i}{g_i + \alpha_i}\right)^{2\lambda} \tag{20.6}$$

Where λ is an integer and α_i denotes a tolerance number.

To solve Eq. (20.5), an iterative method is used. At each optimization iteration, j, the design parameters are given by the vector $x^{(j)}$. A search direction vector, $d^{(j)}$, is calculated, and the estimates of the design parameters are obtained as

$$x^{(j+1)} = x^{(j)} + s_j d^{(j)} \tag{20.7}$$

where the line search parameter, s_j, corresponds to the minimum value of Q in the direction of $d^{(j)}$. This line search can be solved using a combination of a golden-section algorithm and a local quadratic fitting technique. The search parameter s_j is usually limited to the range given by

$$0 \leq s_j \leq \frac{s_{\max}}{100} s_j^* \tag{20.8}$$

where s_j^* is the largest possible step size for the line search at the current iteration, j, and $s_{\max}$ is the maximum line search step size.

20.2.3 Joint Modeling

The components are updated and validated with the test data in the first step. The second step is to model the joints and update the joints parameters for the assembly of the valid components using the measurement data tested from the assembled structure. In this paper, thin layer elements are employed to model the joints. For simplicity, the thin layer elements are often formulated using a linear elastic constitutive relation, which generally has the form

$$\begin{Bmatrix} \sigma_{xx} \\ \sigma_{yy} \\ \sigma_{zz} \\ \sigma_{xy} \\ \sigma_{yz} \\ \sigma_{xz} \end{Bmatrix} = \begin{bmatrix} c_{11} & c_{12} & c_{13} & c_{14} & c_{15} & c_{16} \\ & c_{22} & c_{23} & c_{24} & c_{25} & c_{26} \\ & & c_{33} & c_{34} & c_{35} & c_{36} \\ & & & c_{44} & c_{45} & c_{46} \\ & & & & c_{55} & c_{56} \\ sym & & & & & c_{66} \end{bmatrix} \begin{Bmatrix} \varepsilon_{xx} \\ \varepsilon_{yy} \\ \varepsilon_{zz} \\ \varepsilon_{xy} \\ \varepsilon_{yz} \\ \varepsilon_{xz} \end{Bmatrix} \tag{20.9}$$

Where σ and ε represent the stress and strain respectively, and c_{ij}, $i, j = 1,\ldots,6$, represents the stiffness at each degree of freedom. The behavior of the joint is governed mainly by the normal and shear stiffness, whilst the coupling terms have a secondary effect [16]. Hence the coupling between the normal and shear stiffness will be neglected. Thus, the constitutive relation further simplifies to

$$\begin{Bmatrix} \sigma_{xx} \\ \sigma_{yy} \\ \sigma_{zz} \\ \sigma_{xy} \\ \sigma_{yz} \\ \sigma_{xz} \end{Bmatrix} = \begin{bmatrix} c_{11} & & & & & \\ & c_{22} & & & & \\ & & c_{33} & & & \\ & & & c_{44} & & \\ & & & & c_{55} & \\ & & & & & c_{66} \end{bmatrix} \begin{Bmatrix} \varepsilon_{xx} \\ \varepsilon_{yy} \\ \varepsilon_{zz} \\ \varepsilon_{xy} \\ \varepsilon_{yz} \\ \varepsilon_{xz} \end{Bmatrix} \quad (20.10)$$

The off diagonal elements are 0 and there are only 6 parameters which represent the normal and shear stiffnesses to be identified.

20.3 Case Study: Model Updating of a Jointed Aero-Engine Casing

In order to demonstrate the effectiveness of the two-step strategy for updating complex structures, a joint assembly of two aero engine casings is used. First, models of two single casings are updated individually with modal test data to reduce the possible modeling errors of each casing. Then, the assembled model of both casings with bolts is updated with emphasis on updating the joint parameters (Fig. 20.2).

20.3.1 Aero-Engine Casings and Modal Test

The investigated physical casings are shown in Fig. 20.4. Both individual casings are made of two separate parts (the upper and the lower bodies) that are connected together with spot welds. Also, there are 36 identical bolt holes distributed at the flange of the casings, which are used to assemble the casings, and 4 mounting bases placed at the upper part to fix attachments. The FE model for each individual casing is developed using the ANSYS software tool. Modeling the spot weld, shown in Fig. 20.3, is the most difficult task since there are many local effects such as geometrical irregularities, residual stresses and defects due to the welding process, which are not taken into account in the FE modeling. Furthermore, the casing contains many spot welds and modeling each of them in detail would require a major computational effort. Thus, the welded regions are simplified to rigid connections. In order to save computational effort, the holes and the mounting bases are also removed from the FE model. The FE model of the rear casing is show in Fig. 20.4, and is formed using ten-noded solid elements. The model has 32,827 elements and 66,247 nodes. The Young's modulus of the casing is 180 GPa and the mass density is $7{,}920\,\mathrm{kg/m^3}$.

Free-free boundary conditions were employed for the measurements by supporting the casings with elastic bands. A roving hammer test was chosen for the force excitation. Figure 20.5 shows that the optimum suspension locations of the rear casing are within the blue regions, where is not easy to suspend the structure. Hence the casing was suspended by the small flange. Figure 20.6 shows the optimum transducer locations for the rear casing. Clearly the best transducer locations are in the middle region of the casing. In order to have sufficient spatial resolution to correlate with the finite element model, a test model with 128 DOFs was established for the single casing, shown in Fig. 20.7. Two reference accelerometers were placed at orthogonal directions. The frequency range of interest is 0–1,000 Hz and the number of spectral lines was set to 3,200, which provided sufficient resolution.

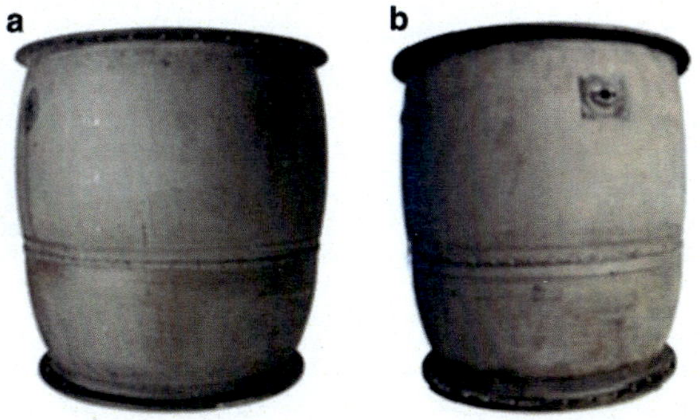

Fig. 20.2 The aero-engine casings. (**a**) Front casing and (**b**) rear casing

Fig. 20.3 The spot weld

Fig. 20.4 The FE model of the rear casing

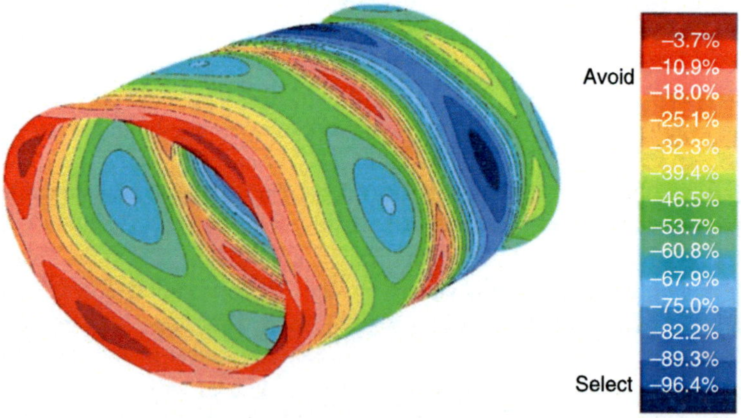

Fig. 20.5 The optimum suspension locations for the rear casing

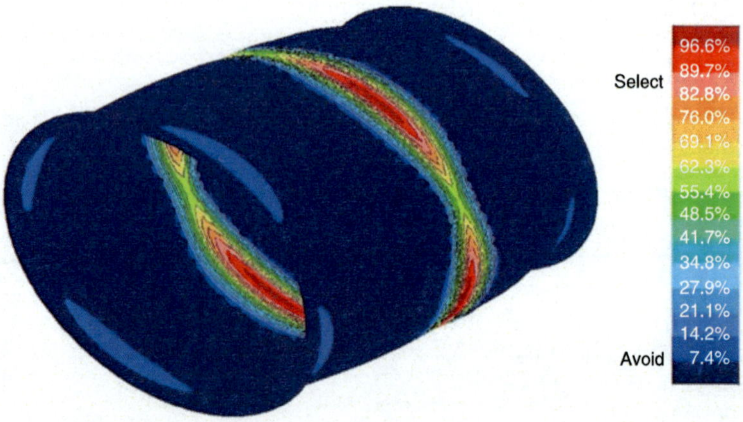

Fig. 20.6 The optimum transducer locations for the rear casing

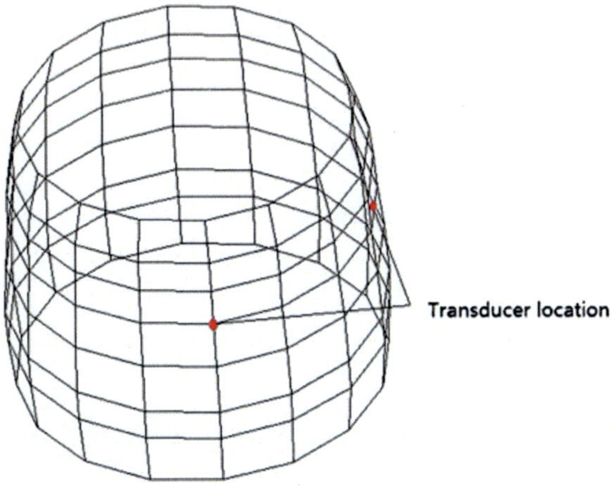

Fig. 20.7 The test model showing the impact and accelerometer locations

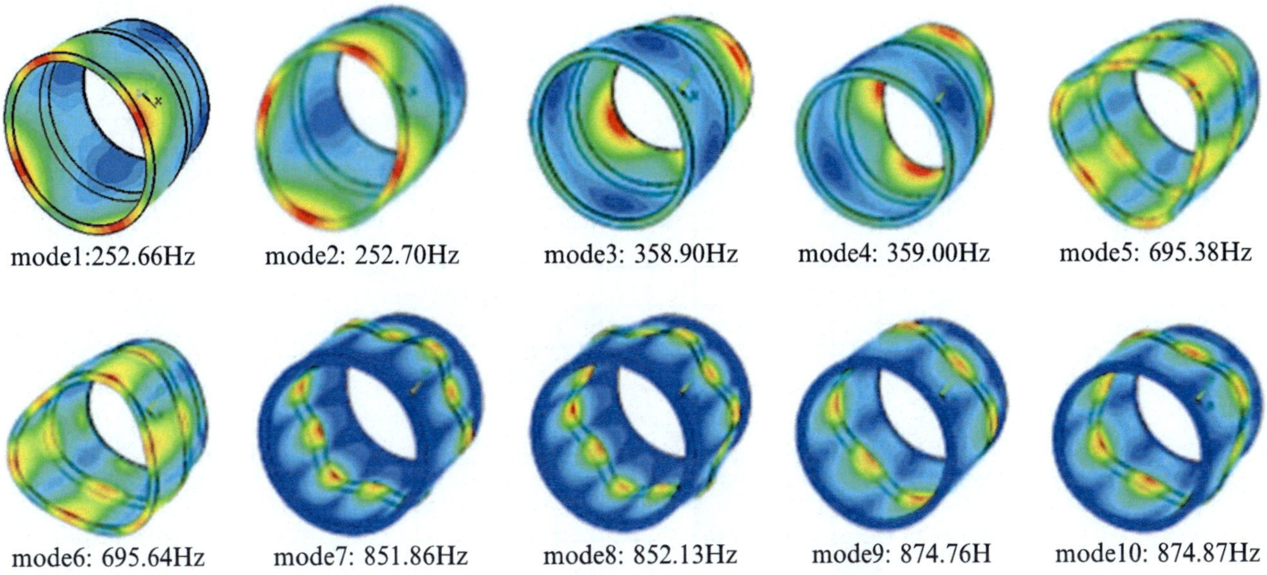

Fig. 20.8 The predicted natural frequencies and mode shapes for the rear casing

20.3.2 Updating of the Casing Components Using Test Data

The predicted natural frequencies and mode shapes for the front and the rear casings are calculated from the FE models. The modes of the rear casing are shown in Fig. 20.8. It can be seen that the natural frequencies appear in close pairs since the casing is almost axisymmetric.

The initial mode pairing for the individual casings are given in Tables 20.1 and 20.2 and the corresponding SCO matrix is visualized in Figs. 20.9 and 20.10. It can be seen that the modal properties from the initial FE models show significant deviations from those of the test models. Many of the pairs of modes have been swapped and there is some coupling between modes 9 and 10 of the front casing. Also, modes 9 and 10 of the real casing are in the wrong sequence. To improve these unsatisfactory correlations, the suggested model updating procedure is applied.

The choice of the updating parameters is critical to improve the modeling of the casing. Physical properties of the elements such as Young's modulus and mass density could be chosen as updating parameters. As a general rule, only parameters that the response is sensitive to should be selected, otherwise the updating process will be ill-conditioned since insufficient information is available to estimate the parameters accurately. Among these parameters, only those having physical meaning should be updated, otherwise the updated model may reproduce the data used for updating but might not accurately predict the response of the casing under different conditions [11]. In order to determine the most sensitive parameters, an eigenvalue

Table 20.1 The initial correlation between the predicted and measured modal data of the front casing

Test mode no.	FE mode no.	Test freq./Hz	Initial freq./Hz	Error %	Initial SCO/%
1	2	229.44	241.43	5.23	89.0
2	1	230.11	241.65	5.01	97.0
3	3	360.72	357.25	−0.96	81.6
4	4	361.36	357.34	−1.11	79.2
5	5	609.42	631.63	3.64	87.9
6	6	611.31	632.95	3.54	83.2
7	8	659.02	662.18	0.48	58.6
8	7	661.83	660.06	−0.27	90.6
9	10	675.51	691.72	2.40	58.8
10	9	677.57	690.47	1.90	57.0

Table 20.2 The initial correlation between the predicted and measured modal data of the rear casing

Test mode no.	FE mode no.	Test freq./Hz	Initial freq./Hz	Error/%	Initial SCO/%
1	1	247.20	252.66	2.21	94.6
2	2	248.26	252.70	1.79	70.7
3	4	361.27	359.00	−0.63	94.7
4	3	362.85	358.90	−1.08	92.9
5	6	684.11	695.64	1.68	97.0
6	5	686.02	695.38	1.36	96.7
7	8	731.93	852.13	16.42	76.4
8	7	733.84	851.86	16.08	82.9
9	13	771.46	956.28	23.96	72.2
10	14	773.79	956.57	23.62	65.5

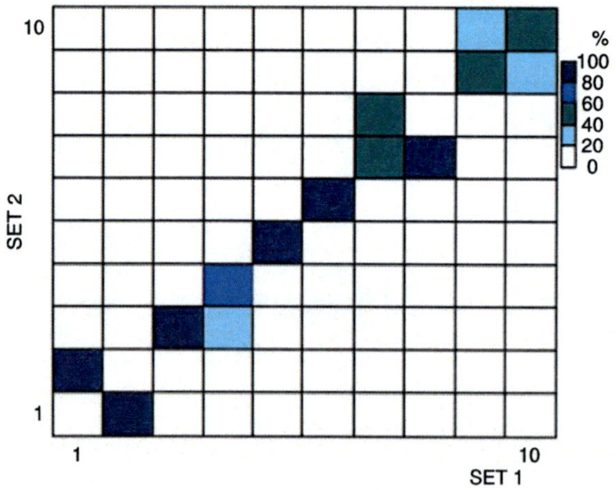

Fig. 20.9 The initial SCO for the front casing

sensitivity analysis to changes in the stiffness and mass of element were carried out. Figure 20.11 shows that the welded connections regions and the top and bottom flange were more sensitive than other model regions. In the FE model, the holes on the front and rear flange were removed and the welded connections was simplified as rigid connection. Hence, it is consistent that the main modeling errors in the FE model are located at these three regions. As the mass change of the FE model, compared with the actual structure, is very small, it was decided to select the Young's modulus of the flanges and the welded connection regions as the updating parameters, as shown in Fig. 20.12.

The objective function is formulated using the first ten natural frequencies, defined as

$$\min \sum_{i=1}^{n} W_i (1 - \omega_i^e / \omega_i^a)^2 \tag{20.11}$$

where ω_i^e and ω_i^a are experimental and analytical eigenvalues, respectively. The weights W_i are set to one.

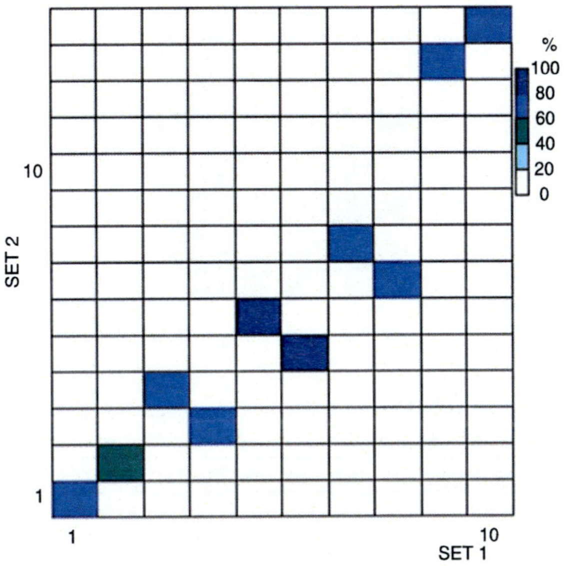

Fig. 20.10 The initial SCO for the real casing

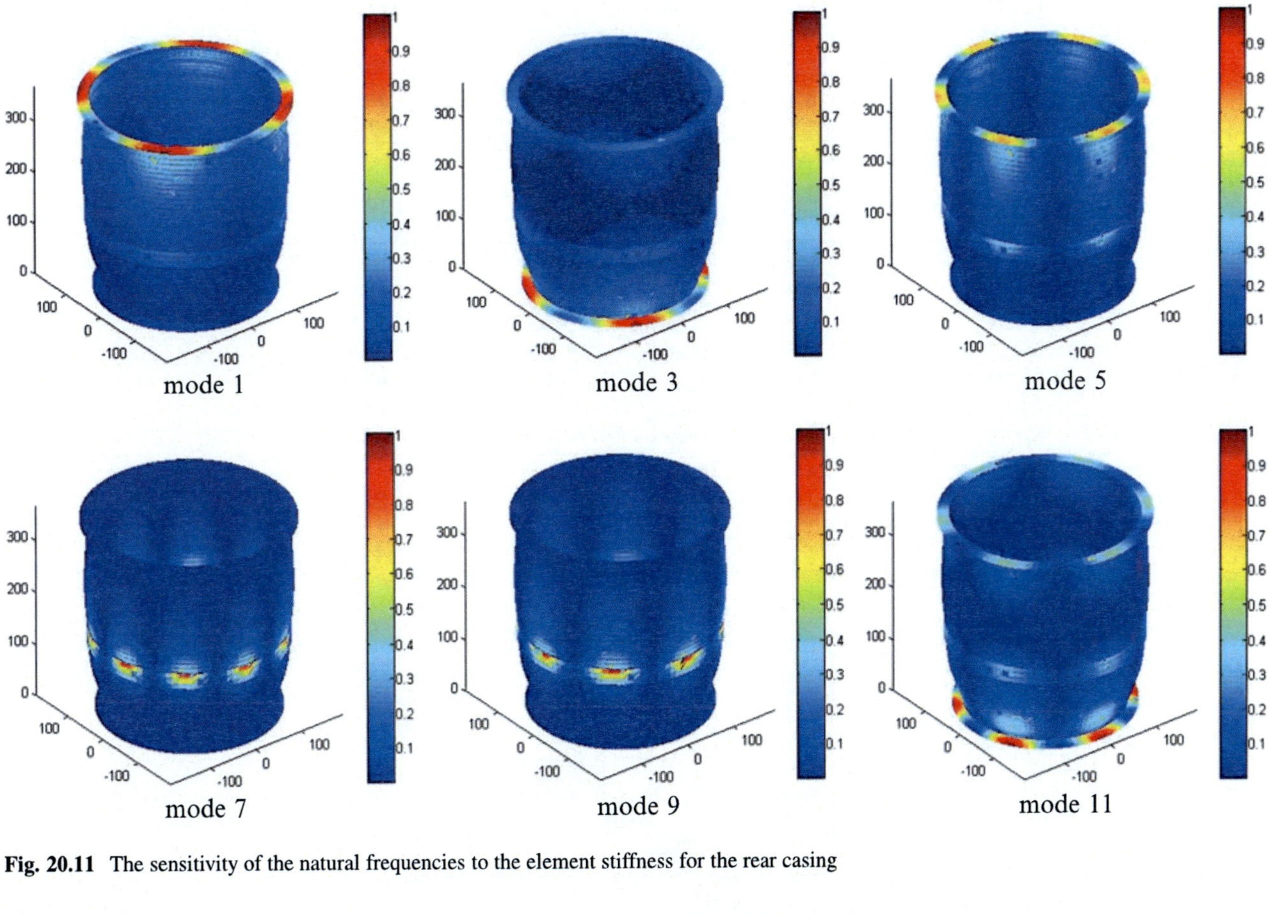

Fig. 20.11 The sensitivity of the natural frequencies to the element stiffness for the rear casing

The updated results for the two individual casing are presented in Tables 20.3 and 20.4. It can be seen that the dynamic behavior of both the front and rear casings have been improved significantly after updating. The highest absolute natural frequency error is reduced from 5.32 to 1.41 % for the front casing and from 23.96 to 1.26 % for the rear casing. The updated SCO is shown in Figs. 20.13 and 20.14 respectively. Although the SCO values have no significant improvement, all of the modes are corrected to the right sequence. For the updated models, all of the Young's moduli at the three regions become smaller, which reflects that the simplifications in the initial FE model resulted in increased stiffness. The updated Young's

Fig. 20.12 The updating regions for the rear casing

Table 20.3 The updated results for the front casing

Test freq./Hz	Updated freq./Hz	Initial error/%	Updated error/%	Updated SCO/%
229.44	232.29	5.23	1.24	89.5
230.11	231.45	5.01	0.58	96.5
360.72	359.91	−0.96	−0.22	80.7
361.36	360.19	−1.11	−0.32	78.4
609.42	618.08	3.64	1.42	87.8
611.31	618.71	3.54	1.21	82.6
659.02	653.08	0.48	−0.90	59.4
661.83	652.51	−0.27	−1.41	90.6
675.51	673.02	2.40	−0.37	64.0
677.57	673.77	1.90	−0.56	62.4

Table 20.4 The updated results for the rear casing

Test freq./Hz	Updated freq./Hz	Initial error/%	Updated error/%	Updated SCO/%
247.20	248.66	2.21	0.59	94.9
248.26	248.71	1.79	0.18	72.5
361.27	360.66	−0.63	−0.17	94.6
362.85	360.53	−1.08	−0.64	92.5
684.11	684.67	1.68	0.08	97.7
686.02	684.30	1.36	−0.25	96.9
731.93	738.88	16.42	0.95	79.2
733.84	738.65	16.08	0.66	80.6
771.46	761.73	23.96	−1.26	72.7
773.79	765.14	23.62	−1.12	66.2

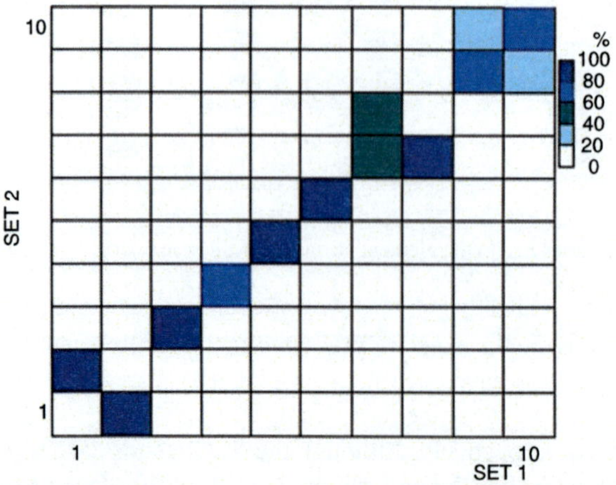

Fig. 20.13 The updated SCO for the front casing

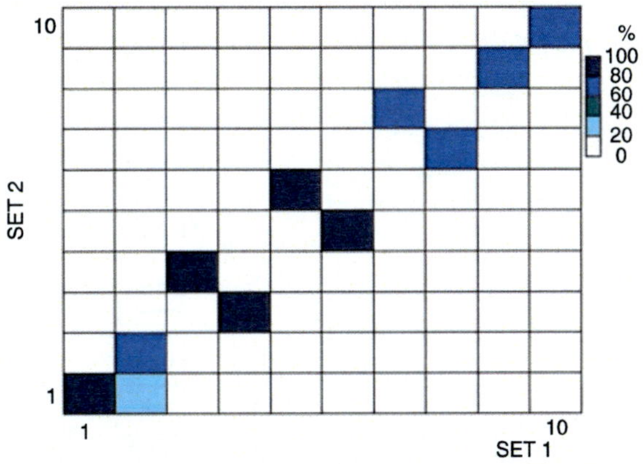

Fig. 20.14 The updated SCO for the rear casing

Fig. 20.15 The joined casing

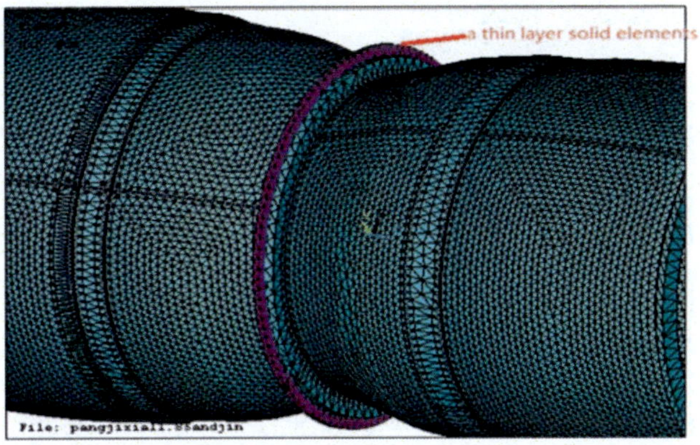

Fig. 20.16 The FE model of the joint

moduli of the top flange, bottom flange and the welded regions are 182, 169, and 105 GPa for the front casing respectively, and 180, 174, and 100 GPa for the rear casing. Note that the Young's moduli in the welded regions have a significantly decrease, which indicts that the spot weld significantly weakens the structure's stiffness.

20.3.3 Updating of the Joint Assembly Using Test Data

With the updated individual casing models, the jointed casing model updating can be performed with the emphasis on the joints. The assembly is fabricated from components by using 36 M6 bolted joints, shown in Fig. 20.15.The thin layer elements, which have material properties that may be adjusted, are used to model these bolted joints, shown in Fig. 20.16.

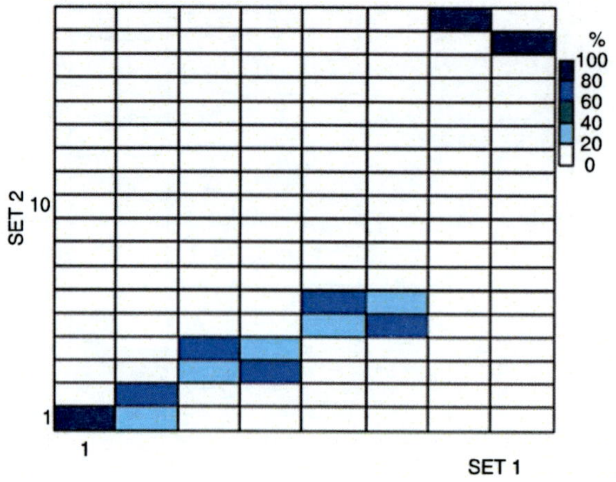

Fig. 20.17 The initial SCO for the assembled casing

Table 20.5 The initial correlation between the predicted and measured modal data of the joined casing

Test mode no.	FE mode no.	Test freq./Hz	Initial freq./Hz	Error/%	Initial SCO/%
1	1	238.21	251.44	5.55	83.2
2	2	238.53	251.92	5.61	75.3
3	4	272.68	317.23	16.34	69.8
4	3	274.45	316.13	15.19	68.0
5	6	385.72	468.70	21.51	75.2
6	5	386.81	468.36	21.08	60.3
7	18	598.55	847.64	41.62	87.8
8	17	599.75	843.89	40.71	88.0

Table 20.6 The updated results of the assembled casing/4 Nm

Test freq./Hz	Updated freq./Hz	Initial error/%	Updated error/%	Updated SCO/%
238.21	241.43	5.55	1.35	84.3
238.53	242.10	5.61	1.50	77.5
272.68	277.94	16.34	1.93	92.1
274.45	278.21	15.19	1.37	90.8
385.72	376.82	21.51	−2.31	80.5
386.81	376.42	21.08	−2.69	68.5
598.55	597.06	41.62	−0.25	88.8
599.75	594.50	40.71	−0.88	75.1

The assembled casing was tested 3 times with different torque levels of 3, 4, and 5 Nm for the tightness of the bolts. The initial correlation between the predicted and measured modal data of the joined casing with the torque level of 4 Nm is presented in Table 20.5 and the initial SCO is shown in Fig. 20.17. Clearly the initial frequency error is very significant; the maximum frequency error is 41.62 % and modes 7 and 8 are in the wrong sequence.

As the behavior of the joint is governed mainly by the normal and shear stiffnesses, it was decided to select the elastic and shear moduli of the joints as updating parameters. A permissible range of variation for the parameters was defined during the updating. Since the joints usually introduce a local softening effect and sharply reduce the normal and shear stiffnesses, the upper bound for the variation of the updating parameters was set to the initial value and the lower bound was set to 0.000001 of the initial value. The updated results are presented in Table 20.6. The maximal frequency error is reduced from 41.62 to −2.69 %. The SCO values in Fig. 20.18 show little improvement. After updating, the normal and shear stiffnesses have been reduced to 300 and 50 MPa, which shows a sharp decrease in the stiffnesses of the structure due to the joints.

Tables 20.7 and 20.8 give the updated results of the assembled casing with torques 3 and 5 Nm, respectively. The frequency deviations in both cases, between the updated model prediction and the measurement test, are within 3 %. The updated normal and shear stiffnesses are 240 and 52 MPa for the casing joined with 3 Nm and 400 and 57 MPa for the casing joined with 5 Nm. This shows that the torque of the joints has a significant impact on the normal stiffness.

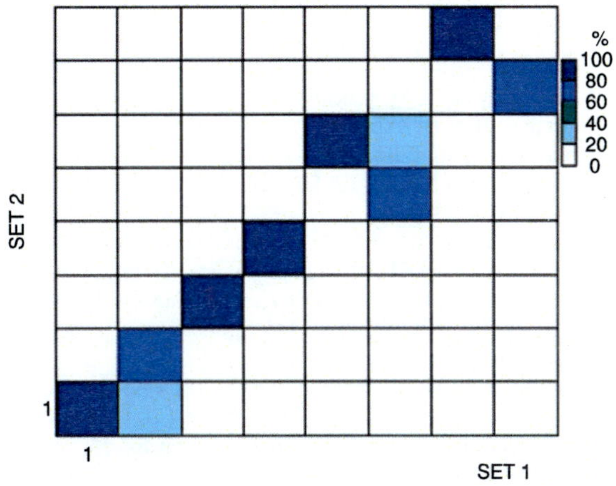

Fig. 20.18 The updated SCO for the assembled casing

Table 20.7 The updated SCO for the assembled casing/3 Nm

Test freq./Hz	Updated freq./Hz	Initial error/%	Updated error/%
238.45	241.16	5.45	1.14
238.74	241.85	5.52	1.30
272.16	276.18	16.56	1.48
272.76	276.44	15.90	1.35
385.50	376.55	21.58	−2.32
386.47	376.19	21.19	−2.66
576.38	574.51	47.06	−0.32
577.15	574.04	46.22	−0.54

Table 20.8 The updated SCO for the assembled casing/5 Nm

Test freq./Hz	Updated freq./Hz	Initial error/%	Updated error/%
238.53	241.76	5.41	1.35
238.92	242.41	5.44	1.46
275.18	280.45	15.28	1.92
277.26	280.76	14.02	1.26
386.86	376.73	21.15	−2.62
387.47	377.17	20.88	−2.66
628.13	620.14	34.95	−1.27
630.84	622.77	33.77	−1.28

20.4 Conclusions

Model updating has been performed on a joined aero-engine casing. The thin layer elements are suggested to model the bolted joints and this model was successfully updated for three different torque levels. The final updated FE model for the aero-engine casing is able to predict natural frequencies and mode shapes that are very close to the measured ones. Therefore, the jointed casing model may be reliably used for the analysis of the whole engine. The parameters of the joint model used in this paper are heavily dependent on the experimental measurements. Parameters of the model have to be identified using the test data. Thus, a big challenge is to analyze the dynamic characteristics of the structure accurately at the design stage. Truly predictive models of the joints require further investigation.

Acknowledgements The financial support of the National Natural Science Foundation of China (Project No. 51175244)C Research Fund for the Doctoral Program of Higher Education of China (Project No. 20093218110008) and the Priority Academic Program Development of Jiangsu Higher Education Institutions (PAPD) are gratefully acknowledged. SC Ma also acknowledges the support of Fundamental Research Fundsfor the Central Universities of China (Project No. kfjj20110204).

References

1. Boswald M, Link M, Schedlinski C (2005) Computation model updating and validation of aero-engine finite element models based on vibration test data. IFASD, Munich
2. Mottershead JE, Friswell MI (1993) Model updating in structure dynamics: a survey. J Sound Vib 167(2):347–375
3. Mares C, Mottershead JE, Friswell MI (2006) Stochastic model updating: part 1—theory and simulated example. Mech Syst Signal Process 20(7):1674–1695
4. Baruch M, Bar-Itzhack Y (1978) Optimal weighted orthogonalisation of the measured modes. AIAA J 16:346–351
5. Berman A, Nagy EJ (1983) Improvement of a large analytical model using test data. AIAA J 21:927–928
6. Baruch M (1984) Methods of reference basis for identification of linear dynamic structures. AIAA J 22:561–564
7. Zang C, Ewins DJ (2009) Model validation of structural dynamics in the aero-engine design process. Front Energy Power Eng China 3(4):480–488
8. Ahmadian H, Mottershead JE, James S, Friswell MI, Reece CA (2006) Modelling and updating of largesurface-to-surface joints in theAWE-MACE structure. Mech Syst Signal Process 20:868–880
9. Mottershead JE, Friswell MI, Ng GHT, Brandon JA (1996) Geometric parameters for finite element model updating of jointsand constraints. Mech Syst Signal Process 10(2):171–182
10. Li W-M, Hong J-Z (2011) New iterative method for model updating based on model reduction. Mech Syst Signal Process 25:180–192
11. Zapicoa JL, Gonzaleza MP, Friswell MI, Taylor CA, Crewe AJ (2003) Finite element model updating of a small scale bridge. J Sound Vib 268:993–1012
12. Ibrahim RA, Pettit CL (2005) Uncertainties and dynamic problem of bolted joints and other fasteners. J Sound Vib 279:857–936
13. Ahmadian H, Jalali H (2007) Generic element formulation for modelling bolted lap joints. Mech Syst Signal Process 21:2318–2334
14. Mayera MH, Gaul L (2007) Segment-to-segment contact elements for modelling joint interfaces in finite element analysis. Mech Syst Signal Process 21:724–734
15. Bograd S, Reuss P, Schmidt A, Mayer M (2011) Modeling the dynamics of mechanical joints. Mech Syst Signal Process 25:2801–2826
16. Ahmadian H, Ebrahimi M, Mottershead JE, Friswell MI (2002) Identification of bolted-joint interface models. In: 27th international seminar on modal analysis, Katholieke Universiteit Leuven, Leuven, 16–18 Sept 2002, pp 1741–1747
17. Imamovic N (1998) Validation of large structural dynamics models using modal test data. PhD thesis, Imperial College, London
18. Swanson Analysis Systems Inc. (2004) ANSYS user's manual version 8.0, Houston

Chapter 21
Experimental Modal Analysis and Modelling of an Agricultural Tire

F. Braghin, F. Cheli, S. Melzi, S. Negrini, and E. Sabbioni

Abstract Traction properties of agricultural tires on deformable grounds are strongly dependent on the interaction between tread elements with the soil, which is in turn function of parameters like tread design and inflating pressure. The present paper presents a numerical model aimed at investigating tire performance and soil compaction under different conditions, considering effects due to tangential and normal compliance of the terrain surface as well as tread design and inflating pressure. In particular, the ground has been modelled as a deformable springs layer and the tire as a flexible ring reproducing the tread geometry. Longitudinal and vertical dynamics of the tire are taken into account. Moreover, effect of tire carcass deformation was included considering the in-plane eigenmodes of the tire identified through a series of experimental tests.

Keywords Tire model • Tire flexibility • Deformable soil • Modal identification • Numerical analysis

21.1 Introduction

Experimental [1–3] and numerical [4] analysis carried out in recent years on off-road tires, evidenced the role of inflating pressure on several aspects of tire performance. In particular, inflating pressure deeply influences the shape and the dimension of the contact area together with the position and sinkage of tread lugs with respect to the soil. As a consequence, inflating pressure has a strong impact on the distribution of both normal and tangential contact stresses. These factors determine the performance of a tire [5] in terms of: traction force and traction efficiency [6], soil compaction [7] and vibrations transmitted to the tractor.

The work carried out by the authors aims at setting up a numerical model able to describe the interaction between tire and soil a tread level, taking into account soil characteristics, effect of tire flexibility (in particular associated with inflating pressure) and tread geometry. On this purpose, a model developed in recent years [8] by the authors was adopted. The model allows to take into account the soil deformation and the 3D design of the tread pattern. In addition, with respect to [8], the model was further developed including the description of tire structure flexibility, introduced by means of a modal superposition approach.

The paper is organized as follows: the first section is focused on the description of the numerical model developed to investigate the interaction between tire and deformable soil; the second part of the paper describes the experimental campaign aiming at identifying the in-plane eigenfrequencies and eigenmodes of the tire. The last part of the work reports some results of a numerical analysis, showing the effect of inflating pressure on traction performance, riding comfort and soil compaction.

F. Braghin• F. Cheli • S. Melzi • S. Negrini • E. Sabbioni (✉)
Dipartimento di Ingegneria Meccanica, Politecnico di Milano, via La Masa 1, 20156, Milano, MI, Italy
e-mail: francesco.braghin@polimi.it; federico.cheli@polimi.it; stefano.melzi@polimi.it; silvia.negrini@mail.polimi.it; edoardo.sabbioni@polimi.it

T. Simmermacher et al. (eds.), *Topics in Model Validation and Uncertainty Quantification, Volume 5*,
Conference Proceedings of the Society for Experimental Mechanics Series 41,
DOI 10.1007/978-1-4614-6564-5_21,

21.2 Numerical Model

Evaluation of tire performance was carried out by means of a numerical model of the interaction between tire ad soil. The model describes the motion of the tire in the vertical plane and considers the flexibility of the carcass and the 3D geometry of the tread pattern design. The soil model allows to describes the normal and tangential contact stresses associated with the relative motion between soil and tire.

21.2.1 Structure Model

The tire structure model describes the dynamics of the tire in the vertical plane. In particular the tire hub presents one single degree of freedom (d.o.f.) represented by its vertical displacement. The tire is assumed to be mounted on a tractor and, according to its weight distribution, part of the total mass of the tractor is assigned to the tire hub. Longitudinal speed of the hub and slippage are imposed, that is longitudinal displacement and rotation of the hub are regarded as imposed inputs.

The flexibility of the tire carcass is reproduced by means of a modal superposition approach. The equations of motion for the tire structure thus can be written as:

$$\begin{bmatrix} m_{hub} & & & \\ & m_1 & & \\ & & \ddots & \\ & & & m_n \end{bmatrix} \begin{Bmatrix} \ddot{z}_{hub} \\ \ddot{q}_1 \\ \dots \\ \ddot{q}_n \end{Bmatrix} + \begin{bmatrix} 0 & & & \\ & r_1 & & \\ & & \ddots & \\ & & & r_n \end{bmatrix} \begin{Bmatrix} \dot{z}_{hub} \\ \dot{q}_1 \\ \dots \\ \dot{q}_n \end{Bmatrix} + \begin{bmatrix} 0 & & & \\ & k_1 & & \\ & & \ddots & \\ & & & k_n \end{bmatrix} \begin{Bmatrix} z_{hub} \\ q_1 \\ \dots \\ q_n \end{Bmatrix} = \begin{Bmatrix} Q_{hub} \\ Q_1 \\ \dots \\ Q_n \end{Bmatrix} \tag{21.1}$$

Where m_j, r_j, k_j represent the modal mass, the equivalent damping and stiffness of the modal coordinate q_j respectively. As previously mentioned, m_{hub} is the mass of the hub and collects also part of the mass of the tractor. The terms Q_{hub} and Q_j introduce the effect of external forces (weight and contact forces) acting on the system. External forces actually couple the equations of motion.

21.2.2 Tread Model

Though the proposed model analyzes the dynamics of the tire in the vertical plane, a complete 3D description of the tread design is included. The outer surface of the tire is divided into N_{el} elements through a rectangular grid; the grid presents different spacing along circumferential and transversal direction. As an example, Fig. 21.1 shows the rectangular grid superposed to a picture of the tire analyzed in this work. Each element of the grid characterizes the local geometry of the tread considering the area of four surfaces: front, rear, top and base. Front and rear surfaces represent the total lateral surfaces of one or more tread lugs included in one element of the grid; in particular, normal contact stresses may contribute to the development of the total thrust. Base and top surfaces are instead normal to the radial direction and respectively represent the extremity of the tread lugs and the tread surface between lugs. In these tread regions, the main contribution to total traction force is associated with tangential stresses.

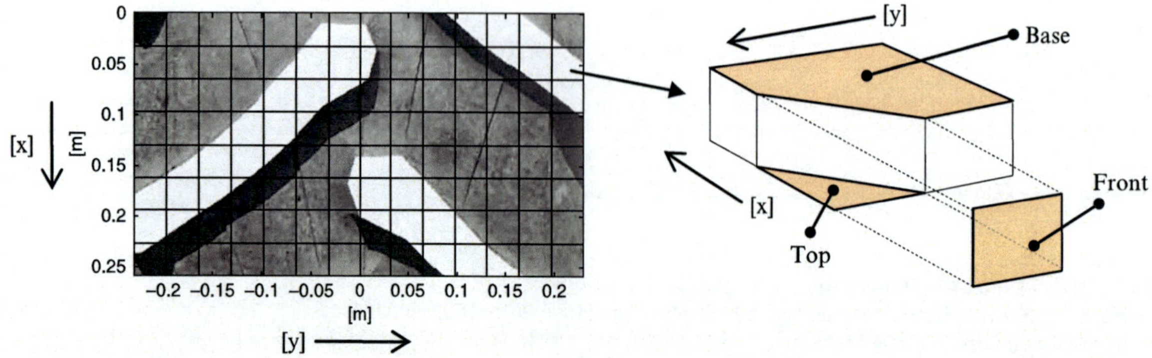

Fig. 21.1 Discretization of the tread surface and properties of the grid element

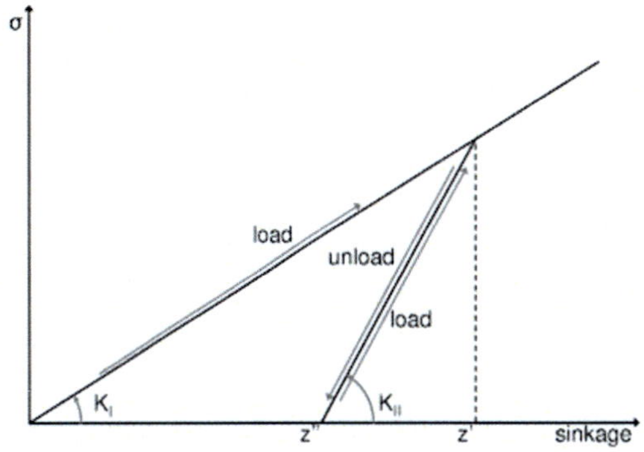

Fig. 21.2 Normal contact stress as function of the history of load cycles

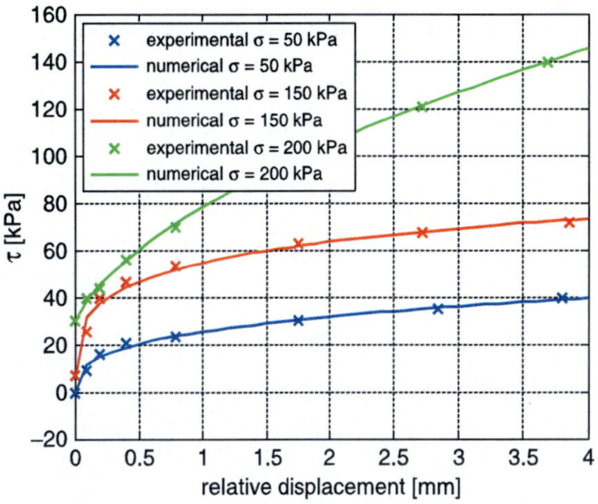

Fig. 21.3 Relation between tangential contact stress and relative displacement between tread and soil

21.2.3 Soil Model

The soil model is made up of a layer of springs whose compression, caused by the sinkage of the tire, gives rise to a normal contact stress σ: the stress is assumed to be hydrostatic, i.e. even though sinkage is computed along a vertical axis, the resulting pressure acts along three directions [9]. The relation between sinkage and normal stress depends on the time history of the deformation of the terrain. Figure 21.2 describes the adopted model: the load phase follows a linear relation whose slope KI is determined according to the experimental tests carried out with a standard penetrometer. The model takes into account the increase of the soil stiffness associated with its compaction: assuming that a soil has been compressed at z′, the unload phase from z′ to z″ and further load phases back to z′ are described through a different linear relation whose slope KII is assumed to be five times higher than KI.

As far as the tangential contact stress is concerned, a direct shear test has been carried out on a soil sample, to measure the shear stress as a function of the relative displacement between soil layers and the normal stress [10]; thus an analytical relation has been stated to fit the experimental data.

As it is shown in Fig. 21.3, the tangential contact stress τ increases along with the normal contact stress σ and the relative displacement of the soil, until a value of 4 mm of displacement is reached, at this point the soil is supposed to fail.

21.2.4 Generalized Forces

Recalling Eq. (21.1), this paragraph is devoted to the determination of the terms Q_{hub} and Q_j which account for the effect of external forces (weight and contact forces) acting on the system d.o.f.s. Figure 21.4 depicts the absolute position of point C_{ik}, geometrical center of the k−th area (A_{ik}) of the i−th element of the grid (Fig. 21.1); each element is in fact characterized by four surfaces (top, base, front, rear) so that k varies between 1 and 4. Figure 21.4 reports also the position of the tire hub (point O) and the distance between C_{ik} and O, named R_{ik}. If point Cik happens to be below the undeformed soil level, it will be characterized by a sinkage s_{ik}.

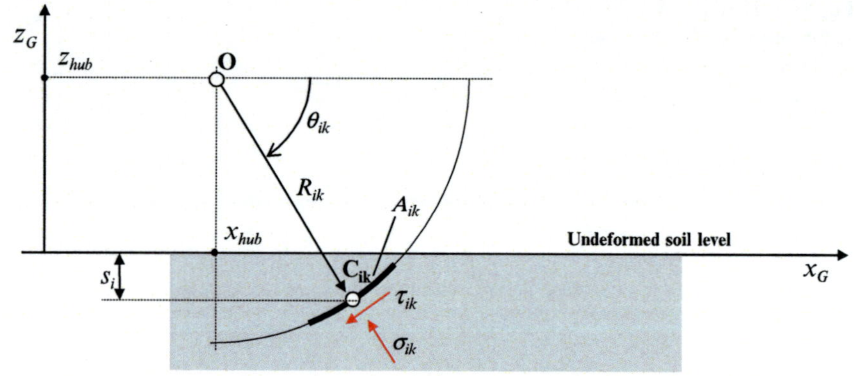

Fig. 21.4 Position of the center of the k-th area of the i-th element of the grid

The absolute position of C_{ik} can be computed as follows:

$$\begin{cases} x_{cik} = x_{hub} + R_{ik}\cos\theta_{ik} \\ z_{cik} = z_{hub} - R_{ik}\sin\theta_{ik} \end{cases} \tag{21.2}$$

Considering the flexibility of the tire structure, the radius R_{ik} will be function of the modal coordinates, according to the expression:

$$\begin{cases} x_{cik} = x_{hub} + \left(R_0 + \sum_{j=1}^{N} \Phi_{ik,j}q_j\right)\cos\theta_{ik} \\ z_{cik} = z_{hub} - \left(R_0 + \sum_{j=1}^{N} \Phi_{ik,j}q_j\right)\sin\theta_{ik} \end{cases} \tag{21.3}$$

In Eq. (21.3) of the term $\Phi_{ik,j}$ is the element of $j-$th eigenvector corresponding to the location of C_{ik}; modal coordinates actually represents variation of the radius of the tire along the circumference.

Considering the vertical position of C_{ik}, it is possible to compute the sinkage s_{ik} and the value of the contact pressure σ_{ik} acting on the $k-$th area of the $i-$th element. According to the procedure described in the previous paragraph, value of the tangential stress τ_{ik} can also be found. The virtual work of the external forces acting on the $k-$th area of the $i-$th element(Aik) is given by:

$$\delta L_{ik} = [-m_{ik}g + (\sigma_{ik}\sin\theta_{ik} - \tau_{ik}\cos\theta_{ik})A_{ik}]\,\delta z_{cik} - (\sigma_{ik}\cos\theta_{ik} + \tau_{ik}\cos\theta_{ik})A_{ik}\delta x_{cik} \tag{21.4}$$

In Eq. (21.4) m_{ik} is the mass associated with the $k-$th area of the $i-$th element. Substituting Eq. (21.3) in Eq. (21.4), the virtual work can be expressed as:

$$\begin{aligned} \delta L_{ik} = {} & [-m_{ik}g + (\sigma_{ik}\sin\theta_{ik} - \tau_{ik}\cos\theta_{ik})A_{ik}]\left(\delta z_{hub} - \sum_{j=1}^{N} \Phi_{ik,j}\delta q_j \sin\theta_{ik}\right) \\ & - (\sigma_{ik}\cos\theta_{ik} + \tau_{ik}\cos\theta_{ik})A_{ik}\left(\delta x_{hub} + \sum_{j=1}^{N} \Phi_{ik,j}\delta q_j \cos\theta_{ik}\right) \end{aligned} \tag{21.5}$$

Considering also the mass of the hub itself, the value of the generalized forces due to vertical displacement of the tire hub is represented by:

$$Q_{hub} = -m_{hub}g - \sum_{i=1}^{Nel}\sum_{k=1}^{4} m_{ik}g + \sum_{i=1}^{Nel}\sum_{k=1}^{4} (\sigma_{ik}\sin\theta_{ik} - \tau_{ik}\cos\theta_{ik})A_{ik} \tag{21.6}$$

Also the value of Q_j can be derived

$$\begin{aligned} Q_j = {} & \sum_{i=1}^{Nel}\sum_{k=1}^{4} \Phi_{ik,j}\sin\theta_{ik}m_{ik}g - \sum_{i=1}^{Nel}\sum_{k=1}^{4} (\sigma_{ik}\sin\theta_{ik} - \tau_{ik}\cos\theta_{ik})A_{ik}\Phi_{ik,j}\sin\theta_{ik} + \\ & - \sum_{i=1}^{Nel}\sum_{k=1}^{4} (\sigma_{ik}\cos\theta_{ik} + \tau_{ik}\cos\theta_{ik})A_{ik}\Phi_{ik,j}\cos\theta_{ik} \end{aligned} \tag{21.7}$$

Since σ_{ik}, and thus τ_{ik}, are functions of the sinkage, which is itself function of both the vertical displacement of the hub and of the modal coordinates, the equations of motion are coupled by the contact forces.

21.3 Identification of Modal Parameters

To include the tire deformation in the model, a modal identification has been performed on a tire of size 460/85R34. As the numerical model considers only in-plane forces, only the radial modes have been studied, considering two different inflation pressures.

The inflation pressure exerts a high influence on the tire dynamic response, determining the tire stiffness; the first inflation pressure inflation is 1.0 bar, which is generally used during field operation, the second inflation pressure inflation is 1.6 bar, preferable on public asphalt roads.

During the experimental tests modes in the range 10–60 Hz have been identified, whose shapes are shown in Fig. 21.5. The first mode corresponds to the expansion and contraction of the tire carcass, the second mode is the rigid displacement of the tire, and the third, fourth, and fifth mode correspond to carcass shape deformation, presenting respectively two, three and four lobes.

The modes shapes are poorly influenced by the inflation pressure, which has instead a significant effect on the tire stiffness.

Table 21.1 shows frequency and damping factor of the modes previously described as a function of the inflation pressure. Due to the stiffness increase with the pressure, the modes frequency increases and the damping factor decreases, proportionally to the mode order.

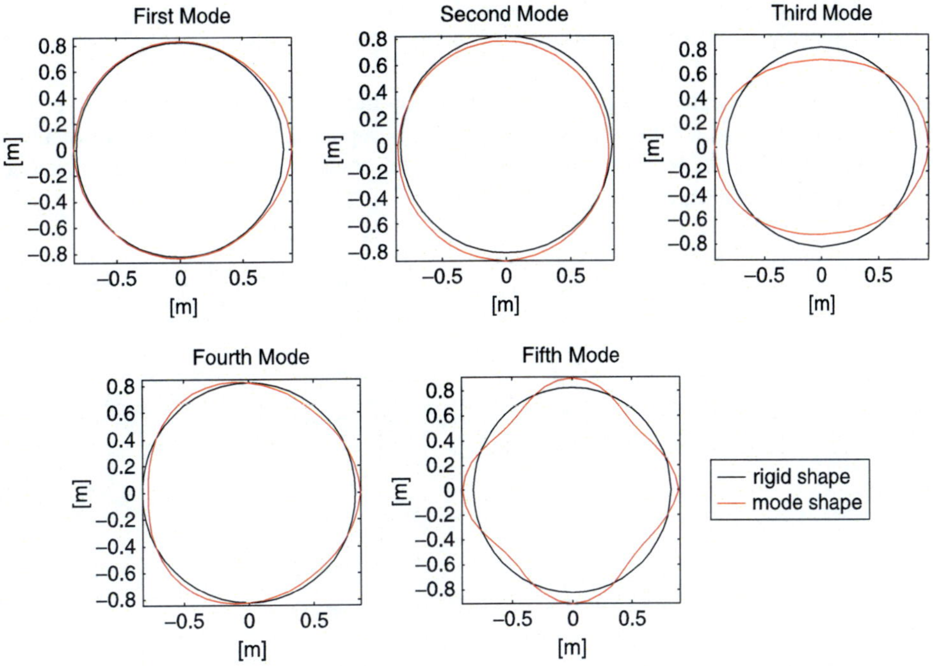

Fig. 21.5 Shapes of the five modes experimentally indentified

Table 21.1 Frequency and damp factor of the identified modes

	1.0 bar		1.6 bar	
	Frequency (Hz)	Damp. factor (%)	Frequency (Hz)	Damp. factor (%)
0 lobe	11.2	5.8	11.8	5.8
1 lobe	26.5	4.4	26.1	2.2
2 lobes	39.3	3.1	42.4	3.6
3 lobes	46.3	1.3	51.9	2.8
4 lobes	56.3	4.8	62.2	3.8

21.4 Simulation Results

The influence of the inflation pressure on tire performances has been studied, considering the tire moving at a speed of 7 km/h, with tire slip varying from 5 to 30%. Tire slip is defined as:

$$\varepsilon = \frac{\omega R - v}{\omega R} \tag{21.8}$$

The tire performance has been evaluated considering three aspects:

1. The traction force developed at the contact interface,
2. The soil compaction after tire passage, whose influence is crucial on roots growth
3. The driver comfort, which has been evaluated considering the standard deviation of the vertical acceleration at the hub,

Fig. 21.6 shows the traction force as a function of the tire slip it is possible to observe the increase of traction force with slip. Althrough the model does not consider the compaction of the soil between the treads it is able to account for the traction force increment at high slip values. This behaviour is reproduced into the model by introducing a coefficient to reduce forces generated on rear surface of lugs to about half of the front ones.

The traction force supplied is barely influenced by the tire inflation pressure; in fact, in the proposed model, contact forces are mainly related to the mean sinkage, which is quite similar for both the considered inflation pressures as it is possible to infer from Fig. 21.7. It is in fact worth noting that influence of inflation pressure on tire torsional stiffness is not included into the model.

On the contrary the inflation pressure have a severe influence on the driver comfort, as shown in where the standard deviation of the vertical acceleration as a function of the tire slip is reported. The model does not allow to evaluate the vertical acceleration at the driver's seat, as it is not modeled, hence the standard deviation of the vertical acceleration at the hub has been considered as comfort index.

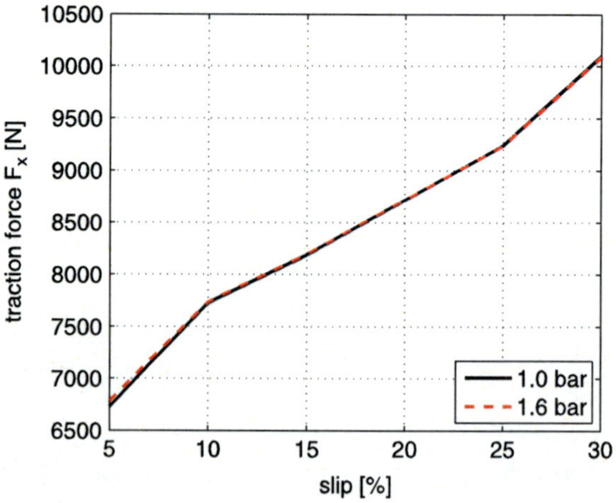

Fig. 21.6 Longitudinal traction force available at the hub as a function of the slip, for two different inflation pressures

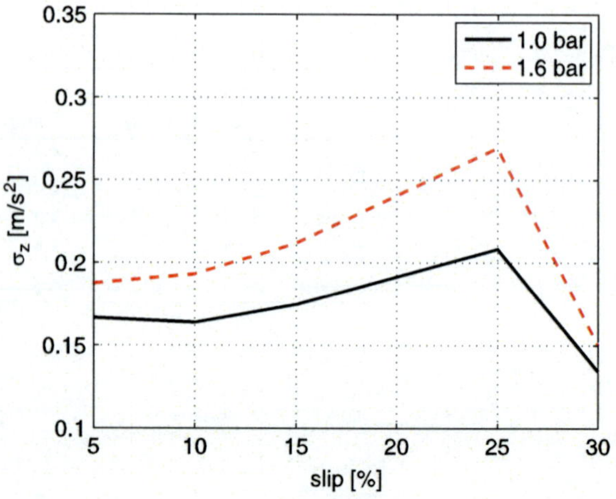

Fig. 21.7 Standard deviation of the hub vertical acceleration

Fig. 21.8 Soil compaction after tire passage as a function of slip and inflation pressure

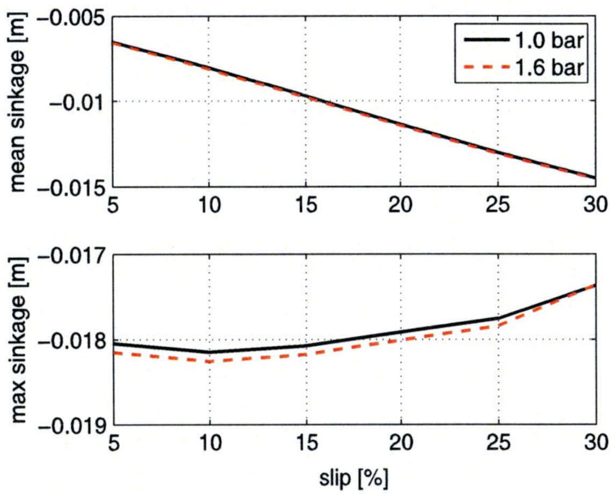

A lower inflation pressure dramatically decreases the standard deviation of the vertical acceleration at the hub, supplying a better filtering of the disturbance, as shown in Fig. 21.7.

The soil sinkage has been considered to evaluate the compaction level of the soil after the tire passage. The simulation results are shown in Fig. 21.8, where the mean value and the maximum value of the sinkage are shown as a function of the slip.

The mean value is related to the total vertical force on the tire, and it is almost the same for both the considered inflation pressures as the vertical load on the tire, due to the tractor load, is the same; moreover it is possible to observe the sinkage increase with the slip value, as the tire rotation digs the soil.

As far as the maximum sinkage is concerned, it is possible to observe that the inflation pressure affects its value thus reducing the soil compaction.

21.5 Conclusion

In this paper a tire soil interaction model allowing to evaluate the performance of a tractor tire riding on a deformable soil was presented, also including tire flexibility.

This model is suitable for estimating the traction force, the driver comfort and the vertical compaction of the soil's layers.

Influence of inflation pressure was analysed through numerical simulations. An increase in inflation pressure involves a larger value of the maximum soil sinkage, corresponding to a larger soil compaction, Moreover an increase in inflation pressure involves a rise of the standard deviation of the vertical acceleration at the hub, which can be assumed to be related to the driver comfort. Thus, as expected, a lower inflation pressure can improve comfort.

Furthermore this model can be used to carry out the optimal design of the tread pattern, since it considers the tire geometry. A sensitivity analysis of the influence of tire characteristics on performances will be provided in future developments of the work.

Acknowledgements Authors would like to thank ISMA (Istituto Sperimentale per la Meccanizzazione Agricola) for the support offered during the experimental tests.

References

1. Mohsenimanesh A, Ward SM (2010) Estimation of a three-dimensional tyre footprint using dynamic soil–tyre contact pressures. J Terramech 47:415–421
2. Jun H, Way TR, Löfgren B, Landström M, Bailey AC, Burt EC, McDonald TP (2004) Dynamic load and inflation pressure effects on contact pressures of a forestry forwarder tire. J Terramech 41:209–222
3. Nguyen VN, Matsuo T, Inaba S, Koumoto T (2008) Experimental analysis of vertical soil reaction and soil stress distribution under off-road tires. J Terramech 45:25–44
4. Schjønninga P, Lamandé M, Tøgersenb FA, Arvidssonc J, Keller T (2008) Modelling effects of tyre inflation pressure on the stress distribution near the soil–tyre interface. Biosyst Eng 99:119–133

5. Garciano LO, Upadhyaya SK, Jones RA (2010) Measurement of soil parameters useful in predicting tractive ability of off-road vehicles using an instrumented portable device. J Terramech 47:295–305
6. Lyasko Modest I (2010) How to calculate the effect of soil conditions on tractive performance. J Terramech 47:423–445
7. Hamzaa MA, Anderson WK (2005) Soil compaction in cropping systems. A review of the nature, causes and possible solutions. Soil Tillage Res 82:121–145
8. Braghin F, Melzi S, Negrini S, Sabbioni E (2011) A 2D model for tractor tire-soil interaction: evaluation of the maximum traction force and comparison with experimental results. SAE technical paper, SAE
9. Bekker MG (1956) Theory of land locomotion: the mechanics of vehicle mobility. The University of Michigan Press, Ann Arbor
10. ASTM Standard D3080/D3080M (2011) Standard test method for direct shear test of soils under consolidated drained conditions. ASTM International, West Conshohocken. doi:10.1520/D3080_D3080M-11, www.astm.org

Chapter 22
International Space Station Modal Correlation Analysis

Kristin Fitzpatrick, Michael Grygier, Michael Laible, and Sujatha Sugavanam

Abstract Modal analyses, model validations and correlations are performed for the different configurations of the International Space Station (ISS). Three Dedicated Thruster Firings (DTF) tests were conducted during ISS Stage ULF4; this paper will focus on the analysis and results of the DTF S4-1A, which occurred on October 11, 2010. The objective of this analysis is to validate and correlate analytical models used to verify the ISS critical interface dynamic loads.

During the S4-1A Dedicated Thruster Firing test, on-orbit dynamic measurements were collected using four main ISS instrumentation systems along with a Russian high rate sensor; Internal Wireless Instrumentation System (IWIS), External Wireless Instrumentation System (EWIS), Structural Dynamic Measurement System (SDMS), Space Acceleration Measurement System (SAMS) and Internal Measurement Unit (IMU). ISS external cameras also recorded the movement of one of the main solar array tips, array 1A.

Modal analyses were performed on the measured data to extract modal parameters including frequency, damping, and mode shape information. Correlation and comparisons between test and analytical frequencies and mode shapes were performed to assess the accuracy of the analytical models for the configuration under consideration. Based on the frequency comparisons, the accuracy of the mathematical model is assessed and model refinement recommendations are given.

Keywords Correlation • Modal analysis • Photogrammetry • International space station • Structural dynamics

22.1 Introduction

The on-orbit construction of the International Space Station (ISS) began in November 1998, and was completed in July of 2011. The ISS has been designed to operate for at least 15 years to conduct science and engineering projects. To maintain its structural integrity during its construction and life span, structural loading distributions have been rigorously analyzed through numerical simulations and included in the design of the structure and its mission operations [2, 15, 16]. The accuracy of such analysis results is directly affected by the integrity of structural dynamic mathematical models and estimated input forces.

On-orbit dynamic math models of ISS configurations are generated by combining component math models. Each component model is required to be correlated with ground test data. However, it is expected that on-orbit math models will still contain modeling inaccuracies due to differences in boundary conditions, mass distributions, and gravitational fields [7, 15]. Uncertainty factors are used to compensate for inherent inaccuracies in the math models and the estimated input forces [19]. The latter ISS configurations will have greater uncertainties due to the accumulation of component model inaccuracies. Large uncertainties would restrict the ISS mission operations. This problem may be alleviated by correlating on-orbit math models using test data measured in space. On-orbit testing of earlier ISS configurations, with ground testing

K. Fitzpatrick • M. Laible (✉)• S. Sugavanam
The Boeing Company, 3700 Bay Area Blvd., Houston, TX, USA
e-mail: michael.r.laible@boeing.com

M. Grygier
NASA, Johnson Space Center, Houston, TX, USA
e-mail: Michael.S.Grygier@nasa.gov

T. Simmermacher et al. (eds.), *Topics in Model Validation and Uncertainty Quantification, Volume 5*,
Conference Proceedings of the Society for Experimental Mechanics Series 41,
DOI 10.1007/978-1-4614-6564-5_22, © The Society for Experimental Mechanics, Inc. 2013

Table 22.1 Configuration of ISS stage ULF4 S4-1A DTF

ISS stage ULF4 on – orbit event	Date	SM Aft	MRM1 Nad	DC1 Nad	MRM2 Zen	Stbd SARJ angle	Port SARJ angle	Stbd HRS angle	Port HRSangle
S4-1A DTF	10/11/10	39P	23S	37P	24S	195	75	30	45
ISS stage ULF4 on – orbit event	BGA 3B angle	BGA 1B angle	BGA 1A angle	BGA 3A angle	BGA 4A angle	BGA 2A angle	BGA 2B angle	BGA 4B angle	
S4-1A DTF	279	70	279	70	270	90	270	90	

of new hardware components, will lead to the verification of later, more complex configurations. This "phased configuration verification" allows the use of the same uncertainty factors in predicting structural dynamic loads for all configurations.

This paper summarizes the on-orbit modal test and the related modal analysis, model validation and correlation performed for the ISS Stage ULF4, DTF S4-1A, October 11, 2010, GMT 284/06:13:00.00. The objective of this analysis is to validate and correlate analytical models with the intent to verify the ISS critical interface dynamic loads and improve fatigue life prediction.

For the ISS configurations under consideration, on-orbit dynamic responses were collected with Russian vehicles attached and without the Orbiter attached to the ISS. ISS instrumentation systems that were used to collect the dynamic responses during the DTF S4-1A included the Internal Wireless Instrumentation System (IWIS), External Wireless Instrumentation System (EWIS), Structural Dynamic Measurement System (SDMS), Space Acceleration Measurement System (SAMS), Inertial Measurement Unit (IMU) and ISS External Cameras.

Experimental modal analyses were performed on the measured data to extract modal parameters including frequency, damping and mode shape information. Correlation and comparisons between test and analytical modal parameters were performed to assess the accuracy of models for the ISS configuration under consideration. Based on the frequency comparisons, the accuracy of the mathematical models is assessed and model refinement recommendations are given.

Section 22.2 of this report presents the math model used in the analysis. This section also describes the ISS configuration under consideration and summarizes the associated primary modes of interest along with the fundamental appendage modes. Section 22.3 discusses the details of the ISS Stage ULF4 DTF S4-1A test. Section 22.4 discusses the on-orbit instrumentation systems that were used in the collection of the data analyzed in this paper. The modal analysis approach and results used in the analysis of the collected data are summarized in Sect. 22.5. The model correlation and validation effort is reported in Sect. 22.6. Conclusions and recommendations drawn from this analysis are included in Sect. 22.7.

22.2 Math Models and Dynamics

An ISS math model was created for the configuration of the ISS Stage ULF4 DTF S4-1A. This math model was generated from collections of the latest ground test verified component models. The component models are represented by Finite Element Models (FEMs) that also include internal and external Component Mode Synthesis (CMS) models. In addition, the analyzed model incorporated, as closely as possible, the actual on-orbit boundary and interface conditions. The post-flight modeling effort also attempted to account for the actual array orientations that were recorded during the event. The model used in this analysis are linear and do not account for non-linearity that may be present.

The configuration of the ISS Stage ULF4 DTF S4-1A is outlined in Table 22.1. The configuration includes a Progress docked to the SM Aft and DC1 Nadir ports, and a Soyuz docked to the MRM1 Nadir and MRM2 Zenith ports. Figure 22.1 illustrates the vehicle and element configuration for this event. The specific SARJ and BGA angles are presented in Table 22.1 and depicted in the math model MSC/PATRAN^{TM} view in Fig. 22.2. The ISS solar array, 1A, that was recorded during the S4-1A DTF is also labeled in that figure.

Table 22.2 is the latest on-orbit ISS mass properties compared to the NASTRAN model. The associated global system and truss modes (up to 5.0 Hz) are summarized in Table 22.3. The mode descriptions in the table were determined from kinetic energy distributions and mode shape animation using MSC/PATRAN^{TM}.

22.3 On-Orbit Flight Testing: Dtf s4-1A

The purpose of the ISS Stage ULF4 S4-1A DTF test was to excite primary load inducing modes which would be recorded by the ISS instrumentation systems. The test was conducted on October 11, 2010, GMT 284/06:13:00.00 and was comprised of two firings. Firing 1 was an ISS Yaw firing designed to excite dynamic modes in the ISS XY plane, though other modes

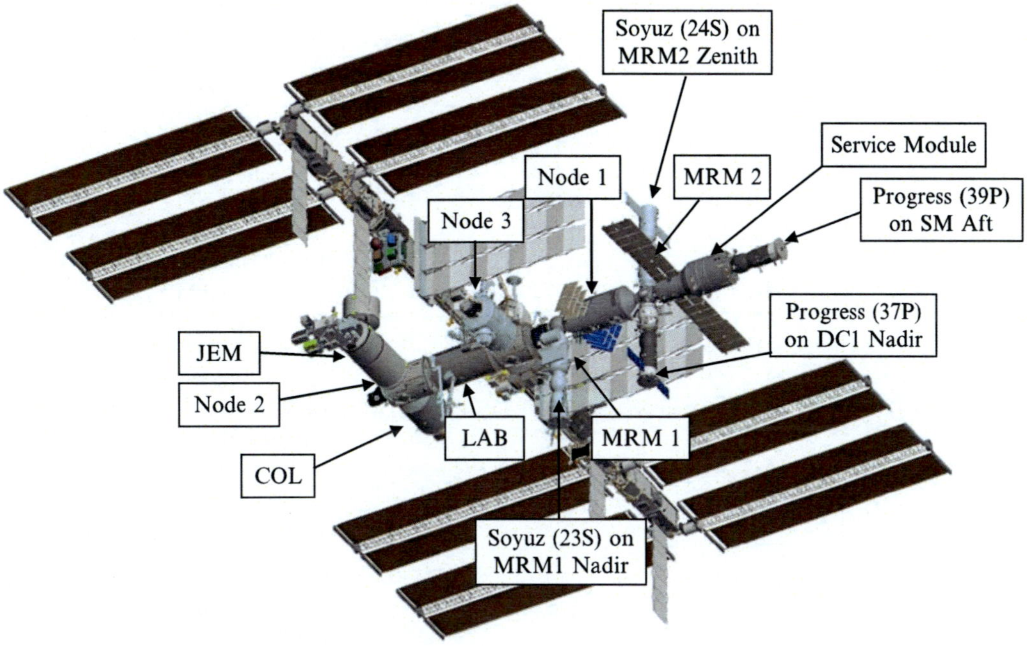

Fig. 22.1 ISS stage ULF4 dedicated thruster firing S4-1A configuration

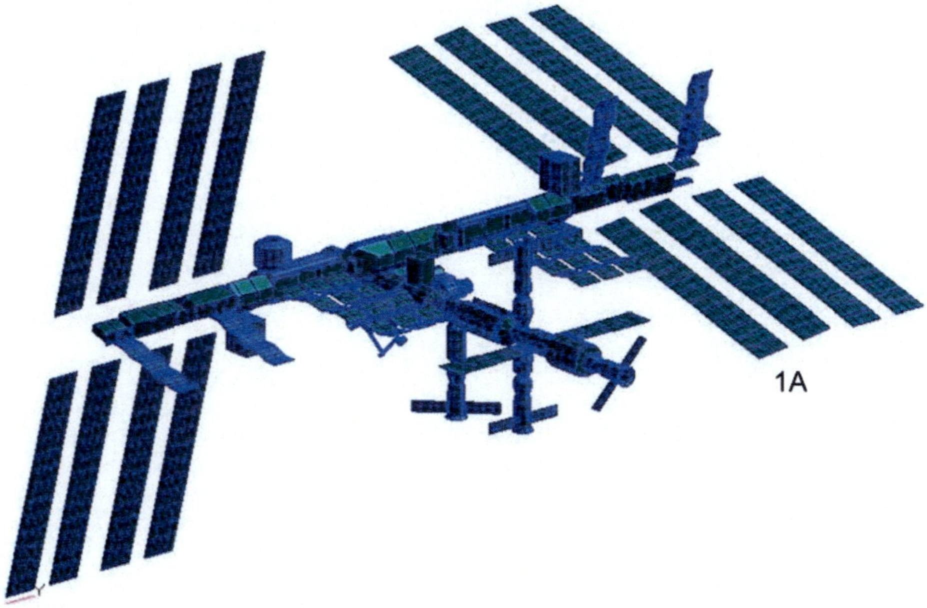

Fig. 22.2 ISS stage ULF4 S4-1A DTF math model PATRAN view

Table 22.2 Mass properties: ISS stage ULF4 S4-1A DTF configuration

	24S_AR ISS + 37P+23S+39P+24S after 24S docking						
		Center of gravity (in)			Moments of inertia (Slug – ft^2)		
Description	Mass (lbs)	X	Y	Z	Ixx	Iyy	Izz
VIPER properties	827,126	−160.9	−38.4	130.0	84,599,810	50,464,111	127,527,980
Loads model fv84ds	789,386	−161.6	−34.9	127.4	85,752,622	47,993,160	123,591,383
% diff	−4.8 %	0.5 %	−10.1 %	−2.0 %	1.3 %	−5.1 %	−3.2 %

Table 22.3 Mode descriptions: ISS stage ULF4 S4-1A DTF math model

Mode #	Freq. (Hz)	Mode description
10	0.061	Array OP/module XZ
17	0.092	Array IP/station XY
36	0.112	Station XY (slight)/array OP
44	0.114	Station XY
57	0.150	HRS out-of-plane
83	0.173	Truss XY
84	0.197	HRS IP/module XZ/truss XYZ
102	0.225	Truss YZ/module XZ
131	0.272	Station XY (JEM XY)
142	0.285	Station XZ
171	0.362	SM array IP/COL-JEM YZ
175	0.368	Truss YZ/module TOR/HRS torsion
211	0.385	HRS torsion
245	0.447	Modules-JEM YZ/truss XYZ
254	0.494	Station torsion X/truss YZ
256	0.513	Station torsion X – JEM XYZ/HRS accordion/ truss XYZ
276	0.553	Module XY/STBD truss XY/HRS accordion
286	0.578	HRS accordion
288	0.584	PHRS torsion/RSA tor X/truss YZ
297	0.637	EPS torsion
302	0.665	Station XZ
313	0.720	Truss XY bending
339	0.793	Module XY/Node 3 YZ/EPS IP
352	0.840	Module RSA YZ
353	0.843	Node 3 YZ/RSA XZ
355	0.888	Station XY/Soyuz MRM1 OP/EPS IP
359	0.928	HRS IP
368	0.994	Truss XZ/module XZ
371	1.030	Module XY (N3/N1/AL)
382	1.099	RSA XZ
443	1.174	Station XY/JEM EF
484	1.251	JEM XY/Node 3 XY/COL YZ
486	1.310	Module XZ
504	1.425	RSA XZ/COL XY/JEM EF XYZ
511	1.508	JEM YZ/APM YZ/RSA XZ
540	1.633	Station XY
549	1.672	Truss XY/station XYZ/RSA YZ
553	1.703	Station XY
625	1.834	SPDM MB2
646	1.956	Truss XY bending/more...
661	1.983	P3 torsion
674	2.037	Truss XY/SPDM MB2
723	2.171	Airlock YZ/SM XY
728	2.190	Station XY
803	2.434	Truss COL N2 JEM YZ/SM XY
819	2.560	Module XZ
833	2.754	SM XY/airlock YZ
834	2.785	SM XY/airlock YZ/PHRS IP
921	3.458	MRM YZ/N2 N3 airlock XY/S3 P3 torsion
931	3.722	Station XZ
935	3.769	Truss XYZ
951	4.001	Station XYZ
960	4.103	US LAB XY/Node 2 XY
979	4.264	EPS IP 2nd
1,003	4.564	Truss XYZ/station XY
1,008	4.703	Module XY

Table 22.4 ISS stage ULF4 S4-1A DTF firings

S4-1A Array	Event	Description	Thruster firing sequence	Duration (S)	Start time (S)	Stop time (S)
		Free drift			0.0	200.0
	Firing 1	Yaw	SM 12, 13, 14, 18, 19, 20	1.0	200.0	500.0
	Firing 2	Pitch	SM 21, 22, 23, 27, 28, 29	0.6	500.0	700.0
		Return to attitude				

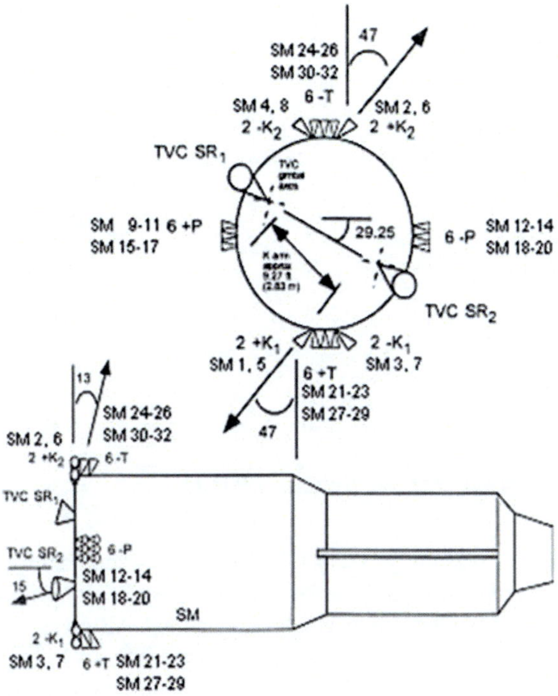

Fig. 22.3 Service module thruster diagram

were also excited due to complexity of the ISS structure. Firing 2 was an ISS Pitch firing designed to excite dynamic modes in the ISS XZ plane, which also excited other modes due to the complexity of the ISS structure. Both test firings used ISS Service Module (SM) thrusters. The Service Module is a Russian Module located at the ISS aft, see Fig. 22.1. A breakdown of each firing set is shown in Table 22.4. A diagram of the Russian Service Module thrusters is shown in Fig. 22.3.

22.4 On-Orbit Instrumentation Systems

22.4.1 Overview

During the ISS Stage ULF4 S4-1A DTF, structural responses were measured and recorded with a variety of instrumentation systems that include: External Wireless Instrumentation System (EWIS), Internal Wireless Instrumentation System (IWIS), Structural Dynamic Measurement System (SDMS), Space Acceleration Measurement System (SAMS), Inertial Measurement Unit (IMU) and ISS External Cameras. A detailed description of each of these systems is given in the following sections.

Photogrammetry data, from the ISS External Cameras, was taken of the 1A solar array during the S4-1A DTF. The data consists of a displacement time history response of two tracked points at the end the array. This data allows for an independent analysis of the solar array wing's modal parameters. Having a better understanding of the on orbit array mode frequencies and damping allows for less uncertainty to be placed on array modes during loads analysis which feed into reducing solar array position constraints during flight operations.

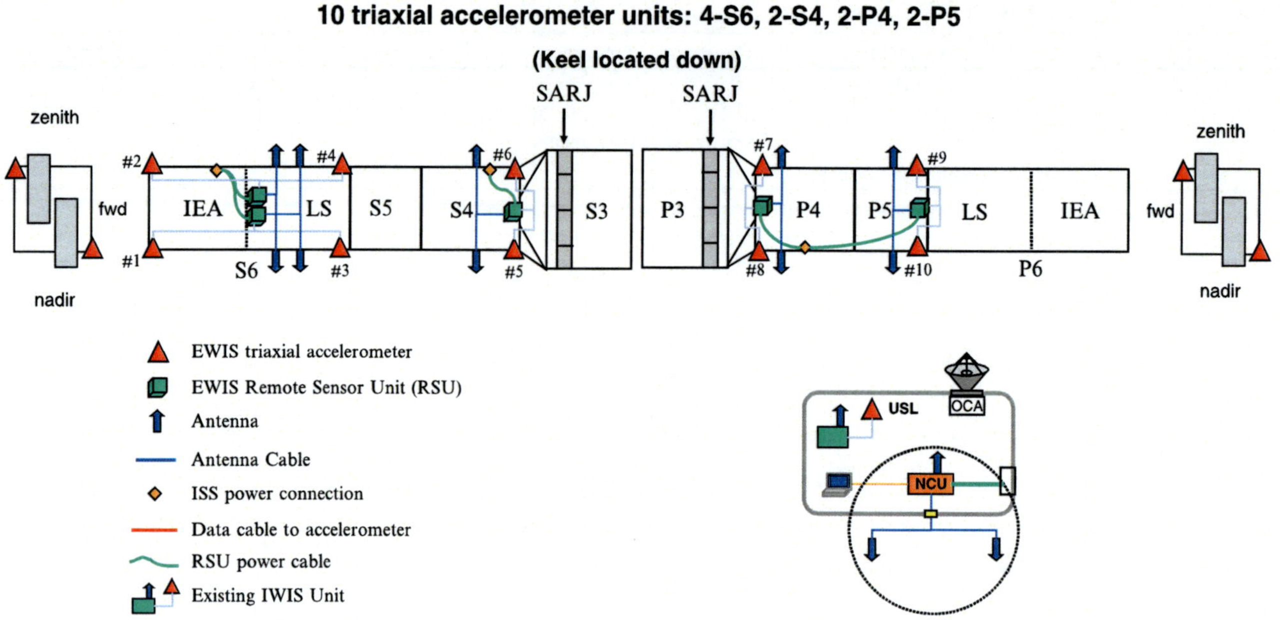

Fig. 22.4 EWIS architecture

22.4.2 *External Wireless Instrumentation System (EWIS)*

The ISS truss segments outboard of the solar array rotary joint (S4, S6, P4 and P5) were outfitted with External Wireless Instrumentation System (EWIS) accelerometers and Remote Sensor Units (RSUs) prior to launch [5]. The EWIS Network Control Unit, which sends user commands to the RSUs, resides inside the LAB pressurized module and two NCU antennas were externally mounted to the LAB via EVA. The EWIS architecture is shown in Fig. 22.4. The EWIS System has both a Continuous Mode and a Scheduled Mode for data acquisition. The Continuous Mode aspect of EWIS records a 2 min window of accelerometer data when a prescribed acceleration threshold is reached; such that 30 s prior to the threshold exceedance and 90 s after the threshold exceedance is stored for download. The Continuous Mode also collects data and sorts cycle counts into 200 μg bins for the time period in between Scheduled Mode data takes, which is utilized for structural life assessment. The Scheduled Mode allows for a user to command EWIS accelerometers to record at a set sampling frequency for a prescribed length of time. The EWIS RSUs store the accelerometer data until a data take has completed. The RSUs then transmit the data to the EWIS NCU via Remote Frequency (RF). The EWIS NCU then passes the data through a cable to a Space Station Computer (SSC) where it is stored until it is downloaded to the ground via KU-band. Due to communication programming issues to the outboard truss only the accelerometers on the S4 and S6 IEA truss segments recorded data during the ISS Stage ULF4 S4-1A.

22.4.3 *Internal Wireless Instrumentation System (IWIS)*

The Internal Wireless Instrumentation System (IWIS) operates within the pressurized ISS modules [8]. The IWIS hardware consists of Remote Sensor Units (RSUs) each connected by a cable to a single Triaxial Accelerometer (TAA). There is one IWIS RSU that also connects to eight strain gages. The Triaxial Accelerometers and strain gages receive their power from their connected RSU, which gets its power via a Green or Cobalt power brick connected to the Russian or US power system, respectively. The EWIS NCU, which commands and receives data from the EWIS RSUs, also commands and receives data from the IWIS RSUs via RF communication. The IWIS hardware configuration during the ISS Stage ULF4 S4-1A DTF is, shown in Fig. 22.5, comprised of the EWIS NCU) six (6) Remote Sensor Units (RSU), six (6) triaxial accelerometer units, eight (8) strain gages, and accelerometer mounting plates.

Similarly to EWIS Scheduled Mode programming, IWIS RSUs are programmed prior to an on orbit event with a prescribed duration and sampling frequency. IWIS RSUs do not have a Continuous Mode they will only record data if preprogrammed to do so. After a programmed data take is complete the measured data is stored in the IWIS RSU until it

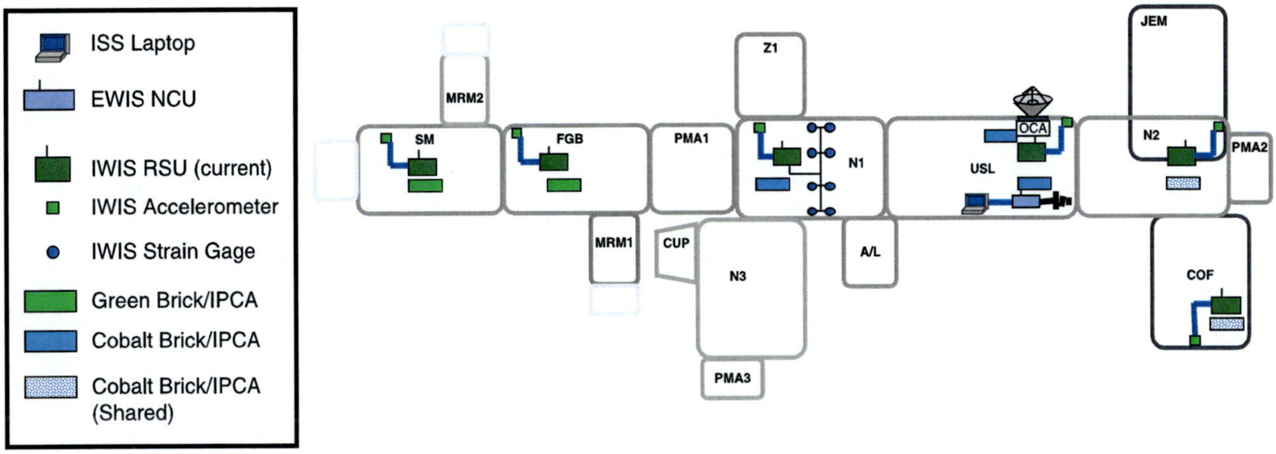

Fig. 22.5 IWIS sensor Configuration during ISS stage ULF4 S4-1A DTF

receives a command from the EWIS NCU to download via RF to the EWIS NCU. Once the EWIS NCU has received the measured data it is passed to a SSC and later downloaded via KU-band, in the same manner as EWIS TAA data.

The IWIS accelerometers are temporarily attached to mounting plates that are installed on primary ISS structure. The mounting plates are attached to the structure using a combination of adhesive and grey tape. The IWIS strain gages are permanently mounted to the US Node 1 radial port struts. New IWIS components are scheduled to be implemented on orbit in early 2013, which will allow accelerometer data to be recorded in MRM 1 and MRM1. Also, and IWIS TAA and RSU will be installed in the MLM prior to its launch and connection to station.

22.4.4 Structural Dynamic Measurement System

The Structural Dynamic Measurement System (SDMS) was developed by The Boeing Company for NASA-JSC. The SDMS is intended to measure on-orbit dynamic responses of the ISS Inboard Truss Segment and module-to-truss structure (MTS) struts. SDMS is comprised of 33 accelerometers, 38 strain gage bridges, and 2 signal conditioning units (SCUs). The accelerometers and strain gages are externally mounted on the five segments of the main inboard truss between the Solar Alpha Rotary Joints (SARJs).

The SDMS accelerometers are proof-mass type and their locations are shown in the schematic of the five inboard truss segments shown in Fig. 22.6. The accelerometers are mounted in groups of one, two, and three, on the truss primary structure. Each strain gage-bridge uses four strain-gages to form a four-active-arm bridge circuit. A total of 152 strain gages were used to generate all 38 strain-gage bridges. Each strain gage bridge generates a single strain measurement. The general locations of the strain gage bridges are shown in Fig. 22.6. Electrical power is provided to the accelerometers and strain gages by two SCUs. The SCUs are also used to amplify, filter, and digitize the signal output by the accelerometers and strain gage bridges. Sensor data is stored on a memory buffer before it is downlinked directly to the ground by telemetry. The SDMS can be fully operated by commands up-linked from the ground.

The SDMS has a fixed sampling rate and anti-alias filtering of 40 and 7.5 Hz, respectively. The SDMS system is capable of recording approximately 10.5 min of data. The memory buffer is circular so that if over 10.5 min of data is recorded, the data will be overwritten gradually starting from the buffer's beginning. When data recording stops (and over 10.5 min of data was collected), the last time step of the data collection immediately precedes the first time step that hasn't been overwritten. A more comprehensive discussion on the SDMS hardware can be found in [13].

22.4.5 Space Acceleration Measurement System (SAMS)

The Space Acceleration Measurement System (SAMS-II), Fig. 22.7, provides a continuous measurement of the ISS vibratory acceleration environment from 0.01 to 300 Hz using a distributed, configurable set of tri-axial accelerometers. The accelerometers are housed inside module racks, secondary structure, instead of on the main structure like IWIS. The sensors

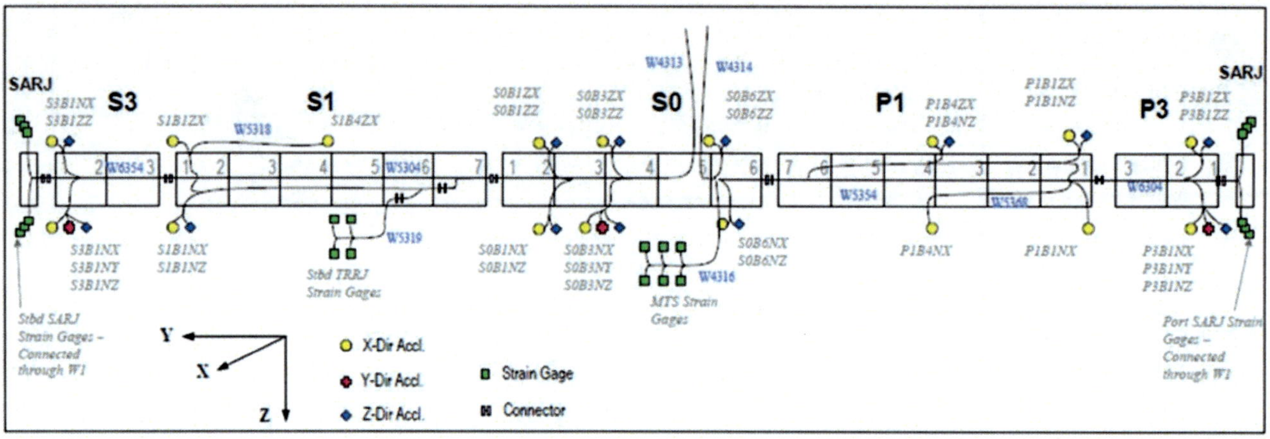

Fig. 22.6 SDMS accelerometer and strain gage locations

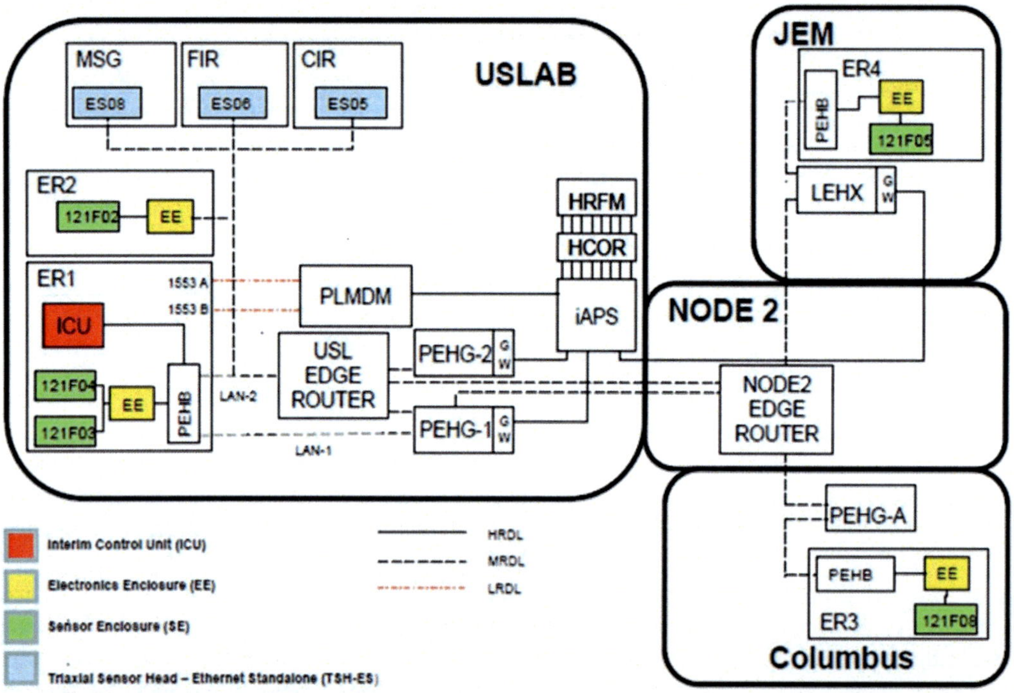

Fig. 22.7 SAMS ISS hardware

are in racks in order to provide microgravity data for a variety of science projects, rack system analysis and for the overall ISS microgravity environment. Though these sensors are housed inside module racks, the frequency of interest for ISS loads modal correlation <5 Hz, has been found to have comparable content as the IWIS sensors on the main structure. The SAMS sensors that were recorded and used for the S4-1A DTF analysis are the US Lab sensors, F03 and F04, and JEM sensor, F05.

22.4.6 Russian Inertial Measurement Unit (IMU)

The IMU-D (Inertial Measurement Unit) is located in the MRM1 (Mini Research Module), a Russian Pressurized Module connected to the FGB (Functional Cargo Block). The sensor samples at a rate of 2,400 Hz. This sensor can measure acceleration +/-10 mg with a frequency range between 0.01 to 50 Hz. The data for a single event is recorded in three audio files, one file for each coordinate direction. The data is downloaded to a Russian ground site and delivered to US NASA personnel.

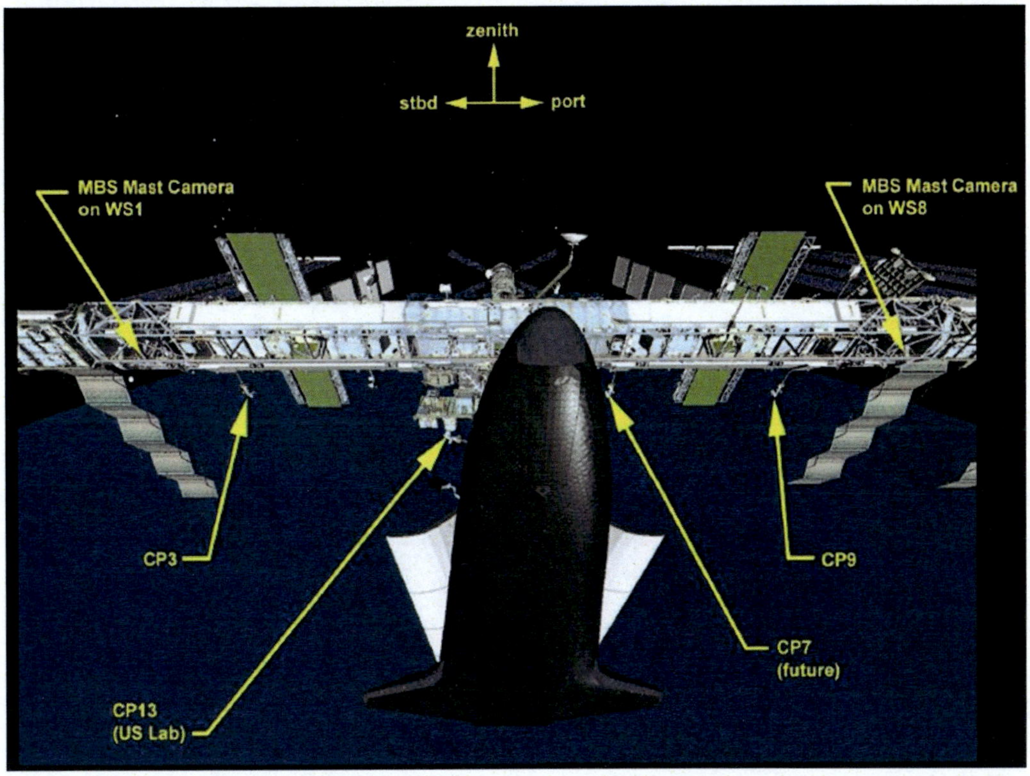

Fig. 22.8 ISS cameras used for structural dynamics testing

22.4.7 ISS Photogrammetric System

The ISS photogrammetric system uses EVTCG cameras outside of the ISS structure. The photogrammetric system is intended to be a non-contact instrument to record the dynamic deflections of low frequency space structures such as solar arrays, EPS radiators, and antennas. The Image Science and Analysis Group at NASA-JSC perform the photogrammetric processing of the recorded video footages to generate deflection time response histories. Daylight testing is the optimum condition for photogrammetric processing. Ideally, photogrammetric data processing should involve video footages from at least two video cameras to give a more accurate three-dimensional perspective of deflections. However, single camera approximate processing is also possible using some assumptions on the nature of the deflections. The current ISS video system and data processing method, Fig. 22.8, offer a time history with a sampling rate of 15 or 30 Hz and a resolution of 0.1 in.

22.5 Modal Analysis

22.5.1 Overview

The test data from all instrumentation systems was prepared for analysis. Time history data was converted, filtered, and plotted as needed. The power spectrum density (PSD) of this data was calculated and plotted to investigate its modal content. An attempt was made to synchronize data sets from different instrument system. This task was complicated by the fact that there is no universal time management across the different instrumentation systems. Each system has a different time keeping and standard that is not synchronized.

Modal analysis was then performed on all pre-processed data to determine the structural modal parameters, i.e., frequencies, damping factors, and mode shapes. Modal system identification was also performed separately on the photogrammetry displacement data and MTS strut strain gage data.

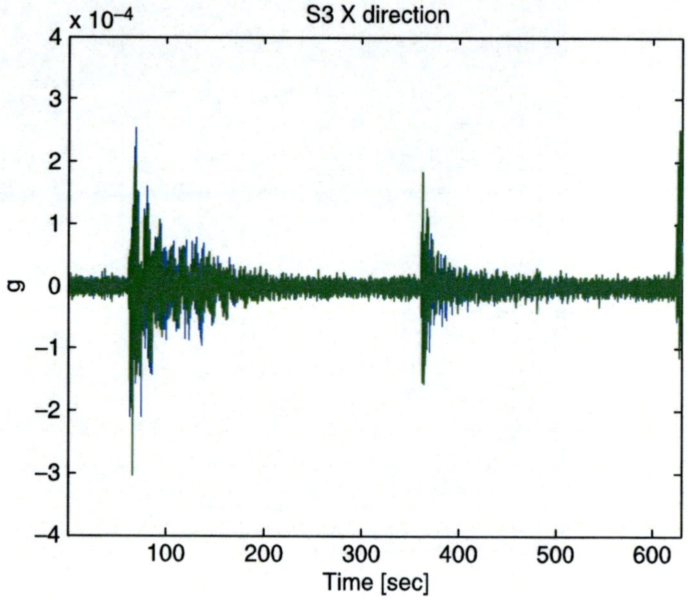

Fig. 22.9 DTF S4-1A S3 ISS X: SDMS: time history

22.5.2 Modal Analysis Procedure

Traditional modal analysis methods using frequency response functions (FRFs) were not used since the input excitation forces were not measured and the duration of free decay data is short. A special modal identification method [17] was used on the accelerometer data, which has been developed for applications to large space structures. It is a time-domain, free-decay method based on the Eigensystem Realization Algorithm (ERA) [14] and a time-domain zooming technique. This method does not require input force measurements and characterizes nonlinearities with a series of linearized modal parameters during the free-decay period.

To utilize the time-zooming technique, each data segment was first detrended to remove the constant and linear biases. The detrended data was filtered by a number of bandpass filters to emphasize different frequency ranges. The filtered data was then decimated to reduce the sampling rate. The intent of this combined filtering and decimation process is to emphasize the frequency content of the data in certain bandwidths. This process is comparable to a frequency-domain zooming technique used in the traditional modal analysis methods. It is again noted that the selected modal analysis process is based on a free-decay method and applied to the free-decay portion of the data sets. The ERA modal extraction is applied to several data time windows varying in length. This is performed to extract the most consistent modal content present in the data. The whole process described herein has been implemented in a Boeing proprietary MATLAB™ based Graphical User Interface (GUI) software entitled "The Boeing Modal Refinement and Identification Tool" (The Boeing MoReID).

The Boeing Test Analysis Correlation Solutions (BTACS), another Boeing proprietary interactive engineering MATLAB™ based GUI, was used to extract modal parameters from the MTS strut strain gage and photogrammetry data. BTACS has a system identification tool that extracts parameters through system realization using the Hankel matrix along with the singular value decomposition method.

22.5.3 Sample Data Plots

The S4-1A Dedicate Thruster Firing Test consisted of two sets of thruster firings. Each set of firings was designed to excite distinct sets of Truss and Module modes. Figures 22.9 through 22.14 show SDMS time history accelerometer data plots for the yaw and pitch firings of the DTF test. Figure 22.9 is SDMS S3 (Starboard Truss Segment) ISS X direction accelerometer data, which shows the Yaw firing (first firing at $t = 60\,s$) has higher magnitude in the ISS X direction than the pitch firing (second firing at $t = 360\,s$), as expected. Figures 22.12 through 22.14 are the plots of the SDMS S0 (Center Truss Segment) ISS Z direction accelerometer data, which show a higher acceleration magnitude during the pitch firing.

Fig. 22.10 DTF S4-1A S3 ISS X: SDMS: PSD

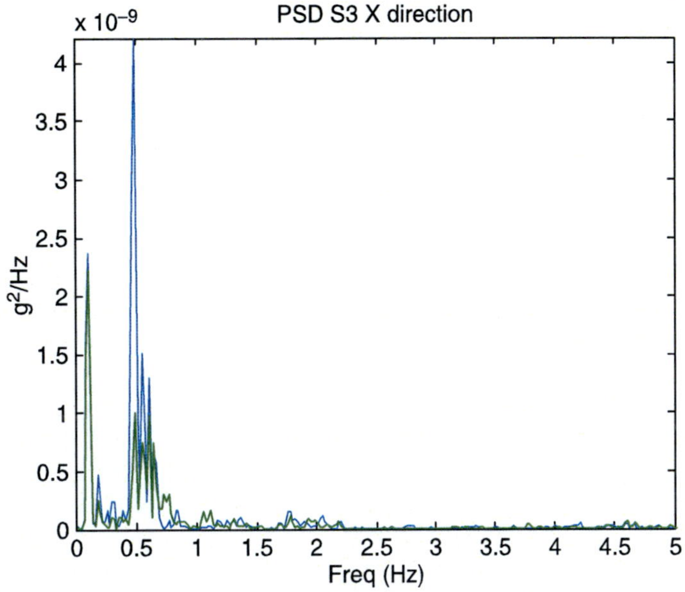

Fig. 22.11 DTF S4-1A: SDMS S3 ISS X: spectrogram

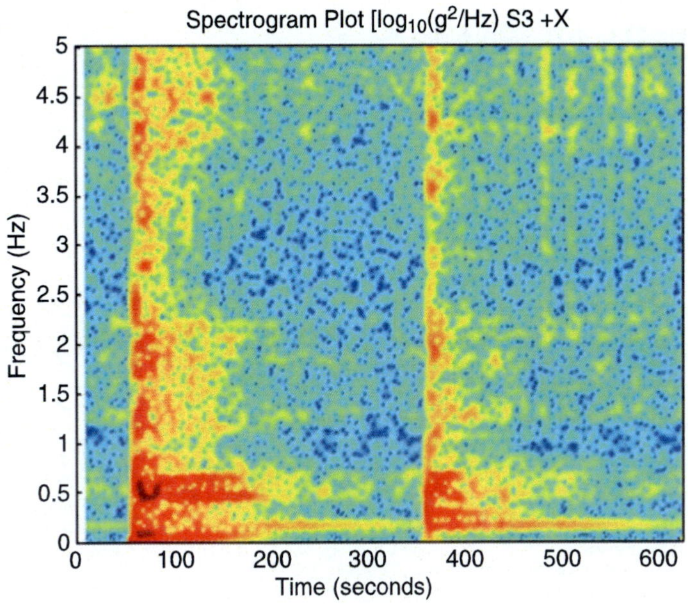

22.6 Model Correlation and Validation

22.6.1 Overview

International Space Station integrated loads and dynamics verification and validation procedures are defined in Ref. [7]. Verification procedures are intended to insure that the on-orbit structure satisfies its structural requirements. Among others, these requirements include the verification that the structure can accommodate on-orbit loads. The model of the structure used in the verification of the on-orbit loads must be validated [19]. One of the requirements in the validation plan states that test and analytical modal frequencies agree within 5 % for primary modes and within 10 % for secondary modes. The model validation plans are intended (i) to prove that the on-orbit models satisfy the validation requirements or (ii) to refine the on-orbit models so that they satisfy the validation requirements.

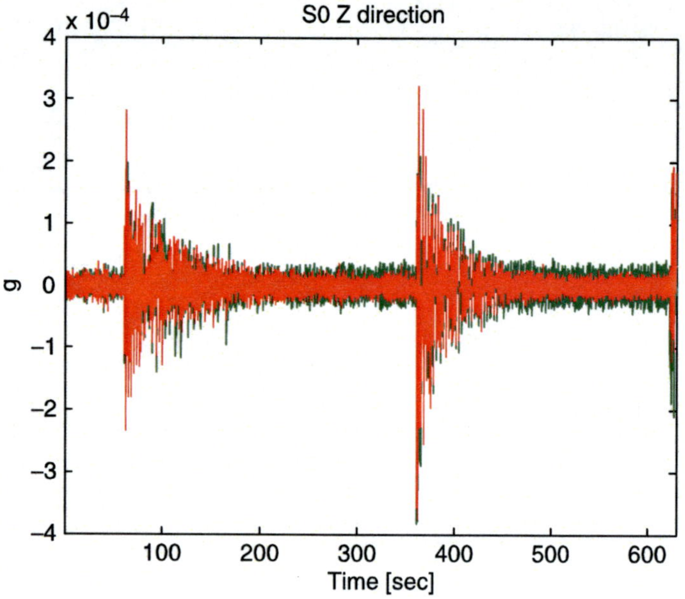

Fig. 22.12 DTF S4-1A: SDMS S0 ISS Z: time history

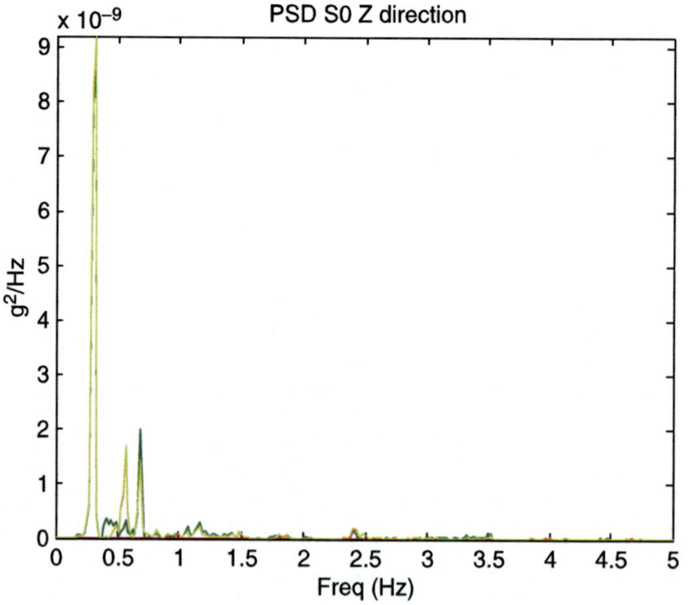

Fig. 22.13 DTF S4-1A: SDMS S0 ISS Z: PSD

22.6.2 Mode Correlation and Modal Analysis

The task of matching test and analytical modes is commonly performed by evaluating the Modal Assurance Criteria (MAC) [21] and the Cross-Orthogonality (XOR) matrices [6]. Both MAC and XOR matrices are indicators that show the level of correlation between test and analytical mode shapes. These indicators are only meaningful with a large number of mode shape measurements spanning a wide spatial distribution. For the ISS, it is desirable to have two sensors in each ISS module; at least one sensor at each side of an interface. This would aid in defining axial and bending modes. Furthermore, the addition of three sensors placed in each ISS module would aid in defining ISS torsion modes. Such criteria are only partially met by modes extracted from the IWIS, SDMS, EWIS, SAMS and IMU-D dynamic measurement systems.

In the computation of the XOR matrix, a reduced mass matrix, having degrees-of-freedom (DOFs) consistent with the measurement DOFs, is needed. A reduced mass matrix for this particular problem requires a significant reduction from over

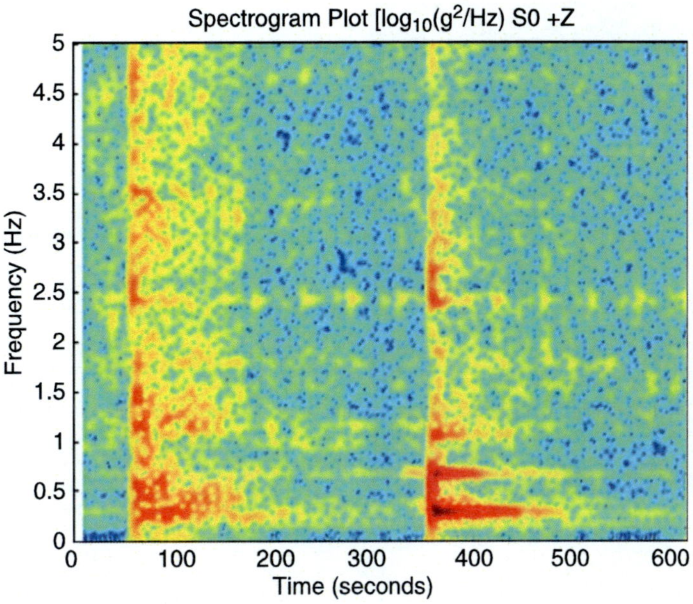

Fig. 22.14 DTF S4-1A: SDMS S0 ISS Z: spectrogram

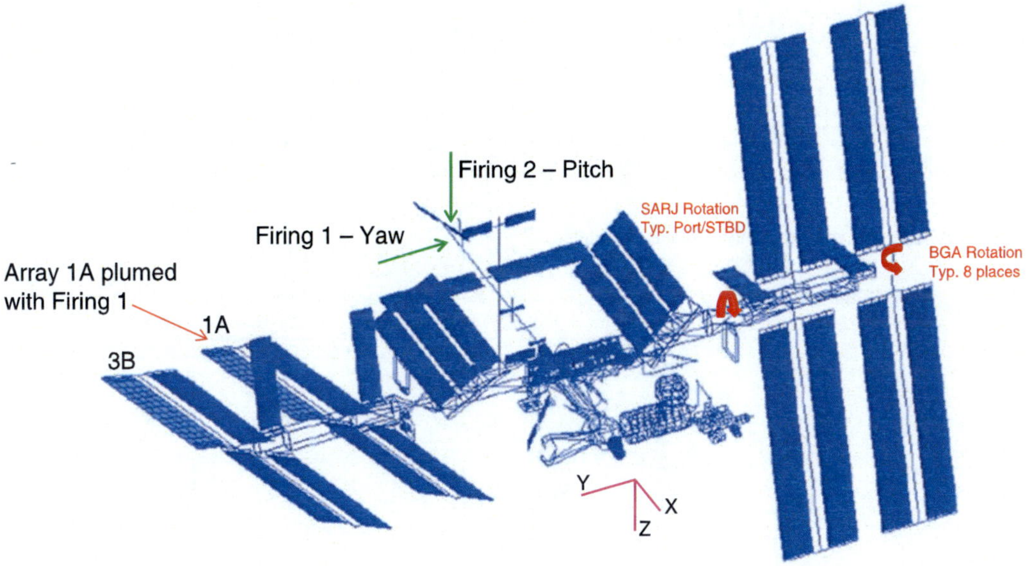

Fig. 22.15 NASTRAN system model S4-1A DTF

several thousand DOFs (residual model) to less than 75 DOFs and was unavailable for this study. The computation of the MAC matrix does not require the use of a reduced mass matrix. Thus, test-analytical mode correspondences based on the MAC was used for the modes extracted from the ISS sensor systems.

It should be noted that the damping was very difficult to estimate with the type of on orbit test that was conducted and the type of data that was available. In order to estimate modal damping with high confidence, free decay data created from a test with several input sites is required. The S4-1A DTF had one input site, the SM thrusters, it is not feasible to conduct an on-orbit dynamic test on the ISS with multiple input sites and nominal on-orbit dynamic events (i.e., vehicle dockings, undockings) do not have the characteristics of this ideal data.

The analytical model and test data both exhibit high modal density above 1.5 Hz. Also, the data exhibits very low modal amplitude above 2.0 Hz. Therefore, mode correlation above 1.5 Hz was very difficult.

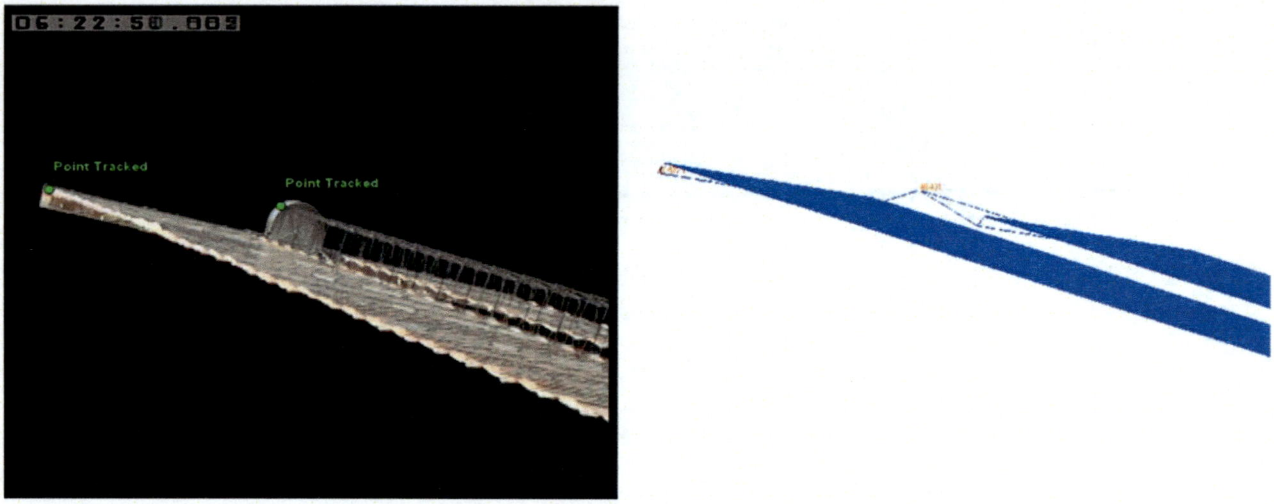

Fig. 22.16 Tracked points of 1A, *Left*: camera image, *Right*: PATRAN™ image

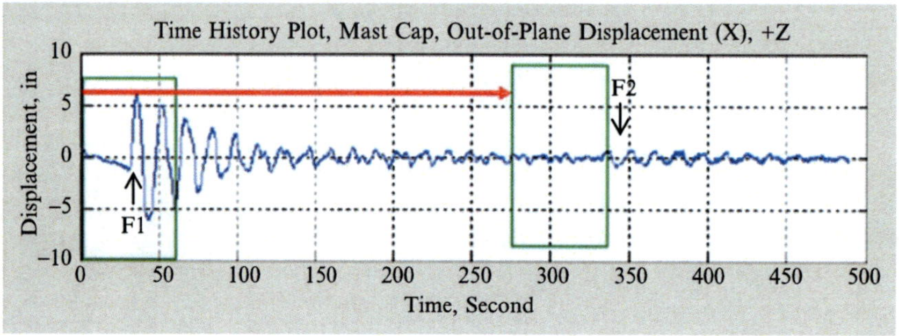

Fig. 22.17 Time history, 1A array mast cap, out-of-plane, detrended

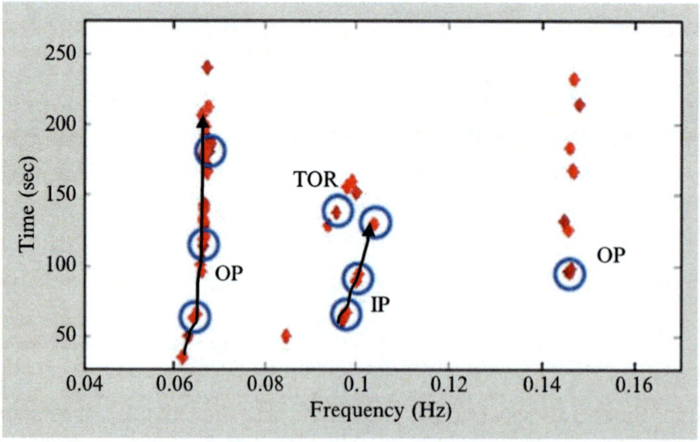

Fig. 22.18 BTACS Sys ID result for 1A photo-g data of S4-1A DTF F1

The photogrammetry data that was collected on the 1A solar array wing, attached to starboard truss segment S4, using ISS ETVCG cameras was analyzed independently from the accelerometer data. The Image Science and Analysis Group at NASA JSC received the analog video and, using their image processing software, created discrete displacement time history data sampled at 30 Hz. Modal parameters were extracted from this data and was used to assess the frequency and structural damping of the fundamental US PV Array modes.

Table 22.5 BTACS sys ID results/MAC for 1A photo-g data of S4-1A DTF F1

Mode description	On-orbit			Analytical			
	Time	Freq (Hz)	% Damping	Mode #	Freq (Hz)	MAC	% Freq. Dif.
OP	Beg	0.064	8.58	16	0.067	0.975	3.8
–	Mid	0.066	1.90	–	–	0.964	0.7
–	End	0.068	0.34	–	–	0.974	−1.1
IP	Beg	0.097	2.96	20	0.096	0.953	−1.4
–	Mid	0.100	2.05	–	–	0.933	−3.9
–	End	0.104	3.22	–	–	0.975	−7.5
TOR	Constant	0.096	0.81	30	0.1	0.984	4.5
OP	Constant	0.146	0.64	60	0.151	0.955	3.6

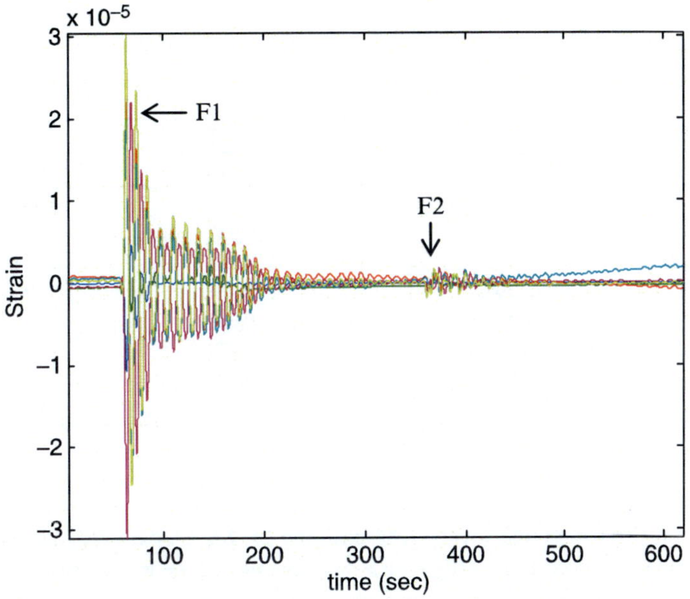

Fig. 22.19 SDMS MTS strain gage time history data S4-1A DTF

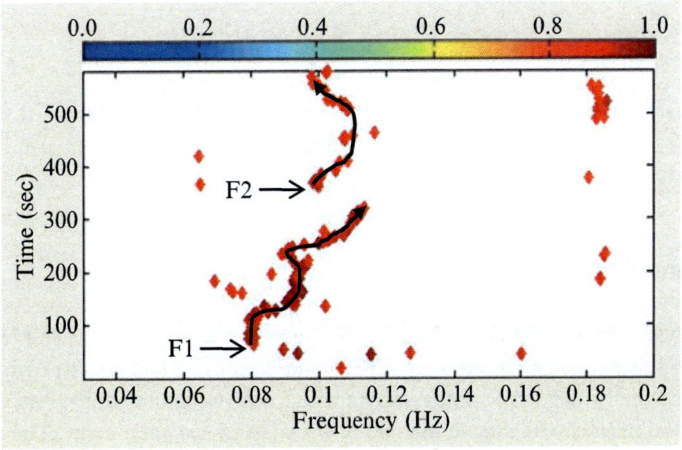

Fig. 22.20 BTACS Sys ID result, EMAC>85% for MTS strain gage data of S4-1A DTF

Table 22.6 BTACS Sys ID results for MTS strain gage data of S4-1A DTF F1

Time (s)	Freq (Hz)	Damping	EMAC (%)
75	0.079	1.26	92.7
99	0.080	1.69	93.7
123	0.082	3.67	94.8
135	0.081	6.82	95.4
159	0.093	5.45	97.1
189	0.094	7.61	96.8
210	0.095	5.22	93.9
216	0.097	5.09	95.8
282	0.109	2.51	94.4
303	0.111	1.77	94.4
318	0.113	1.1	93.4
321	0.114	0.33	91.7

Table 22.7 ISS stage ULF4 S4-1A DTF test/test correlation

				Accel group				
Firing 1 yaw		Firing 2 pitch		All	w/o IMU-D	Modules	Truss	Freq
Mode #	Freq	Mode #	Freq	MAC	MAC	MAC	MAC	% dif
1	0.084	1	0.1	0.977	0.977	0.987	0.969	18.4
2	0.104	1	0.1	0.985	0.987	0.99	0.984	−3.7
3	0.186	2	0.182	0.913	0.918	0.741	0.929	−2.5
8	0.308	5	0.304	0.96	0.957	0.974	0.957	−1.3
11	0.575	7	0.56	0.644	0.737	–	0.837	−2.7
14	0.771	9	0.748	0.68	0.729	–	0.801	−2.9
19	0.948	11	0.912	0.549	0.777	–	0.829	−3.8
24	1.481	14	1.55	–	–	0.622	–	4.7

The SDMS MTS Strut Strain Gage data that was collected during the S4-1A data was also analyzed independently from the accelerometer data. This data was used to investigate a low frequency mode that exhibits nonlinear characteristics. The mode under investigation can range in frequency from 0.08 to 0.11 Hz depending on the amplitude of the response.

The following sections will give results for the photogrammetry array data, the MTS strain gage data and the accelerometer data of the S4-1A DTF.

22.6.3 Stage ULF-4: Dedicated Thruster Firing S4-1A: Results

The analytical model was created in detail to match the ISS configuration during the time of the S4-1A DTF. The visiting vehicles, the solar array rotary joints (SARJ) angles, the array angles and the ISS robotic arm location were all represented in the NASTRAN system model, Fig. 22.15.

22.6.3.1 S4-1A Photogrammetry Results

The Image Science and Analysis Group (ISAG), at NASA JSC received analog video taken during the S4-1A DTF from two ISS ETVCG cameras [18]. The cameras used were CP3 located on the S1 starboard truss segment and CP13 located on the US Lab. The ISAG used their image processing software to track two points at the end of the 1A Solar Array Wing, Fig. 22.16. One point was on the mast cap located at the end of the mast of the solar array. The second tracked point was on the tip of the blanket box at the end of the solar array. The motion of each point was tracked in each video sequence and used to compute the relative displacement of the SAW tip in each axis, defined by the plane of the array during the DTF. Due to the high mathematical correlation between the axial and out-of-plane (OP) motion, the calculations were conducted in a way

Table 22.8 Stage ULF4 S4-1A DTF: test/analysis correlation: accel data

Test data					Analytical data					
Mode #	Freq. (Hz)	Damp (%)	EMAC CMI	F1 – F2 MAC	Mode #	Freq. (Hz)	Freq. Diff (%)	Accel Group	MAC	Mode description
2-F1	0.104	4.011	79/61	0.985	44	0.113	9.3	ISEMLJ	0.862	Station XY/module & truss
1-F2	0.1	2.903	87/69		44	0.113	13.5	ISEMLJ	0.847	"
1-F1	0.084	3.137	94/85	0.977	44	0.113	34.4	ISEMLJ	0.839	"
3-F1	0.186	1.107	83/47	0.931	83	0.172	−7.9	ISEMLJ	0.889	Truss XY bending/module XZ
2-F2	0.182	1.416	94/69	0.931	83	0.172	−5.5	ISEMLJ	0.933	"
3-F2	0.221	2.626	89/32		102	0.225	1.6	SE	0.731	Module & truss XZ
					102	0.225	1.6	ISE	0.669	"
7-F1	0.259	1.36	83/48		125	0.271	8.1	ISEMLJ	0.782	Station XY (JEM XY)
								IMLJ	0.873	"
8-F1	0.308	0.394	90/25	0.96	142	0.285	−7.5	ISEJ	0.789	Station XZ
5-F2	0.304	1.249	97/23		142	0.285	−6.3	ISEJ	0.811	"
6-F2	0.404	1.464	82/3		175	0.368	−9	ISEMLJ	0.705	Truss YZ/module TOR X/HRS Torsion
7-F2	0.56	1.639	76/19	0.742	254	0.494	−11.7	S	0.823	Station TOR X/truss XYZ
11-F1	0.575	1.157	84/58		254	0.494	−14	S	0.746	"
8-F2	0.684	1.13	84/58		303	0.666	−2.6	I	0.885	RSA module XZ/COL JEM OB Truss YZ
14-F1	0.771	0.484	87/37	0.741	313	0.719	−6.7	IS	0.83	RSA modules XY/COL JEM YZ/truss XYZ
9-F2	0.748	0.318	83/54		313	0.719	−3.9	IS	0.811	"
17-F1	0.853	0.627	80/57		354	0.86	0.8	IS	0.903	RSA module YZ/truss XYZ
16-F1	0.827	1.014	90/71		339	0.793	−4.2	IS	0.748	Station XY/EPS OP
18-F1	0.906	1.331	80/18	0.777	354	0.86	−5.1	IS	0.749	RSA module YZ/truss XYZ
11-F2	0.912				354	0.86	−5.7	S	0.869	"
14-F2	1.55	0.411	81/40		511	1.508	−2.7	I	0.83	COL JEM YZ/mocules XZ second
25-F2	2.901	0.856	76/44		833	2.78	−4.2	IML	0.763	Module XZ (third bending)/truss XY
56-F2	4.572	0.31	84/17		962	4.109	−10.1	SE	0.703	US LAB N2 XY/RSA XZ/truss XYZ
57-F2	4.556	0.06	81/49		962	4.109	−9.8	IS	0.753	"

Note: S-SDMS, I-IWIS, E-EWIS, M-IMU-D, L-SAMS Lab, J-SAMS JEM

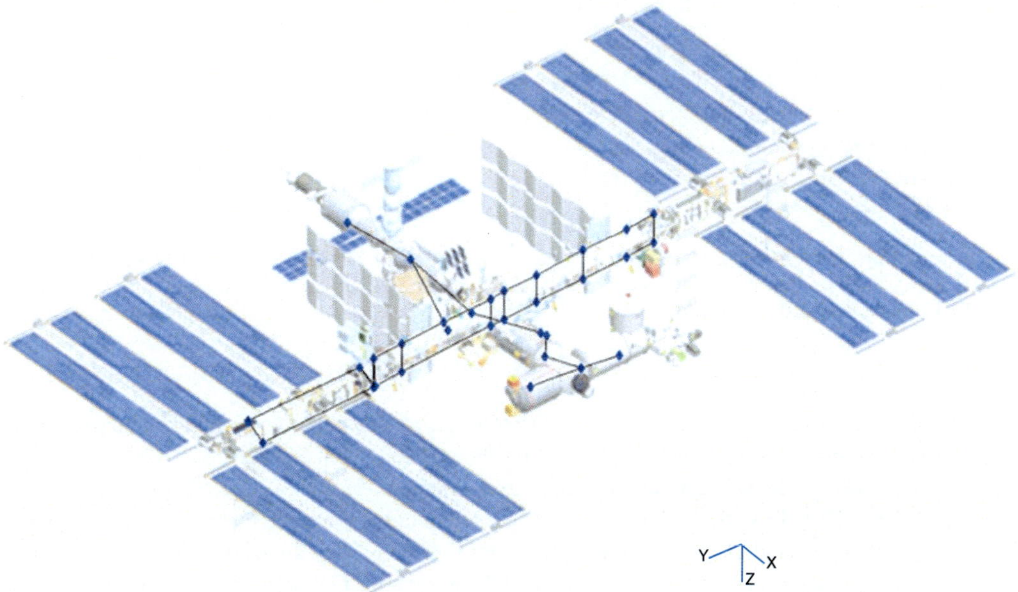

Fig. 22.21 Sensor location mode shape diagram

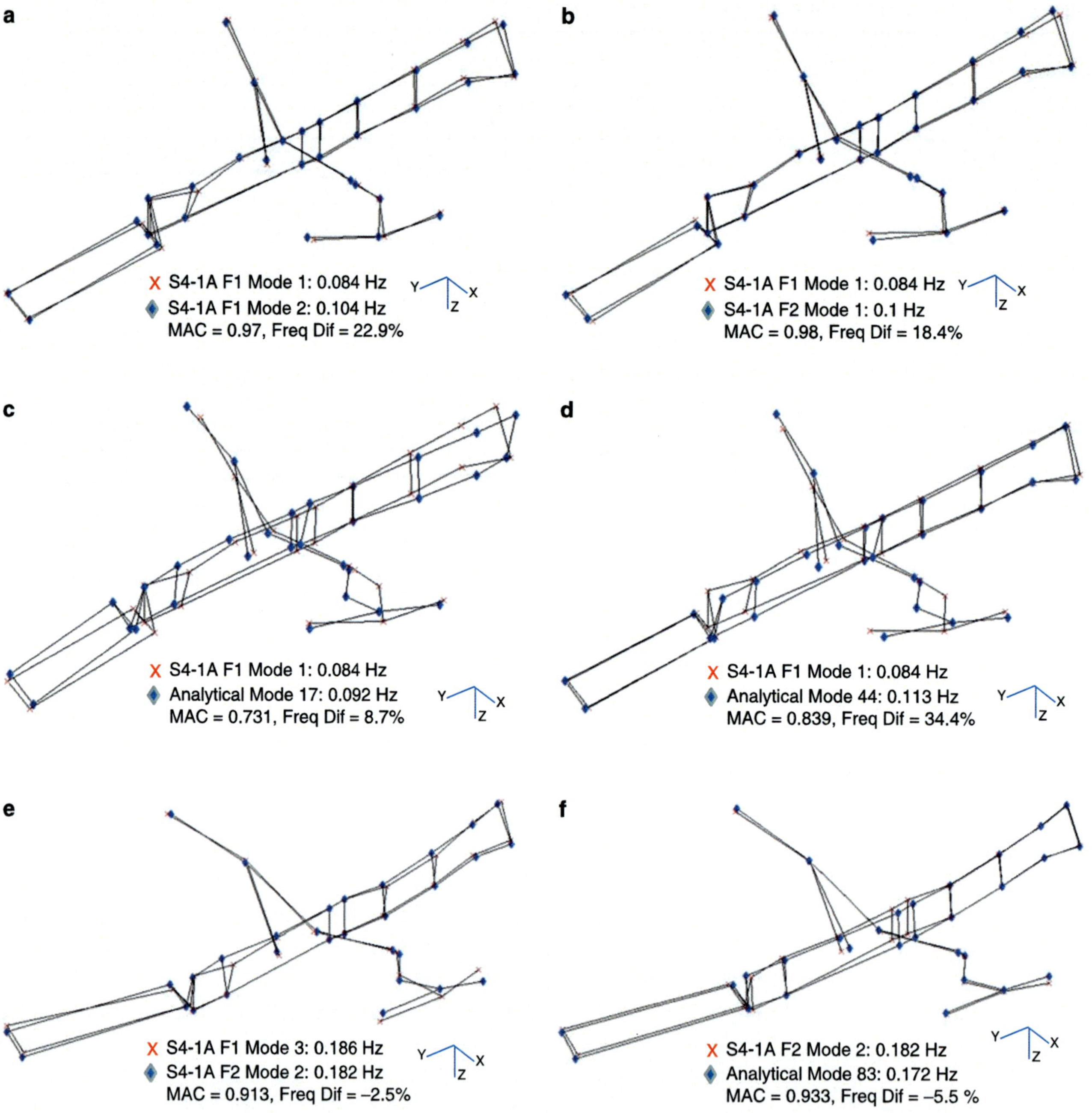

Fig. 22.22 Mode shape overlays, test/test and test/analytical (**a** through **f**). (**a**) S4-1A F1 Mode 1 vs. S4-1A F1 Mode 2, (**b**) S4-1A F1 mode 1 vs. S4-1A F2 mode 1, (**c**) S4-1A F1 mode 1 vs. analytical mode 17, (**d**) S4-1A F1 mode 1 vs. analytical mode 44, (**e**) S4-1A F1 mode 3 vs. S4-1A F2 mode 2, and (**f**) S4-1A F2 mode 2 vs. analytical mode 83

which constrains the axial position to a fixed value of 0. This constraint is acceptable given that the motion of the array in the axial direction is significantly less than the in-plane (IP) or out-of-plane (OP) motions.

The displacement time history data was detrended and the modal parameters were extracted and analyzed using the system identification tool of BTACS. The BTACS system identification tool includes a method that extracts modal parameters within a set window over a prescribed period of time within the data set, Fig. 22.17. The first firing set, Yaw direction, plumed the 1A array and provided the best data set for identifying the modal parameters. The second firing set, Pitch direction, was also analyzed but did not provide as high a quality results as the data from the first firing. The first two modes identified in the Yaw firing data showed an increase in frequency during the free decay of the array motion. The frequency shift overtime

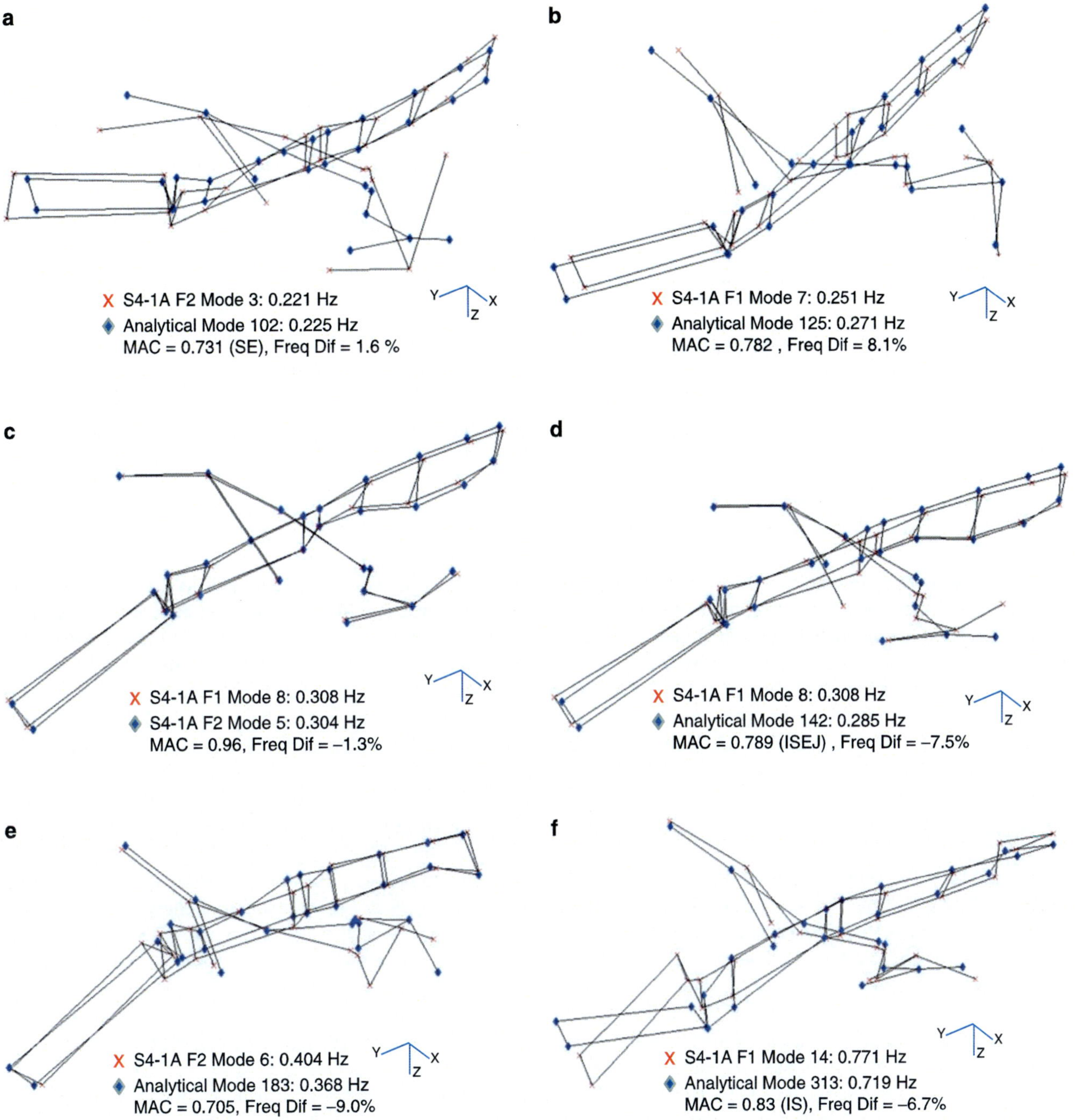

Fig. 22.23 Mode shape overlays, test/test and test/analytical (**g** through **l**). (**g**) S4-1A F2 mode 3 vs. analytical mode 102, (**h**) S4-1A F1 mode 7 vs. analytical mode 125, (**i**) S4-1A F1 mode 8 vs. S4-1A F2 mode 5, (**j**) S4-1A F1 mode 8 vs. analytical mode 142, (**k**) S4-1A F2 mode 6 vs. analytical mode 183, and (**l**) S4-1A F1 mode 14 vs. analytical mode 313

can be seen in the BTACS system ID window in Fig. 22.18. The comparison results between the extracted modal parameters from the data and the analytical modal parameters are summarized in Table 22.5. The first OP mode had a frequency range from 0.064 to 0.068 Hz. The first IP mode had a frequency range of 0.097–0.104 Hz. There was also an array torsion mode identified at 0.096 Hz which is closely spaced with the first IP mode and these modes appear to interact with each other. There was also a second OP mode identified at 0.146 Hz. The results of the S4-1A photogrammetry analysis, along with the analysis conducted on other arrays, will aid in reducing the restrictions in the solar array constraint matrices; used by ISS ground personnel to select appropriate array angles to park solar arrays prior to a dynamic on orbit event (i.e. vehicle docking).

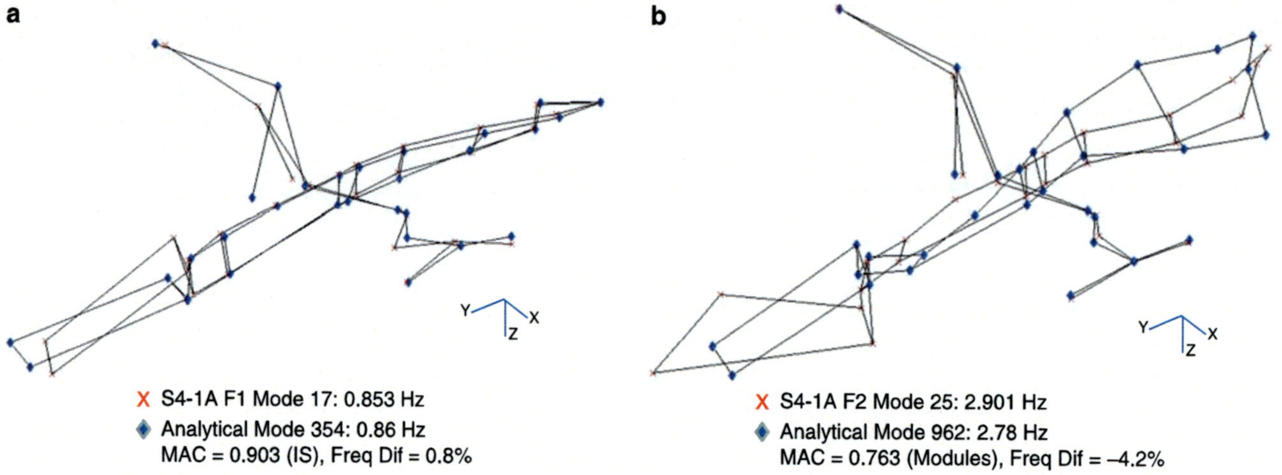

Fig. 22.24 Mode shape overlays, test/test and test/analytical (**m** through **n**). (**m**) S4-1A F1 mode 17 vs. analytical mode 354 and (**n**) S4-1A F2 mode 25 vs. analytical mode 833

22.6.3.2 S4-1A MTS Strut Strain Gage Analysis

In 2011 a maneuver was conducted that produced loads that were 25% higher than predicted loads. An investigation into the event found there was controller-structure-interaction; the controller was amplifying an excited structural mode. The mode in question had previously been identified with on orbit data and was found to correlate well with the shape of the corresponding analytical mode, having a MAC >0.9 repeatedly, but could range in frequency difference from 2 to 18%, even during the same ISS stage. The mode is an ISS XY global mode, where the truss and pressurized modules make a scissor like motion about the module-to-truss structure (MTS) struts. The MTS struts connect the pressurized modules to the truss segments of the ISS. The analytical reconstruction of the event produced high bending loads at the Node 1 to Lab interface, consistent with high MTS strut loads. SDMS MTS Strut strain gage data, for several on orbit events, was analyzed using the BTACS system identification tool which includes a method that extracts modal parameters within a set window over a prescribed period of time within the data set. This section will focus on the results of the data collected during the S4-1A DTF.

The SDMS MTS Strut strain gage time history data, bandpass filtered 0.03–0.2 Hz, for the S4-1A DTF is shown in Fig. 22.19. The first firing is a yaw firing which excited the low frequency ISS XY mode with more energy than the pitch firing that followed. A visual of the frequency results of the system identification analysis is shown in Fig. 22.20. The modal parameters were extracted from 12 time periods throughout the first firing. A modal assurance criterion (MAC) was calculated between each extracted mode. The MAC values between each extracted mode were all above 0.92; confirming that the tool was identifying the same mode over time with a changing frequency. The frequency of the mode shows a dependence on the magnitude of the data that was used for the extraction. The frequency, damping and EMAC value of the extracted modes is presented in Table 22.6: BTACS Sys ID Results for MTS Strain Gage Data of S4-1A DTF F1.

The investigation into the SDMS MTS strut strain gage data, which incorporated several data sets including the S4-1A data, concluded that the mode was nonlinear. Being a nonlinear mode, the frequency of it's free vibration is dependent upon the free vibration amplitude. This amplitude dependent frequency characteristic was seen in the BTACS system identification results where the frequency of the mode changed 44 % over the decay of S4 1A DTF Firing 1. The findings of the MTS Strut Strain Gage data analysis will feed into updating the thruster control algorithm to avoid controller-structure-interaction.

22.6.3.3 S4-1A Model Correlation

Using the Boeing MoReID Global ERA tool, modes were extracted from the combined accelerometer data sets of the S4-1A DTF Firings 1 and 2. Each thruster firing set was treated as a separate on orbit data test. The first step in the correlation effort was to compare the test modes extracted from Firing 1 with Firing 2 with the intent to show the consistency of the modes extracted from each of the dynamic responses. Due to the different types of thruster firings there were some modes that were excited by one firing set and not the other. The MAC was calculated using all of the accelerometers, all the accelerometers minus the IMU-D sensor, using only the pressurized module accelerometers and then using only the truss accelerometers. The MAC was calculated using different sets of accelerometer groups in order to determine the effect each sensor group had on the overall MAC value between the two firing sets; reason being the IMU-D and SAMS sensors have not been used for the purpose of model correlation before. The Firing 1 test mode 1 and 2 are the same mode, extracted twice, at different times within the data set, the overlay of the two mode shapes is presented in Fig. 22.22a. The difference in frequency of this mode is due to the dependence this mode has on free vibration amplitude, as discussed in Sect. 22.6.3.2. In general, the results shown in Table 22.7 show good MAC correlation for the modes that were extracted with a high level of confidence.

The test/analytical correlation for the ISS Stage ULF4 S4-1A DTF configuration is shown in Table 22.8. The test modes that were excited from both firings and showed good correlation among themselves showed good MAC values with the analytical model modes. The modes that were excited well in one firing and not in the other firing, modes with higher frequency and mostly confined to one coordinate plane, also showed good MAC values when compared to the analytical model. The range of frequency differences of the first mode is linked to the nonlinear behavior seen in the on-orbit test mode, where the analytical model is a linear model and does not capture the nonlinear behavior of this mode. A diagram of an extracted mode shape overlaying the ISS structure is shown in Fig. 22.21. That diagram is to aid in visualizing the other mode shape comparisons displayed in Figs. 22.22a through f, 22.23g through l, and 22.24m through n.

22.7 Conclusions

On-orbit structural dynamic data was collected during the ISS Stage ULF4 S4-1A DTF using a variety of instrumentation systems. The main intent of these analyses was to measure dynamic responses of the ISS in order to validate and correlate analytical models. The main instrumentation systems that were utilized during the data collections included Internal Wireless Instrumentation System (IWIS), External Wireless Instrumentation System (EWIS), Structural Dynamic Measurement System (SDMS), Space Acceleration Measurement System (SAMS), Internal Measurement Unit (IMU) and ISS EVTCG Cameras. The photogrammetry data of the 1A Solar Array Wing was used to increase the confidence in the frequency and damping of the fundamental array modes, which will feed into the development of less restrictive solar array constraint matrices. The SDMS MTS Strut strain gage data was used to investigate a low frequency mode whose frequency is dependent on the free vibration amplitude. The results of that study will feed into the update of the thruster control algorithm to avoid controller-structure-interaction. Modal analysis was performed on all measured accelerometer data to extract modal parameters including, frequencies, damping factors, and mode shapes. An analytical math model was developed to simulate the on orbit configuration of the ISS during the time of the S4-1A DTF. The developed models incorporated, as closely as possible, the actual on-orbit configurations including array angles, boundary and interface conditions. The identified test modes were correlated and compared to analytical modes to verify the accuracy of analytical model.

References

1. Costello T (2004) ISS instrumentation to validate the integrated math model. SSCN 008631 revision B, Sept 2004
2. DAC-8 structural loads report: on-orbit transient loads and loads spectra. Boeing report D684-10019-02-01-02, Revision E, Mar 2000
3. Fitzpatrick K, Grygier M, Bartkowicz T (2009) ISS stage 12A post-flight modal analysis, model validation and correlation. In: IMAC-XXVII, Orlando
4. Flight-4A post-flight analysis: modal analysis, model validation and correlation. Boeing MSER memorandum A92-J332-STN-M-MK-2001-0080, June 2001
5. Foster R (2005) EWIS design overview, MSER Presentation, June 2005
6. Guyan RJ (1965) Reduction of stiffness and mass matrices. AIAA J 3(2):380
7. Integrated loads and dynamics verification plan. Boeing report No. D684-10288-01, Rev. B, Aug 1999
8. Internal wireless instrumentation system (IWIS) interface definition document. JSC 28369, Rev. D, Engineering Directorate Avionic Systems Division, NASA-JSC, Oct 2000
9. ISS 11A – 18P post-flight analysis: modal analysis, model validation and correlation. Boeing MSER memorandum ISS-HOU-MSER-SIM-050177, Oct 2005
10. ISS 5A-UF1 post-flight analysis: modal analysis, model validation and correlation. Boeing MSER memorandum J332-2003-0161, June 2002
11. ISS 7P – 5S post-flight analysis: modal analysis, model validation and correlation. Boeing MSER memorandum J332-2003-0163, July 2003
12. ISS stage 12A.1 post-flight analysis: model analysis, model validation and correlation. Boeing Engineering Information Document EID684-13643, Sept 2009
13. ISSA PG-1 On-orbit instrumentation for SDMS, Rev.1. Boeing memorandum A95-J012-SEM-M-9500771R1, May 1995

14. Juang JN, Pappa RS (1985) An eigensystem realization algorithm for modal parameter identification and model reduction. J Guid 8(5):620–627
15. Kim HM, Kaouk M (1998) Final report: mir structural dynamics experiment. The Boeing Company, Contract No. NAS15-10000, Dec 1998
16. Kim HM, Bartkowicz TJ, Van Horn DA (1994) Data recovery and model reduction methods for large structures. Finite Elem Anal Des 16(2):85–98
17. Kim HM, Van Horn DA, Doiron HH (1994) Free-decay time-domain modal identification for large space structures. J Guid 17(3):513–519
18. McNeill SI (2006) Analysis of INC 14 SAW photo-G data. Boeing MSER memo ISS-HOU-MSER-SIM-060212, Dec 2006
19. Space station structural loads control plan. Boeing report No. D684-10019-1, Mar 1994
20. Station development test objective (SDTO) catalog. SSP 50448, Rev. C, NASA-JSC, Mar 2007
21. Targoff WP (1976) Orthogonality check and correction of measured modes. AIAA J 14(2):164–167
22. Wellner N Request photogrammetry of P4 (2A & 4A) solar array tips during SDTO maneuver on GMT 325. Chit ISS 004361, NASA JSC
23. Wilson B (2002) Mission 11A STS-113 reboost#2 IMU data processing results. Lockheed Martin memo 3HEC-ES-2LX-039, Dec 2002

Chapter 23
Numerical Modeling of Vibration Induced Atomization of Liquids

Jesi Ehrhorn and William Semke

Abstract The numerical modeling of vibration induced atomization of liquids validates existing operating parameters of known systems. A computational fluid dynamics analysis is performed which assists in model verification and reveals a critical configuration of driving amplitude and liquid depth that must be fulfilled for atomization. In this configuration the droplet kinetic energy exceeds the fluid resistance energy and the atomization process initiates. Verification and identification of the parameters for atomization of a thin film of water is accomplished.

Existing literature on the operation of vibrating mesh nebulizers does not entirely explain the principles by which these devices atomize liquids. Many previous studies assume a spray or extrusion mode of droplet generation, but it can be demonstrated that the high frequency vibration of these devices is sufficient to produce aerosol droplets. A thin film of liquid vibrated under the correct conditions will produce a fountain of atomized liquid droplets. The formation of standing waves on the surface of a thin film have an oscillating frequency that is half the driving frequency, a wavelength that is equal to a function of the driving frequency, a mean droplet diameter one-third the standing wavelength dimension, and are also dependent of fluid density and interfacial surface tension.

Keywords Nebulizer • Numerical modeling • Atomization • Vibrating mesh • Computational fluid dynamics

23.1 Introduction

This research project is to define the conditions at which atomization initiates for vibrating mesh nebulizers. By defining the conditions at which atomization will occur, it will be possible for future studies to work towards optimizing the design of a vibrating mesh nebulizer in terms of cost and efficiency. This study explains a theory which proposes that a vibrating mesh nebulizer produces an aerosol fountain of atomized particles simply by applying ultrasonic excitation to a thin film of liquid medication. Liquid drugs used for aerosol therapy are typically water-based and thus the properties of the majority of these fluids can be approximated as liquid water at room temperature.

Current literature on the performance of vibrating mesh nebulizers appear to make the claim that the devices operate under a principle similar to an atomizing spray nozzle. To be more precise, this mode of operation would make it necessary for a high-velocity "filament" of liquid medication to be extruded from each orifice and undergo primary and secondary droplet breakup processes before a relatively uniform spray of droplets of appropriate size is realized. In general, this is actually a reasonable theory as it is easy to visualize and a considerable amount of work has been done in the area of atomizing spray nozzles, as well as the particle breakup phenomenon in relation to micron-order droplet sprays. A more simplified theory often referred to as a "micropump" mode of operation, can be described as follows; a continuous body of liquid medication exists between the vibrating actuator and the orifice plate. The actuator's vibration transmits a displacement to the liquid medication, which pumps it through the orifice plate in a periodic fashion.

A new proposed theory detailed in this paper is the ejection of micro-droplets via standing surface waves, commonly referred to as capillary waves. The current study demonstrates that it is quite probable that the primary contributor to the generation of micro droplets is the free surface vibration of a small volume of liquid. Many studies have clearly shown that

J. Ehrhorn • W. Semke (✉)
Department of Mechanical Engineering, School of Engineering and Mines University of North Dakota, Grand Forks, ND 58202, USA
e-mail: william.semke@engr.und.edu

T. Simmermacher et al. (eds.), *Topics in Model Validation and Uncertainty Quantification, Volume 5*, Conference Proceedings of the Society for Experimental Mechanics Series 41,
DOI 10.1007/978-1-4614-6564-5_23, © The Society for Experimental Mechanics, Inc. 2013

thin volumes, or films, of various types of liquids exhibit a very pronounced surface wave phenomenon when a periodic displacement excitation is applied. A few studies have confirmed a relationship between the driving frequency of this excitation, the properties of the working fluid, and the wavelength of the resulting capillary waves. One respected theory has survived for over four decades which relates the observed surface wavelength to the mean diameter of ejected droplets, when the conditions for atomization are met or exceeded. Nearly all of these studies have dealt with working frequencies orders of magnitude below that of current vibrating mesh nebulizers, but it has been stated and some evidence has been provided that the mechanics involved will "scale up" to higher frequencies.

Within the primary focus of this study, which is to elucidate a physical mechanism for micro droplet formation in vibrating mesh nebulizers, is a proposal of the necessary conditions by which atomization will occur. It is shown that a balance between the kinetic energy of a single droplet and the surface tension energy along with a viscous resistance can be met or exceeded in order to produce conditions sufficient for droplet ejection. The methods for testing this hypotheses include analytical calculation based on equations from existing literature and computational fluid dynamics.

23.2 Literature Review

Several works of literature were produced during an era surrounding the 1950s on the potential mechanisms of vibration-induced liquid atomization. Most of these studied focused on finite vibrating liquid films of varying thickness and the ejection of droplets from the peaks of surface waves caused by reaching unstable amplitudes or some other condition sufficient for rupture. In the case of an atomizer with a relatively low flow rate, the surface waves appear to "stand" in place while oscillating. Thus, the term "standing surface waves" is appropriate for the system. A "capillary wave" is technically a wave that travels across a surface (i.e. ripples), but the waves produced by ultrasonic atomizers are often referred to as capillary waves. It is safe to assume that the waves traverse the surface of a small volume of liquid, potentially reflecting at boundaries and experiencing some other forms of bulk motion, albeit at a velocity much lower than that of the wave oscillation and is therefore negligible for most theoretical work.

In the paper "Ultrasonic Atomization of Liquids" by Robert J. Lang [1], published in 1962, one of the first claims of droplet size being directly proportional to capillary wavelength is made. An equation, often referred to as Kelvin's equation for capillary wavelength, relating the two parameters in accordance with the working liquids properties is as follows:

$$\lambda^3 = \frac{2\pi\sigma}{\rho f^2} \tag{23.1}$$

where λ is the surface wavelength in meters, σ is the surface tension coefficient in N/m, ρ is the liquid density in $\mathrm{kg/m^3}$, and f is the surface standing wave frequency in Hz. The equation is reduced and the observed phenomenon of surface vibration frequency being half that of the base driving frequency (F) is taken into account, resulting in:

$$\lambda = \left(\frac{8\pi\sigma}{\rho F^2}\right)^{\frac{1}{3}} \tag{23.2}$$

The use of half the base excitation frequency in determining the capillary wavelength results from the base excitation being transmitted through the fluid body. Each positive-amplitude displacement of the actuator drives each rise of the capillary wave peaks, which are alternating evenly across the surface. This implies that the actuator is at its lowest position when the free surface is between peaks of oscillation, i.e. the surface appears flat. Thus, positive surface oscillation amplitude peaks for either "set" of waves occur on every other positive actuator amplitude peak. The mathematical derivation of this phenomenon is discussed at length in the paper "The Stability of the Plane Free Surface of a Liquid in Vertical Periodic Motion" by Benjamin and Ursell [2], published in 1954; the basis of which is the experimental studies of Faraday [3] and Lord Rayleigh [4], among others.

Further, Lang proposes that the mean diameter of ejected droplets is approximated by the relation:

$$D = 0.34\lambda \tag{23.3}$$

This proportionality constant was obtained via an experiment in which a molten liquid wax is atomized and the generated droplets rapidly cool in air, which are then collected and measured. In its molten state, the liquid wax is similar in fluid properties to oils used in other atomization studies.

Finally, Lang states that particle size tends to increase along with atomization rate and that these results were generated at low atomization rates. The reason for this discrepancy is most likely the result of the conglomeration of smaller droplets into larger droplets immediately after atomization, simply due to the inference that a higher density fog is more likely to have its constituent droplets collide before spreading away from one another.

An aspect of standing surface waves worth considering is if the wave formation or shape is influenced by gravity. It is generally accepted that at a fluid-fluid interface, surface waves shorter than a third of a certain critical wavelength are not significantly influenced by gravity [5]. This critical wavelength can be described as a condition where the effects of surface tension and the acceleration of gravity are equally responsible for the wave motion and shape. The equation for this wavelength is as follows:

$$\lambda_m = 2\pi\sqrt{\frac{\sigma}{g(\rho - \rho')}} \tag{23.4}$$

Here, σ is the interfacial surface tension, ρ is the density of the heavier fluid and ρ' is the density of the lighter fluid For a water-air interface at room temperature, the wavelength at which gravity loses significant influence is a third of 17 mm, approximately 5.7 mm. All systems referenced in the current study do not generate waves this length or longer. Another early work of literature describing the formation of droplets from standing surface waves from a more theoretical standpoint is "Ultrasonic Atomization of Liquids" by Peskin and Raco [6], published in 1963.

Technological advancements in high-speed photography led to further experimental study on ultrasonic atomization in the 1990s. Photographs of initiating ejected droplets were captured in many of these studies, but even today there are practical limitations on camera speed when attempting to capture these effects for very high frequency ultrasonic cases. Thus, the images available are of higher frequency cases than what was possible mid-century and are of much higher resolution, but the scaled-up model approach must still be used. The most important evidence generated from contemporary photographic study is clear sequences of still photos showing the detachment of droplets from the tips of surface waves [7].

The paper "Theoretical and experimental study of the behavior of liquid film free surfaces driven by transverse ultrasonic vibrations" by Sindayihebura and Bolle [8], published in 1995, attempts to shed some light on the evolution of surface wave patterns during atomization. It is stated that resonance of the free surface occurs at some critical amplitude of driving surface excitation, and that the shapes of the surface vibration modes are independent of container boundary shape at any frequency high enough for producing atomized droplets. Another important point is that the early wave formation can generally be approximated linearly, while the progression of these weak surface waves to waves capable of ejecting droplets is a highly nonlinear phenomenon with the droplet ejection itself thought to be a chaotic phenomenon. Theoretical calculation in this paper shows that the excitation of unstable modes of vibration depends upon a dimensionless quantity relating the actuator acceleration to gravity, known as the Froude number (*Fr*). The equation is given as:

$$Fr = \frac{a_o\omega^2}{g} \tag{23.5}$$

where a_o is the driving (actuator) displacement and ω is its angular velocity. So, there is a certain critical Froude number below which the unstable modes cannot be excited. The value of the critical Froude number increases with both fluid viscosity and excitation frequency. It is once again observed that an increase in driving frequency causes the surface wavelength to decrease, further validating past studies.

The phenomenon of wave shapes or configurations as related to the container wall shape is further described in this paper. At relatively low forcing amplitudes, it is very difficult to capture a good photograph of the wave pattern, but the observed shape is similar to that of the container (i.e. a circular container producing axisymmetric wave forms). As the driving amplitude increases, there is a chaotic transition state, then the boundary-independent square and hexagonal arrangements are observed, which tend to be much more clearly defined due to the critical amplitude being reached, and thus the excitation of surface resonance modes. The chaotic transition is most likely the superposition or interference between container dependent and independent wave shapes. After the driving amplitude is increased to show the container-independent orderly wave patterns, further increases result in the onset of atomization. A very interesting hypothesis provided for the approximately square standing wave configuration is the notion that these shapes occur due to the intersection of perpendicular plane wave modes. It follows that some form of polarization exists in the bulk body of water which guides the primary orientation of these two-dimensional waves and that any similar three-dimension system will exhibit the same phenomenon. Thus, careful consideration must be taken when analyzing the system in two dimensions. Finally, it is noted that much more complex surface patterns are observed after atomization commences, which is believed to be largely due to nonlinear and/or chaotic effects originating from boundaries or other localized regions within the fluid body, including additional complex surface disturbances.

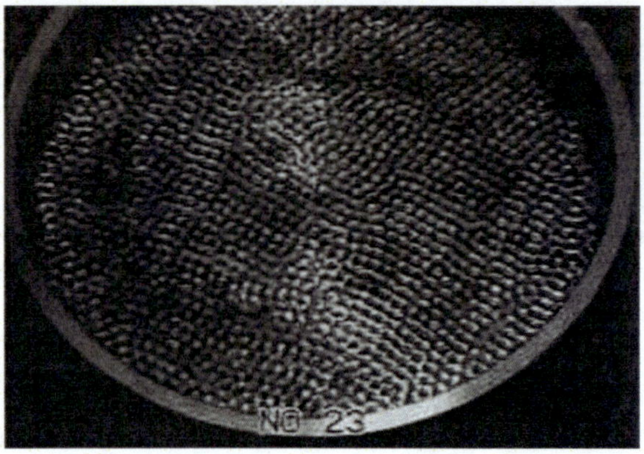

Fig. 23.1 300 Hz vibrating tray setup at atomization inception [7]

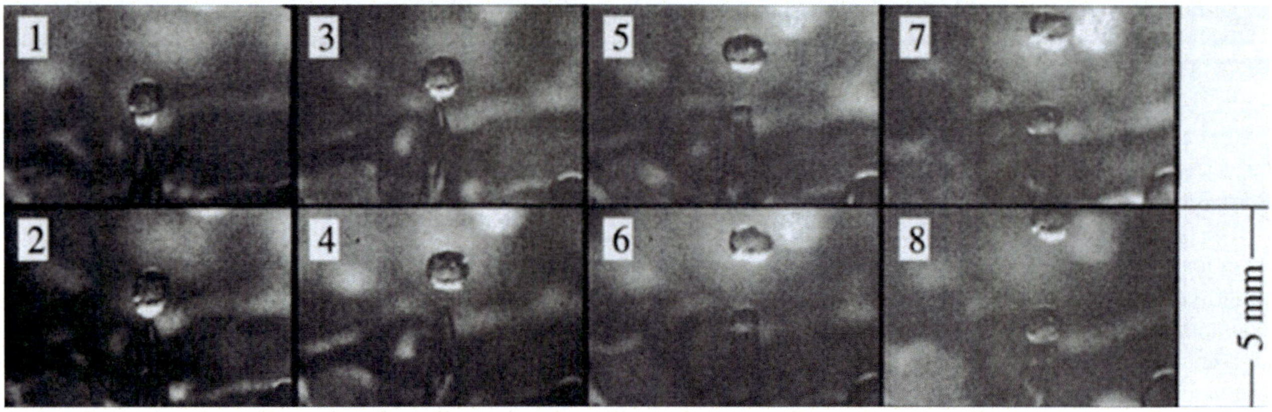

Fig. 23.2 Frame-by-frame droplet ejection from a capillary wave tip [7]

Further contemporary work is carried out by Yule and Al-Suleimani in their paper "On droplet formation from capillary waves on a vibrating surface" [7], published in 1999. The scaled-up system set up by Yule and Al-Suleimani is of great importance to the present work which will be described in detail later. Their experimental setup, shown in Fig. 23.1 consists of a circular tray of liquid water that is approximately 40 mm in diameter. It should be noted that the exact size of the container is not nearly as important as the fluid depth (film thickness), which in this case is 2 mm. The driving amplitude is higher than the critical amplitude for capillary wave formation, and therefore the waves are independent of the container and can be analyzed in terms of expected and observed wavelengths. In this case, the driving frequency is significantly lower than an ultrasonic case at only 300 Hz, but the system is scaled up to provide a more easily observed example of vibration-induced droplet formation. The driving amplitude of vibrating container is 0.1 mm.

Exceptional photographs were taken of a vibrating surface clearly showing droplet formation from capillary waves (Fig. 23.2).

The remainder of the paper attempts to explain why the droplet formation phenomenon is not consistent and orderly, as it is observed that a single wave cell does not continuously produce droplet after droplet. Several oscillations of the surface occur before another droplet is ejected from the unit area, during which time the cell undergoes a form of chaotic sloshing and a redistribution of liquid volume to neighboring cells. It is also observed that the lack of wave momentum which occurs prior to the next expected droplet ejection may cause a "near miss" in which a ligament stretches far enough upward but inevitably falls back downward. In this case, the following oscillation generates an excess of momentum and a very long ligament which may break up into two or three droplets of various sizes. Remembering that the droplet ejection process is expected to scale up and even at these relatively low driving frequencies the described phenomenon are occurring on the order of hundreds of times per second, it is reasonable to assume that this is a good estimate of what causes the distribution of resulting droplet sizes.

Additional support for the calculation of capillary wavelength including a derivation which includes consideration of film thickness can be found in "Visualization and Analysis of Liquid Film Surface Patterns Formed on Ultrasonic Atomisers" by Dobre and Bolle [9], published in 1999. "Theoretical and experimental study of transducers aimed at low-frequency ultrasonic atomization of liquids" by Sindayihebura and Bolle [10] provides more background on this concept, as well as "How Orderly is Ultrasonic Atomization?" by Al-Suleimani, Yule, and Collins [11].

In "Motion of Droplets on Solid Surface Using Acoustic Radiation Pressure" by Alzuaga, Manceau, and Bastien [12], published in 2004, a very unique and important theory in the field of liquid droplet atomization is described. The authors have devised a set of equations for balancing an emerging droplet's kinetic energy with resistant energy from the fluid volume surrounding it. The primary restriction of droplet ejection in liquids such as water and water-based medications is the surface tension energy. This energy can be expressed simply as the surface tension coefficient multiplied by the surface area of a droplet to be ejected, as follows:

$$E_{st} = \sigma S_g \tag{23.6}$$

The resistant energy due to viscosity is computed by first assuming that the flow region for a single droplet is circular in cross section and its velocity distribution is similar to a laminar flow of the same diameter, where after some derivation results in:

$$E_{vis} = 32\pi\mu_o r_o^3 f \tag{23.7}$$

where μ_o is the dynamic viscosity coefficient of water at room temperature, r_o is the droplet radius and f is the surface oscillation frequency. As previously stated, the surface waves of a base-excited liquid film will tend to vibrate at a frequency that is half that of the base excitation frequency. The relationship between these frequencies has been observed experimentally in numerous studies, especially for cases involving lower audible frequencies. The droplet's kinetic energy is simply calculated as:

$$E_v = \frac{1}{2}\rho V_g v^2 \tag{23.8}$$

The quantities of energy of a droplet on the threshold of ejecting from the surface simply balance as:

$$E_v = E_{vis} + E_{st} \tag{23.9}$$

Since water exhibits high surface tension energy in comparison to its unit viscous energy, when calculated the viscous energy can often be left out of the balance because it is generally on the order of 1 % of the surface tension energy. Thus, the balance simplifies to:

$$E_v = E_{st} \tag{23.10}$$

An application study of the droplet-energy balance method for determining the onset of atomization can be found in "Enhanced water removal in a fuel cell stack by droplet atomization using structural and acoustic excitation" by Palan and Shepard [13], published in 2006.

23.3 Analytical Study

All calculations are based on the properties of water at room temperature (about 20–22°c). The surface tension is taken to be 0.072 N/m, the viscosity 0.001 N · s, the density $1{,}000\,\text{kg/m}^3$, and all length, depth, or displacement quantities are expressed as meters for all equations and simulations unless otherwise noted.

The system used for simulation purposes is assumed to be a representation of a wide cylindrical body of liquid water which is vibrated normal to its free surface and is bounded by either no-slip conditions or displacement periodicity on its outer boundaries. The size of the cylindrical body is calculated based on the expected wavelength given by Kelvin's capillary wave equation (Eq. 23.2) which is dependent on the base excitation frequency. Multiples of the wavelength quantities will be used for the size of the CFD simulation models discussed later.

In three dimensions, the shape of the system can be visualized as a relatively thin cylinder with its lower surface on a radial plane which is extruded in its axial direction. Figure 23.3 shows an isometric view of a solid model of the representative system and an outline of the boundary conditions considered. The boundary conditions include the displacement base excitation on the lower circular face as a time-variable sine function, an edge condition on the outer cylindrical face which is dependent on the choice of solver, and a free-surface condition on the upper circular face which is calculated via the Volume of Fluid (VOF) algorithm.

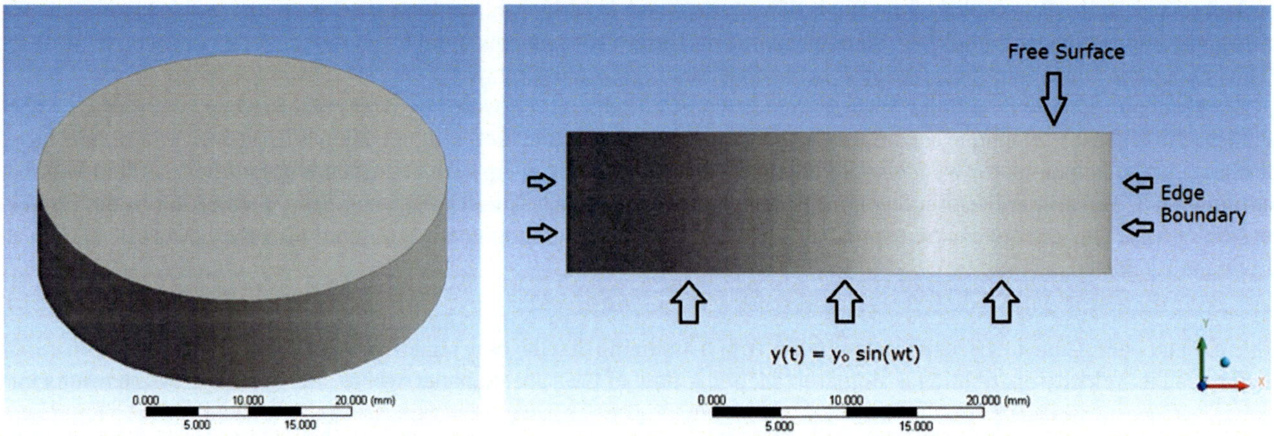

Fig. 23.3 3D CAD model of fluid body, isometric view and overview of CFD boundary conditions, 2D plane view

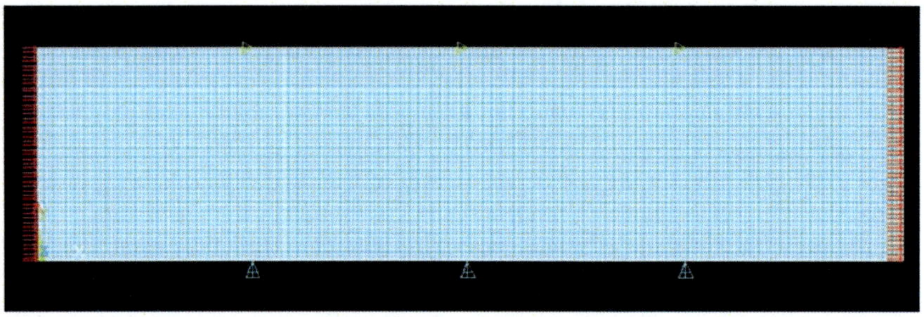

Fig. 23.4 2D planar Flotran model mesh and constraint indicators

A key parameter which is the primary distinguishing factor between the model cases is the operating frequency; the case studied in the proceeding analyses is a representation of the 300 Hz experimental setup of Yule et al. in which the excitation amplitude is clearly defined and the formation of droplets is described and confirmed.

The Flotran Computational Fluid Dynamics (CFD) solver has been included within the ANSYS engineering analysis software package for nearly two decades by the time of this writing. It is able to run simplified systems very efficiently and has been very useful for this study. The Flotran solver includes the option to utilize the Volume of Fluid (VOF) algorithm for analyzing fluid free surfaces. The algorithm is able to track the free surface by including elements that are full, empty, or partially filled. The partially filled elements are treated as polygons which are bounded by at least one completely full element and some combination of partially full or empty elements on the remaining sides.

The polygon is referenced as a volume fraction (denoted by Flotran as VFRC), which is simply the ratio between the size of its filled space and the size of a complete element. An important concept to keep in mind is that for Flotran analyses, only the properties of the working fluid are used for calculation. In more advanced solvers, a second fluid can be introduced as a boundary to the primary fluid. The VOF method is most commonly used for liquid water and gaseous air interactions, so any influence of the latter on the former is negligible and can simply be represented as a constant pressure for most applications.

In a Flotran analysis, the user must choose between 2D and 3D elements based on both the complexity of the model required and the limitations of the software. The primary concern of the analysis required for this study is the behavior of the free surface and because the geometry of the resulting surface waves are expected to be very small in relation to size of the bulk fluid volume, a very small portion or slice can be used effectively to yield appropriate results. Since the liquid volume is a simple film under base excitation which can be approximated as a short cylinder with a uniform surface response, it follows that a simple axisymmetric model can be used. This is of great significance because a 2D model is much easier to constrain and much less computationally demanding. In addition, the VOF option is only available for Flotran's 2D element, which is denoted as FLUID141 by ANSYS. This element provides basic flow data including velocity, temperature, and pressure in addition to free surface tracking.

A 2D axisymmetric model was used for all Flotran calculation. The image capture (Fig. 23.4) shows an example of the necessary mesh density and how the constraints have been placed. The left edge is constrained such that velocity in the

Fig. 23.5 2D planar Flotran model volume fraction location (*red* is filled with water, *blue* is empty space)

x-direction is always zero at this location, which is necessary for an axisymmetric model. The right edge is constrained in the same way, acting as a "free slip" boundary to minimize its effects on the system. The upper edge is constrained as a zero relative pressure boundary and the base is set to oscillate with a displacement in the y-direction.

This model in particular is a 300 Hz excitation case which corresponds to a system having a liquid depth of 2 mm. The expected capillary wavelength is about 2.5 mm for this low frequency system, and the model is sized at 10 theoretical wavelengths (25 mm) wide. The mesh edge length is set at 5 % of the expected theoretical wavelength (125 μm) which is a consideration noted in the ANSYS documentation to be appropriate for capturing the shape of a wave (20 elements per wavelength).

In Fig. 23.5, the lower red-colored region is the initial location of the filled liquid water elements while the upper blue-colored region is simply meshed empty space. A significant amount of empty space is necessary above the free surface so that any fluid motion that occurs during the simulation is contained within the meshed region. In a VOF analysis, any fluid contacting an undefined or simply defined boundary is taken to "spill into the environment" and is lost from the system. In this case, the upper boundary is considered simply defined because it is only a zero relative pressure condition. Applying zero velocity in the y-direction at this edge would contain liquid but would imply a non-slip wall, which is not an accurate portrayal of the real system. The most important observation of this study in terms of CFD analysis is the formation of capillary waves on the surface of the vibrating liquid film.

23.4 Results

The first step in the process of analyzing standing surface waves is to create a model that behaves as expected. Based on photography and sketches from previous work, the general appearance of such a system is known. The model generated showed surface waves forming and oscillating on the surface of a fluid body was a planar rectangular system without axisymmetry specified. The results showing fully developed standing surface waves generated can be seen in Fig. 23.6; in this case, the model is driven at 300 Hz with a base excitation amplitude of 0.1 mm, its depth (later denoted as thickness) is 2 mm, and its width (later denoted as model size) is 10 theoretical wavelengths. At 300 Hz driving frequency, a theoretical wavelength is 2.5 mm and therefore the model is 25 mm wide. The mesh edge length used is 0.125 mm and the time step is 0.1125 s.

Figure 23.6 is an example of a model with fully-developed surface waves; frame-by-frame observation of single wave points on the surface confirmed that the surface of this system was responding at around 120–150 Hz, which is very close to the theoretically expected response of exactly half the driving frequency. Velocity vector plots show the general state of motion of the surface more clearly, since the standing waves are obscured by smaller waves traveling across the liquid surface from the horizontal reflection off outer free-slip boundary. A representative velocity plot is shown in Fig. 23.7.

The time period between these local maximums of velocity and corresponding local minimums of displacement is found. Doubling this value to account for a full wave oscillation period and taking the inverse yields the surface frequency. Further, the expected wavelength for each case is verified by comparing the wave tips to an element plot of the same region. The peak velocity of the surface is also captured.

23.4.1 Amplitude Variable

The interesting case of the 300 Hz excitation system mentioned in the Literature Review has been analyzed in terms of its actuator amplitude. A spike in peak response velocity can be seen at the exact amplitude the previously mentioned study had photographed droplets ejecting at. The data in Fig. 23.8 appears to be averaged at around 0.7 m/s except for a single anomaly.

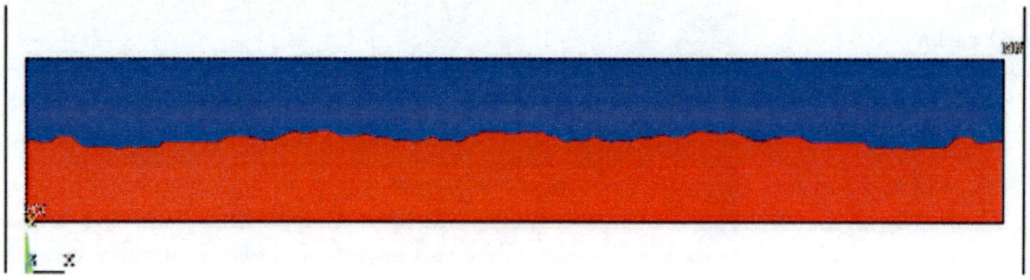

Fig. 23.6 Full development of standing surface waves

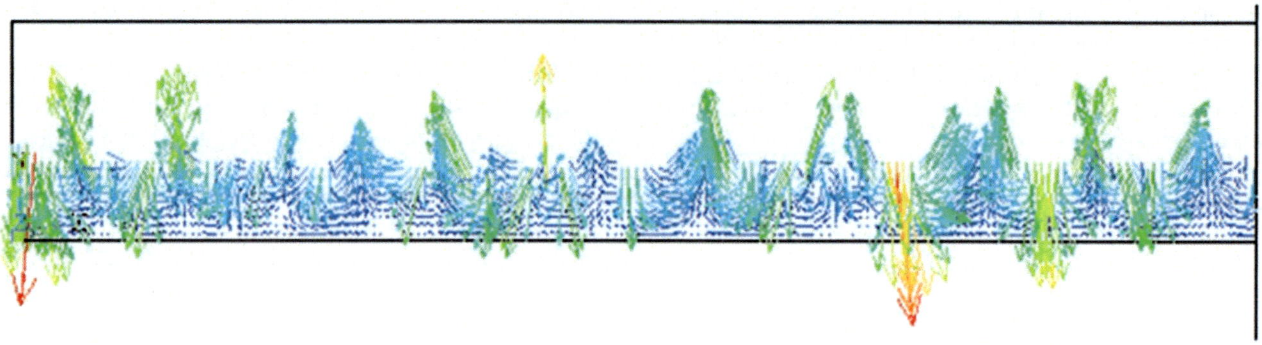

Fig. 23.7 Flotran velocity magnitude vector plot

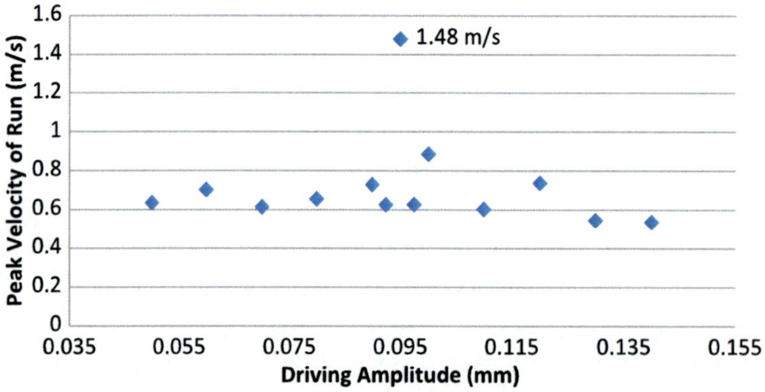

Fig. 23.8 Flotran CFD parameter test, driving amplitude varied, peak velocities plotted for 300 Hz case

A very interesting data point corresponding very closely to the actuator displacement used in the experiments of Yule et al. of 0.1 mm.

The depth of the 300 Hz case was tested around the value reported to have been used by Yule et al. experimentally (2 mm). The increments in Fig. 23.9 are 0.1 mm, except the two additional data points on either side of the spike. Yet again, there is a spike at the experimental depth and no influence otherwise. It is important to note that the 300 Hz case data generated for Fig. 23.9 was conducted at the displacement amplitude of 0.095 mm and that no velocity spike is observed for this depth test using 0.1 mm as the amplitude.

The 300 Hz case was tested to show the influence of driving frequency, with all other parameters remaining at the standard experimental values reported. To do this, the frequency was varied about the 300 Hz mark in 10 Hz increments, with a couple more data points added around the observed spike (Fig. 23.10).

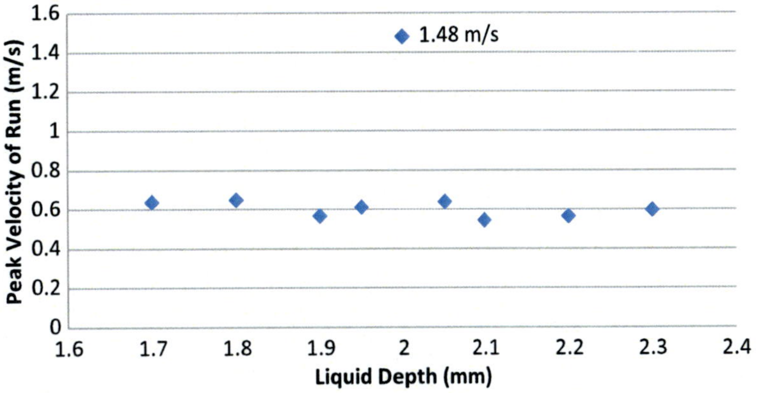

Fig. 23.9 Flotran CFD parameter test, liquid depth varied, peak velocities plotted for 300 Hz case

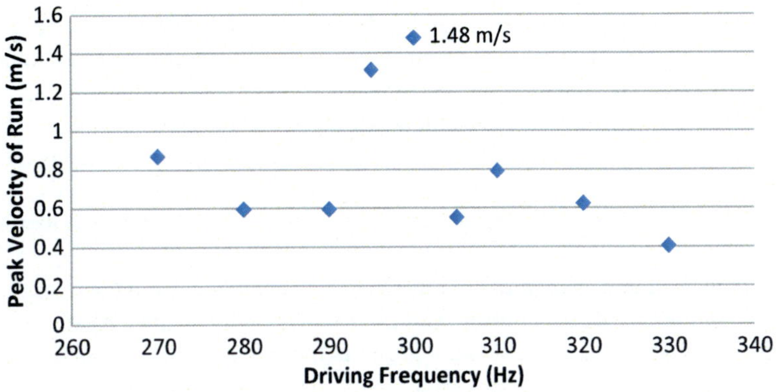

Fig. 23.10 Flotran CFD parameter test, driving frequency varied, peak velocities plotted for 300 Hz case

23.5 Discussion

In order for a vibrating film of liquid to produce atomized droplets, it is necessary that a few conditions are met. The first of these conditions is that the film be deep enough such that the surface can oscillate freely without interference from the actuating base on which the film is held. Second, the actuator amplitude must be great enough to excite the vibrating liquid surface to the point where the peaks of the standing waves have enough kinetic energy to overcome the resistant forces associated with the fluid interface. The quantities of energy required for this threshold are largely determined by the properties of the liquid, with surface tension being much more influential than viscosity for water and water-based liquids (like most medications). Surface tension is the primary resistance to droplets escaping the surface of the liquid medication film and it also plays a major role in determining potential droplet size. Higher surface tension leads to longer standing wavelength, which the mean droplet diameter is proportional to. While raising the droplet size for a given actuating frequency increases droplet kinetic energy, it also increases the resistant energy acting upon it. It follows that the final condition to be met coincides with the previous condition, that the liquid's properties are conducive to generating atomized droplets for the given operating setup which includes film thickness (depth) and actuator amplitude as primary factors.

The vibrating film depth must be sufficient to support surface oscillations; this quantity is generally on the length order of the standing waves produced. This basically means that a wave "aspect ratio" of about one-to-one is necessary in order for the surface to oscillate freely. This means that a fully developed wave's length will correlate closely with its critical peak height, which should be a smaller quantity than the film thickness. Films that are an order of magnitude smaller than the length of an expected standing wave will simply not begin to oscillate, and will "ride along" with the actuator or be flung off entirely in bulk form. If a vibrating film produces atomized droplets at a given depth, increasing this depth beyond a certain threshold will lead to sporadic droplet production and increasing it further will lead to excessive damping in the system which halts the process entirely.

For a given liquid film or body depth, a certain range of actuator amplitude will result in surface oscillation leading to atomized droplets. Below this range, surface vibration may not occur and above it, bulk motion of the fluid body will occur and no atomization will result. It is entirely possible that certain systems will not have critical actuator amplitudes and will

Table 23.1 Energy comparison of peak velocity generated in CFD run to threshold velocity (0.97 m/s) of 300 Hz case, mean droplet diameter of 0.92 mm

Actuator amplitude, mm	Peak CFD velocity generated, m/s	% of required kinetic energy
0.060	0.703	53
0.080	0.656	46
0.090	0.729	57
0.095	1.48	234
0.100	0.886	84
0.120	0.738	58

either fail to produce surface waves, or will simply experience bulk motion of the liquid body. Such systems are beyond the scope of this research.

A distribution of droplet sizes exists in all standing wave atomizers which implies that droplets significant larger than the mean size are produced. As droplet size increases, kinetic energy increases along with resistant energy but kinetic energy increases more quickly due to the nearly spherical shape of an ejecting droplet. Thus, larger droplets on the distribution will tend to initiate the atomization process while destabilizing the entire liquid surface in a sort of chain-reaction.

Under the conditions described by Yule et al., the case of 300 Hz generates a peak velocity of 1.48 m/s using the previously described CFD free-surface method. An exception to the conditions described in the literature is that a 0.095 mm actuator amplitude is used instead of 0.1 mm, but such a small discrepancy is thought to either be experimental error or caused by random variation in the experimental conditions, the reporting of said conditions or their interpretation. Whatever the reason may be, this only represents a difference in the actuator amplitude that is only 0.005 mm or 5% of the measured actuator amplitude.

The most important piece of information to take away from the current study is that of the compelling evidence of a tuned system for each of these cases. The CFD analysis shows that a relatively wide range of actuator amplitude can be used with little change in the peak velocity generated in the system. Of course, the exception is that one particular value of actuator amplitude effectively doubles the velocity generated in the system. This marked increase in kinetic energy occurs at exactly the conditions used for past experimental analysis and only at this specific set of parameters. This set of parameters, which includes the liquid depth, actuator amplitude, and actuator frequency, coincides with a quantity of kinetic energy which significantly surpasses the resistant energy of the liquid surface. Table 23.1 shows values from the 300 Hz actuator amplitude CFD parameter analysis. In this case, the threshold ejection velocity for a droplet of mean diameter (approximately 0.92 mm) is 0.97 m/s. The critical amplitude of this case is highlighted at 0.095 mm which clearly shows the excess kinetic energy resulting from the doubling velocity present in their respective unique parameter configurations.

At the mean droplet diameter, the Flotran CFD analysis shows that the 300 Hz cases generate enough velocity to eject droplets according to the energy balance in Eq. 23.9. These cases are likely to generate a very uniform surface wave pattern and a corresponding droplet distribution that is more tightly centered on the mean droplet diameter than an ultrasonic case due to a reduction in random disorder. The 300 Hz case with a driving amplitude of 0.095 mm is a unique set of conditions that generated 234 % of the required kinetic energy for droplet ejection at the mean droplet diameter. Excesses such as those mentioned may be required to successfully carry the ejected droplet away from the vibrating surface.

23.6 Conclusions

This study validates the process of vibration induced atomization of liquids by relating existing concepts of liquid atomization to the operating parameters of known atomizing systems. By numerically altering these parameters, trends in response conditions can be simulated and examined. The CFD analysis assists in model verification and reveals that there exists some critical configuration driving amplitude and liquid depth must be fulfilled in order for droplet kinetic energy to exceed fluid resistance energy so that the atomization process can initiate. The ability to numerical model and determine this unique set of parameters is an efficient and useful design methodology discovery.

Acknowledgements This research was supported in part by the UND Faculty Research Seed Grant "Medical Drug Delivery using an Aerosol Generating Nebulizer."

References

1. Lang RJ (1962) Ultrasonic atomization of liquids. J Acoust Soc Am 34(1):6
2. Benjamin TB, Ursell F (1954) The stability of the plane free surface of a liquid in vertical periodic motion. Proc R Soc A 225(1163):505–515
3. Faraday, M. (1831) On the forms and states of fluids on vibrating elastic surfaces, Phil. Trans. R. Soc. Lond. 52, 319–340
4. Rayleigh, L. (1883) On the crispations of fluid resting of a vibrating support, Phil. Mag. 16, 50–58
5. Lamb H (1945) Hydrodynamics, 6th edn. Dover, New York, pp 455–460
6. Peskin RL, Raco RJ (1963) Ultrasonic atomization of liquids. J Acoust Soc Am 35(9):1378–1381
7. Al-Suleimani AJ, Yule Y (2000) On droplet formation from capillary waves on a vibrating surface. Proc Math Phys Eng Sci 456(1997):1069–1085
8. Sindayihebura D, Bolle L (1995) Theoretical and experimental study of the behavior of liquid film free surfaces driven by transverse ultrasonic vibrations. Comput Model Free Mov Bound Probl 3:67–74
9. Dobre M, Bolle L (1999) Visualization and analysis of liquid film surface patterns formed on ultrasonic atomisers. sites-final.uclouvain.be. Department of Mechanical Engineering – Université Catholique De Louvain, 5 July 1999. Web
10. Sindayihebura D, Bolle L, Cornet A, Joannes L (1998) Theoretical and experimental study of transducers aimed at low-frequency ultrasonic atomization of liquids. J Acoust Soc Am 103(3):1442–1448
11. Al-Suleimani Y, Yule AJ, Collins AP (1999) How orderly is ultrasonic atomization? yjcorp.co.kr. Department of Mechanical Engineering, UMIST, 5 July 1999. Web
12. Alzuaga S, Manceau J, Bastien F (2005) Motion of droplets on solid surface using acoustic radiation pressure. J Sound Vib 282(1–2):151–162
13. Palan V, Steve Shepard W Jr (2006) Enhanced water removal in a fuel cell stack by droplet atomization using structural and acoustic excitation. J Power Sources 159(2):1061–1070

Chapter 24
Dynamical Modeling and Verification of Underwater Acoustic System

Ahmet Levent Avşar, İstek Tatar, and Cihangir Duran

Abstract Sound generated by different sources in water can be determined by underwater acoustic systems. In order to detect sound in water, the underwater acoustic system should be designed with respect to level and frequency characteristics of sound. Therefore, resonance frequency of the underwater acoustic system is the most important design parameter and it should be obtained by design iterations. During these iterations, resonance frequency of the underwater acoustic system can be found by finite element method with related analysis. Also, this analysis should be verified by experimental techniques. In this study, dynamical design and analysis of underwater acoustic system, which has 5 $\times$ 5 arrays of identical tonpilz transducers and these transducers are attached to the carbon fiber reinforced composite, are detailed. Moreover, this dynamical model of the underwater acoustic system is verified by experimental techniques, which are admittance, transmit voltage response and received voltage sensitivity measurements.

Keywords Underwater acoustic • Transducer • Structural dynamics • Finite element • Admittance measurement

24.1 Introduction

It is very important to transmit the acoustic energy through water for different underwater communication applications [1]. Due to this reason, acoustic transmission devices should be designed carefully. Generally, this type of devices is called as a sonar system and they can be active, passive or both [2]. If the sonar system is active, in other words, it generates underwater sound, it is classified as a projector or transducer. Also, if it listens only to underwater sound, it is named as a hydrophone. The basic working principle of these devices is that they convert pressure, due to sound wave, to electricity. Therefore, special materials are used in these systems, such as piezoelectric or magnetostrictive materials [1]. This types of material are suitable due to their nature, i.e., they can generate electricity because of stress in it or if voltage difference is applied these materials, deformation can occur.

There are different types of transducers or hydrophones. In order to decide which transducer types are suitable for a specific application, property of the acoustic noise in underwater should be known very well. Frequency range and amplitude are two important properties to determine the characteristics of acoustic noise in underwater. Especially, transducer or hydrophone types are decided by considering the frequency range of the acoustic noise. In Fig. 24.1 [3], it can be seen the transducer or hydrophone types with respect to frequency. For lower frequencies, hydroacoustic or variable reluctance type transducer can be used. Tonpilz, cylinder and sphere types are suitable for high frequency applications.

In order to increase the source level of the transmitted acoustic pulse, more than one transducer can be used to form an array [4]. Moreover, the direction of the acoustic pulse can be arranged by constructing an array with identical transducer. Also, if it is wanted to increase the response of the array in receiving mode, hydrophone arrays should be used and signal to noise ratio is improved by these types of arrays [4]. In Fig. 24.2, transducer array and hydrophones line array are illustrated.

A.L. Avşar (✉) • İ. Tatar • C. Duran
Meteksan Savunma (Defense) Industry, Inc., Ankara, Türkiye
e-mail: lavsar@meteksan.com; itatar@meteksan.com; cduran@meteksan.com

T. Simmermacher et al. (eds.), *Topics in Model Validation and Uncertainty Quantification, Volume 5*,
Conference Proceedings of the Society for Experimental Mechanics Series 41,
DOI 10.1007/978-1-4614-6564-5_24,

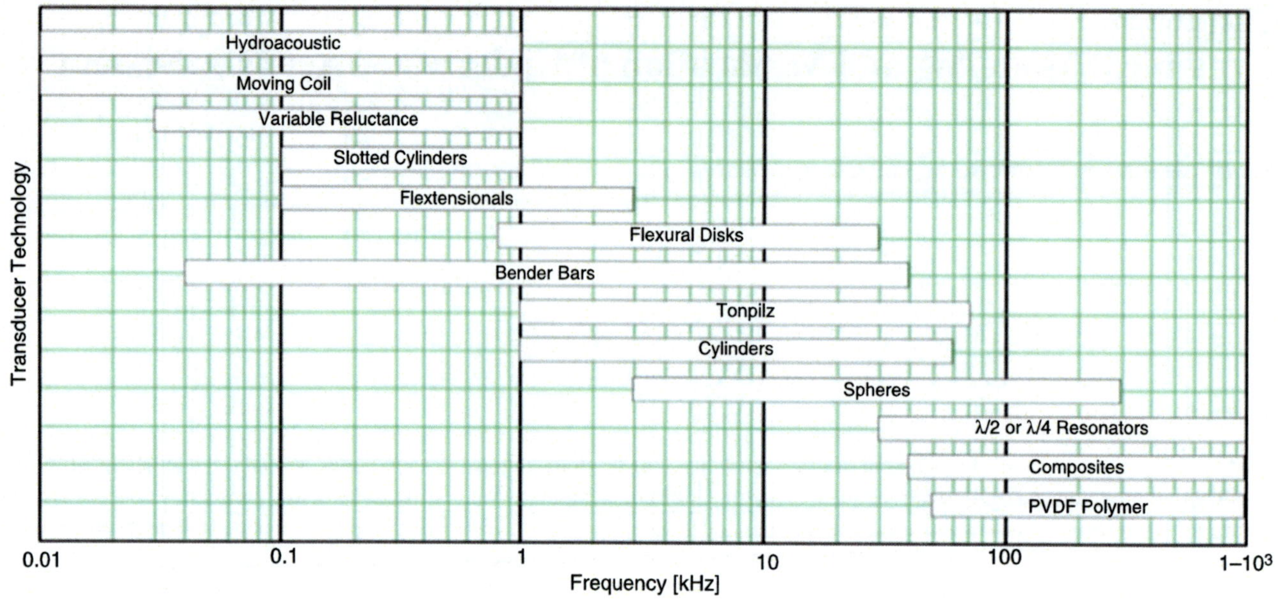

Fig. 24.1 Frequency range of interest and transducer types [3]

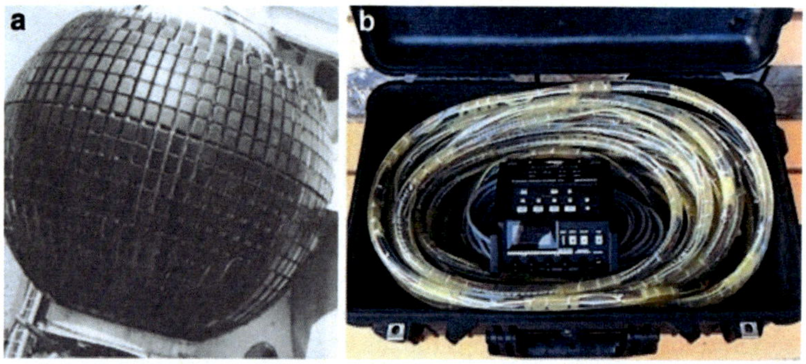

Fig. 24.2 (**a**) Transducer array [4], (**b**) hydrophone line array [5]

In this study, dynamic characteristic of the underwater acoustic system, which is constructed by tonpilz transducers, is investigated. In order to find dynamic characteristics of the system, finite element analysis is conducted in ANSYS Workbench 14.0. After finding the dynamic characteristics of the system, finite element model is validated by special measurement techniques, such as admittance measurement. Moreover, acoustic performance of the system is presented by measuring the transmit voltage response (TVR) and receive voltage sensitivity (RVS).

24.2 Design and Production

24.2.1 Underwater Transducer Type Selection

In order to select the type of transducer, frequency range of interest should be decided. For this application, frequency range is in between 10 and 50 kHz, therefore tonpilz type of transducer is selected and designed. Tonpilz, which is also called as acoustic mushroom, has active materials between rigid radiating head mass and heavy tail mass [6], as shown in Fig. 24.3. Active materials can be piezoelectric or magnetostrictive. The special property of this type of transducer is such that it can be used as both projector and receiver.

During the design phase of tonpilz transducer, different modeling techniques can be used. The basic modeling technique is "Simple Lumped Parameter Model". In this technique, head and tail masses are modeled as a rigid mass and active materials

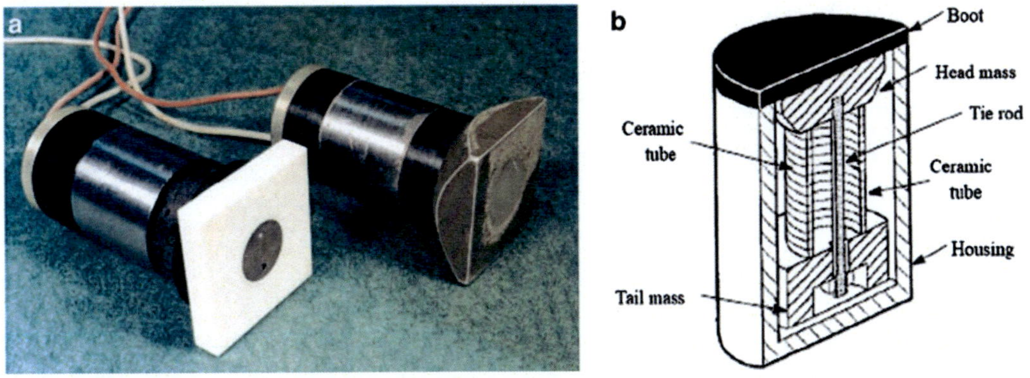

Fig. 24.3 (**a**) Tonpilz transducer, (**b**) parts of tonpilz transducer [7]

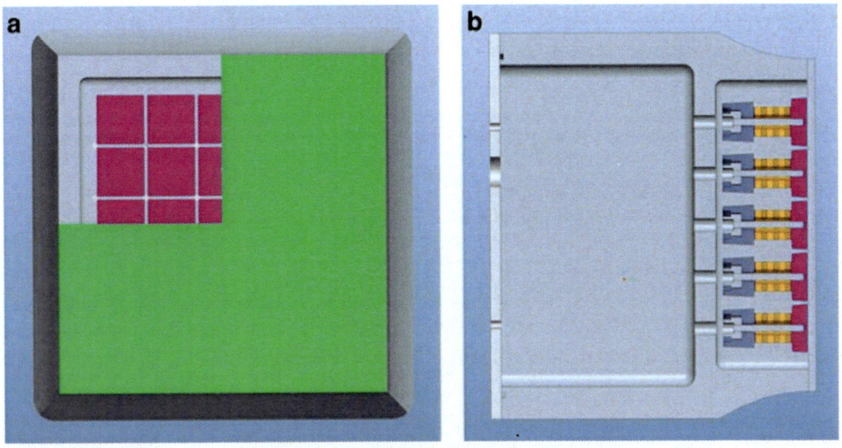

Fig. 24.4 Geometry of the underwater acoustic system, (**a**) front view, (**b**) side and section view

are considered as a stiffness element. Geometrical and material details are ignored by this technique; however, it is very fast and gives first intuitions about the tonpilz transducer properties. The most complex and accurate technique is "Finite Element Modeling". Different geometries, boundary conditions and material properties can be used by this technique. Also, optimization and iteration processes can be run to reach the required properties of tonpilz transducer and arrays.

24.2.2 Underwater Acoustic System

Underwater acoustic system has 5×5 arrays of identical tonpilz type of transducer. These transducers are attached to the composite shell and this shell is glued to the case of the system. The geometrical model of underwater acoustic system can be seen in Fig. 24.4. In the system, transducer case can be considered as rigid with respect to composite shell and transducers, and is assumed that dynamical behavior of the system can be affected only by transducers and composite shell.

Composite shell is made up of carbon fiber reinforced materials. The main property of this type of composite shell is that transverse passion ratio should be lower than 0.1 [8]. If transverse position ratio is lower than 0.1, transducer in the array vibrates like a simple piston, as shown in Fig. 24.5 [9]. Due to this reason, mechanical coupling of the transducer are minimized and acoustic efficiency of the system is increased. Material properties of the composite shell are found by tensile and compression test. During the tensile test, Poisson ration is calculated to be 0.1. Moreover, this shell has isotropic behavior and Young's Modulus is approximately 60 GPa.

During the production of the transducer, piezoceramic disks are combined by epoxy glue to make the active stacks. Active stacks consist of 6 PZT-4 piezoceramics. After production of the active stacks, head and tail mass of the tonpilz transducer are attached to it. Finally, tonpilz transducers are produced and production process is validated by experimental methods, such

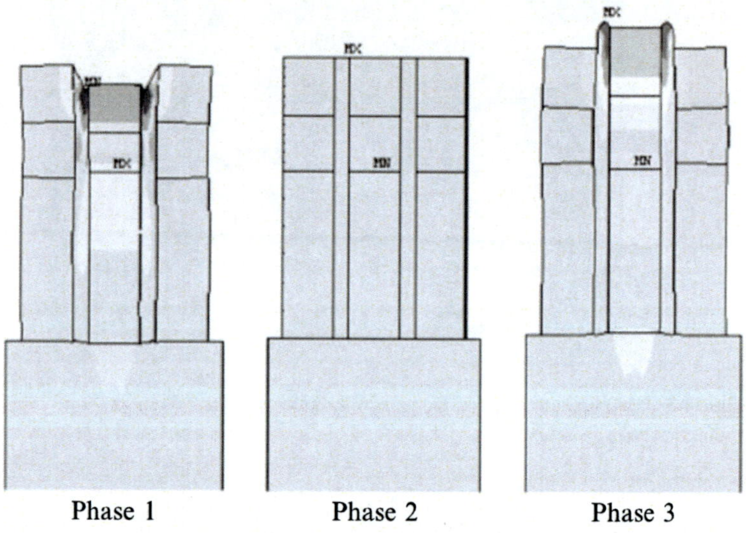

Fig. 24.5 Deformation shape of the transducer with composite shell [9]

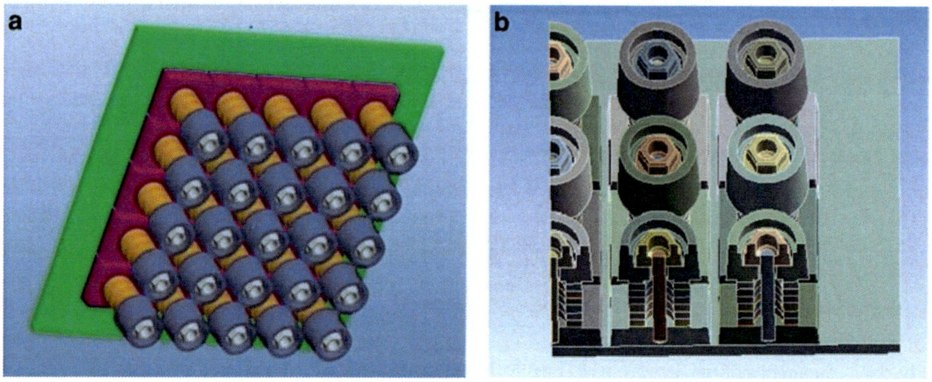

Fig. 24.6 (**a**) Transducer array, (**b**) finite element geometry of the system

as admittance measurements. After obtaining the tonpilz transducer, 25 transducers are aligned and glued on the composite shell. Finally, other components of the underwater acoustic system are assembled together and sealed to prevent leakage of water.

24.3 Finite Element Model of Underwater Acoustic System

The finite element model of the 5 × 5 array is constructed in ANSYS Workbench 14.0 [10]. In this model, only composite shell and 25 transducers are taken into account; however, other components are not considered due to rigidity in the frequency range of interest. The geometrical model of the 5 × 5 array can be seen in Fig. 24.6. Due to the symmetry only quarter part of this array is used in ANSYS Workbench 14.0 to decrease the simulation time, Fig. 24.6.

Mesh of quarter part of the array is formed in ANSYS Workbench 14.0. Piezoceramics are meshed by SOLID 226 [10] and SOLID 186 [10] is used for other components. This mesh has 79,000 nodes and 24,000 elements. After finishing the mesh process, symmetry and fixed support boundary conditions are given in ANSYS (Fig. 24.7). Material properties of the each components of the array are given in Table 24.1.

Finally, harmonic analysis is conducted for the frequency range of interest by using this finite element model. In this analysis, 1 V is given to the middle transducer for driving the system. End of this analysis harmonic response of the system is obtained and frequency response of the system is given in Fig. 24.8. Frequency axis of the frequency response graph is normalized by resonance frequency. Also, deformation response of the system at resonance frequency is shown in Fig. 24.8.

Moreover, harmonic analysis is repeated for the same frequency range by giving 1 V to the all transducers. End of this analysis harmonic response of the system is obtained and frequency response of the system is given in Fig. 24.9. Frequency axis of the frequency response graph is normalized by resonance frequency. Also, deformation shape of the system at resonance frequency is shown in Fig. 24.9.

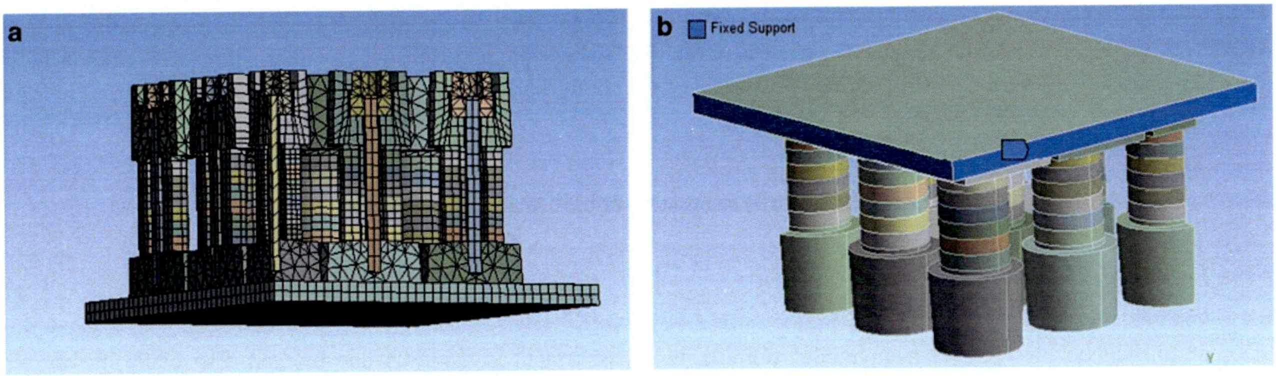

Fig. 24.7 (**a**) Mesh of the system, (**b**) boundary condition, fixed support

Table 24.1 Material properties of the array

Component name	Material
Transducer ceramic	PZT-4
Head mass of transducer	Aluminum
Tail mass of transducer	Steel
Tie rod	Steel
Boot of transducer	Carbon fiber reinforced composite
Housing	Aluminum

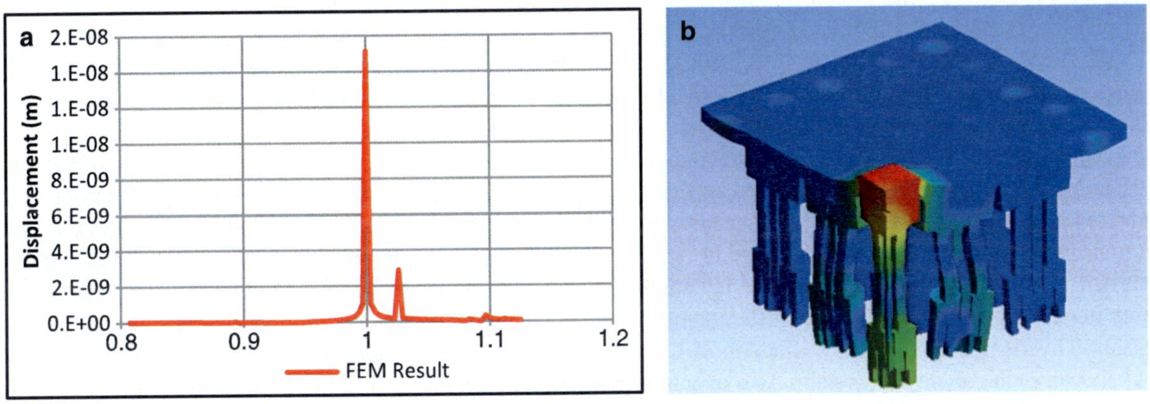

Fig. 24.8 (**a**) Frequency response of the middle transducer, (**b**) deformation shape of the middle transducer

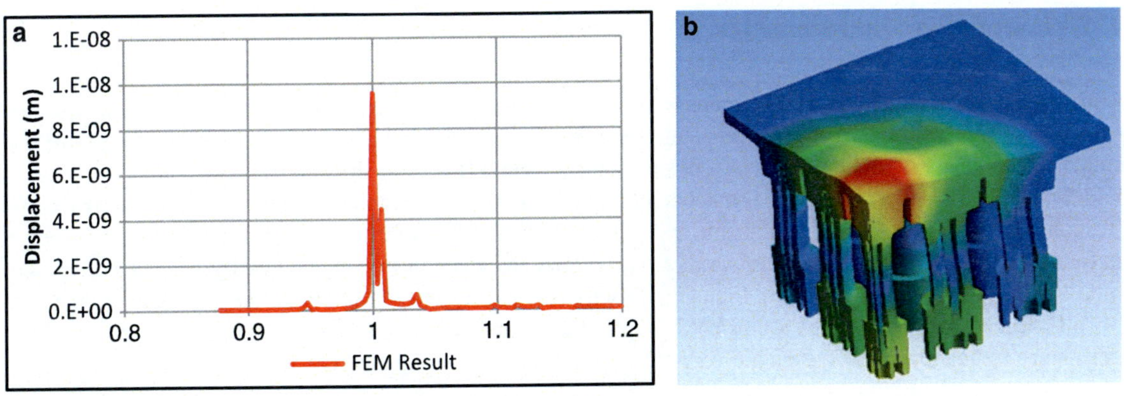

Fig. 24.9 (**a**) Frequency response of the array, (**b**) deformation shape of the array

As seen from Fig. 24.8, due to material properties of composite shell, only one transducer vibrates like as a simple piston. Also if all transducers are drive together, system behaves like a single tonpilz transducer. Other resonance peaks in both Figs. 24.8 and 24.9 are caused by the dynamic properties of the composite shell.

24.4 Validation of Finite Element Model By Experimental Techniques

24.4.1 Admittance Measurement

In order to validate the finite element model, admittance measurement is using as one method. Especially, for micro piezoelectric structures, it is not easy to find resonance frequency by familiar method, such as using accelerometer and shaker. Also it is very hard to find dynamic properties of the high frequency piezoelectric structure by same methods.

If the equation of motion of the simple mechanical system (Eq. 24.1), which has mass, damper and spring, are compared with equation of the electrical series resonant circuit (Eq. 24.2), it is seen that there is very good analogy with electrical and mechanical system [1].

$$m\frac{d^2x}{dt^2}+c\frac{dx}{dt}+kx=F\sin(\omega t) \tag{24.1}$$

where, m is mass, c is damping, k is stiffness, F is force, x is displacement, t is time and ω is frequency.

$$l\frac{di}{dt}+ri+\frac{1}{c}\int idt=V\sin(\omega t) \tag{24.2}$$

where, l is inductance, r is resistance, c is capacitance, V is voltage, i is current, t is time and ω is frequency.

By considering Eqs. 24.1 and 24.2, similarity of these two systems is given in Table 24.2 [1]. Therefore, if admittance of the electrical resonant circuit is obtained, resonance frequency of the system is found easily by using this information. In other words, similar relationship between mechanical mobility and electrical admittance can be seen.

Because of these reasons, admittance measurement is very common method for transducer characterization [11, 12] due to the electrical-mechanical behavior of the piezoelectric material. Impedance analyzer is used to measure the admittance of the piezoelectric structures [13]. By admittance measurement, conductance and susceptance of the structure is obtained, i.e. real part of the admittance is conductance and susceptance is imaginary the part of admittance. Maximum frequency of conductance is given the resonance frequency of the structure [14]. During the admittance measurement in air, only middle transducer is used in one case and also in other case, all transducers are connected in parallel to find total system characteristic. Comparison of the FEM results and conductance measurements are given in both Figs. 24.10 and 24.11.

In both Figs. 24.10 and 24.11, there are different resonance peaks than finite element model in conductance measurements. This is caused by idealization in finite element model, in other words, it is not possible to model real boundary conditions and contact between each parts in the system. Also dynamic characteristics of composite shell affect the results in experimental studies.

24.4.2 TVR and RVS Measurements

The acoustic performance of the underwater acoustic system is measured in water. During the measurement, open water test center, which is in the infrastructure of Meteksan Defense, is used, (Fig. 24.12a). Before measurement, the system is attached to the setup and reference hydrophone is located in front of the system (Fig. 24.12b).

The middle tonpilz transducer in the underwater acoustic system is operated in the frequency range of interest and transmit voltage response of the middle of the composite transducer is obtained. Moreover, total transmit voltage response of the system is found by driving all transducers in parallel. The results are given in Fig. 24.13. Finally, receive voltage sensitivity of the middle transducer and the system is found by driving the reference hydrophone in the frequency range of interest, (Fig. 24.14). Frequency scale of the TVR and RVS graphs are normalized by resonance frequency and as seen from those graphs, resonance frequency of the system are very close to results of the finite element analysis.

Table 24.2 Similarity of the mechanical and electrical systems [1]

Mechanical	Electrical
Mass	Inductance
Compliance	Capacitance
Resistance	Resistance
Velocity	Current
Force	Voltage
Displacement	Charge
Impedance	Impedance

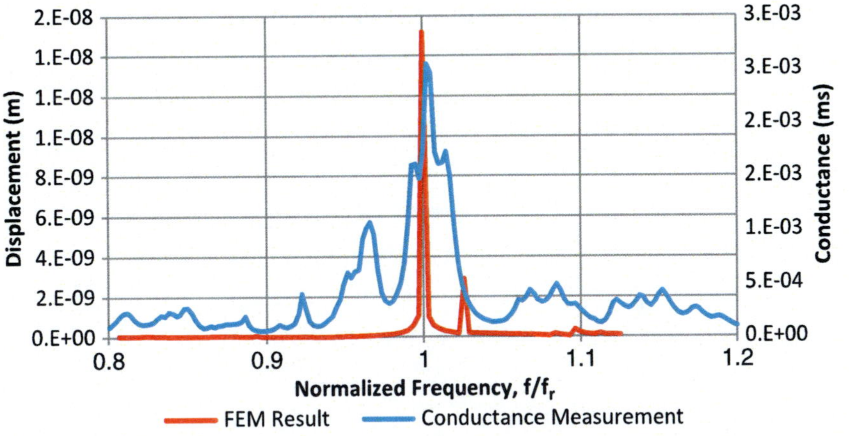

Fig. 24.10 Comparison of the FEM result and conductance measurement in air for the middle transducer

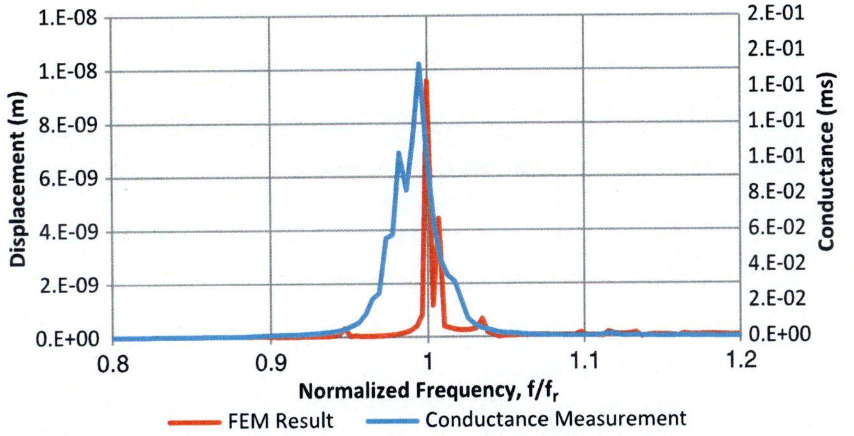

Fig. 24.11 Comparison of the FEM result and conductance measurement in air for the array

24.5 Discussion

Before developing underwater acoustic system, it is very important to understand the system behavior and generally structural dynamics of this system plays critical role. If structural dynamics of the underwater acoustic system is not characterized very well, it is not possible to use such system effectively. Resonance frequency and deformation shape at resonance frequency are very important parameters. Therefore, the resonance frequency should be arranged by requirements of the application for the active underwater acoustic systems. Moreover, the deformation shape of the underwater acoustic system should be piston like movement near resonance frequency to generate acoustic wave in water effectively.

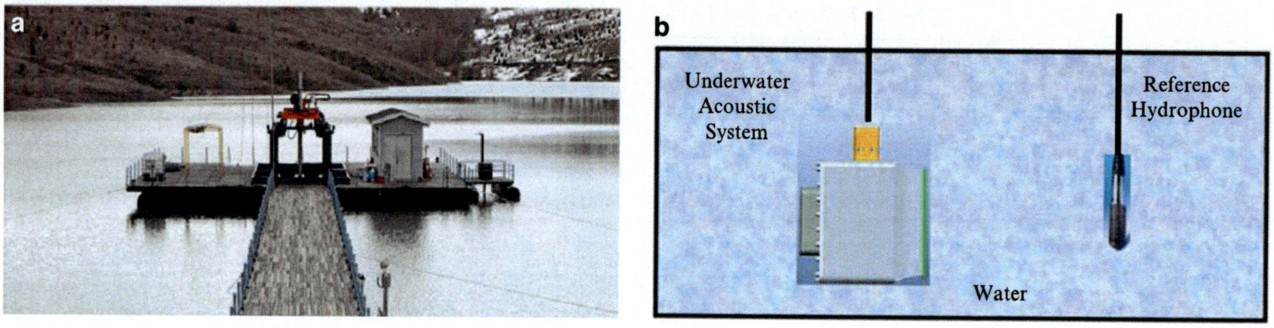

Fig. 24.12 (a) Open water test center, Meteksan Defense, (b) test setup for TVR and RVS measurement

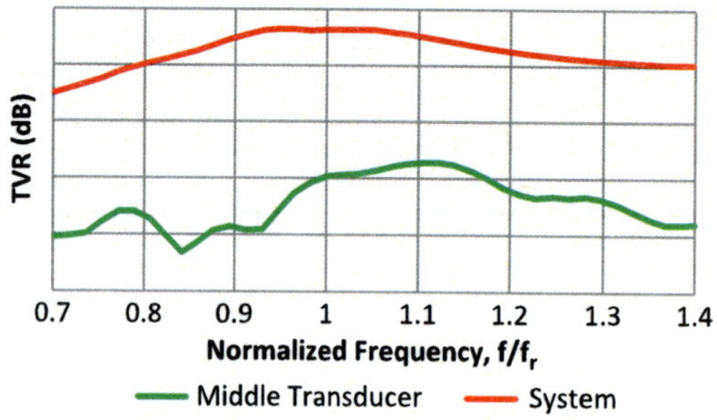

Fig. 24.13 TVR measurement for the middle transducer and array

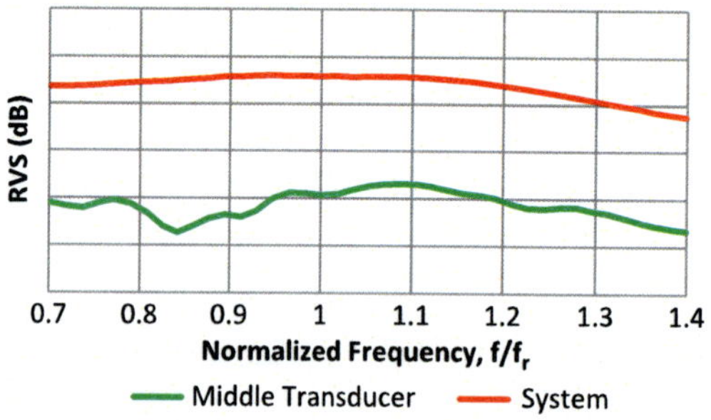

Fig. 24.14 RVS measurement for the middle transducer and array

In this study, dynamic characteristic of the underwater acoustic system is investigated by finite element method and experimental techniques. Finite element method is very fast and reliable method, if all parameters, such as boundary conditions and material properties, are defined very well. It is possible to run trial and error study to determine the required system properties. Transducer geometry, material properties of piezoceramic and composite shell can be found by finite element analysis easily. The presented underwater acoustic system is analyzed in ANSYS Workbench 14.0 and requirement for this system is achieved by iterations in this software. As explained in reference [9], transducers move like a simple piston due to material properties of the composite shell and also system behaves as a single tonpilz transducer if all transducer is drive as parallel. After obtaining the dynamic properties of the system, results of finite element are validated by special measurement techniques for electroacoustic system. One method is admittance or impedance measurement and the resonance frequency of the transducer, which consists of piezoceramic disks, are found by this method. In conductance measurement, there are seen different resonance peaks and this is caused by idealization finite element modeling and dynamic characteristics of the composite shell. Moreover, the underwater acoustic properties of the system are determined by TVR and RVS measurements and resonance frequency of the system also can be seen in these measurements.

References

1. Stansfield D (1991) Underwater electroacoustic transducer, a handbook for users and designers. Peninsula Publishing, Los Altos Hills
2. Urick RJ (1983) Principles of underwater sounds. Peninsula Publishing, Los Altos Hills
3. Cohcran JC (2008) Elements of sonar transducer design
4. Sherman CH, Butler JL (2007) Transducer and arrays for underwater sound. Springer, New York
5. http://www.tonywublog.com/20100609/hydrophone-for-humpback-whale-acoustic-research.html#axzz2AD9YfSyt, page visit: 17 Oct 2012
6. Wilson OB (1988) Introduction to theory and design of sonar transducers. Peninsula Publishing, Los Altos Hills
7. Marshall WJ (1999) Underwater sonar projectors. Wiley encyclopedia of electrical and electronics engineering
8. Kondo T, Kitatuji M (2004) Composite materials and measurement of their acoustic properties. Jpn J Appl Phys 43(5B):2914–2915
9. Kondo T, Kitatuji M (2004) Medical transducer arrays using composite materials for acoustic matching layers. In: IEEE international ultrasonics, ferroelectrics, and frequency control joint 50th anniversary conference, pp 1659–1662
10. ANSYS Workbench 14.0 Help Manuel
11. Sherrit S, Leary SP, Dolgin BP, Bar-Cohen Y, Taskerb RD The impedance resonance for piezoelectric stacks
12. Pardo L, Jiménez R, García A, Brebøl K, Leighton G, Huang Z (2010) Impedance measurements for determination of the elastic and piezoelectric coefficients of films. Adv Appl Ceram Struct Funct Bioceram 109(3):156–161
13. Impedance analyzer: Pv50a, Pv70a. Bandera Electronics Co., Ltd., China
14. http://www.americanpiezo.com/knowledge-center/piezo-theory/determining-resonance-frequency.html, page visit: 17 Oct 2012